Contents

Bibliographic information published by Die Deutsche Nationalbibliothek
Die Deutsche Nationalbibliothek lists this publication in the Deutsche Nationalbibliografie; detailed bibliographic data is available on the Internet at http://dnb.d-nb.de.

A catalogue record for this book is available from the British Library.

Library of Congress Cataloging-in-Publication Data

Names: Crahay, Géraldine, 1986- author.
Title: The hermaphrodite, the effete and the butch : sexual and gender ambiguities in nineteenth-century French narratives / Géraldine Crahay.
Description: Oxford ; New York : Peter Lang, [2022] | Series: French studies of the eighteenth and nineteenth centuries, 1422-7320 ; vol. 40 | Includes bibliographical references and index.
Identifiers: LCCN 2020057126 (print) | LCCN 2020057127 (ebook) | ISBN 9781789976489 (paperback) | ISBN 9781789976496 (ebook) | ISBN 9781789976502 (epub)
Subjects: LCSH: French fiction—19th century—History and criticism. | Gender identity in literature. | Gender nonconformity in literature. | Sex role in literature. | Masculinity in literature. | Literature and society—France—History—19th century.
Classification: LCC PQ637.G35 C73 2022 (print) | LCC PQ637.G35 (ebook) | DDC 843/.7093521—dc23
LC record available at https://lccn.loc.gov/2020057126
LC ebook record available at https://lccn.loc.gov/2020057127

Cover image: Hermaphrodite endormi. Datation : Epoque romaine impériale. Provenance : Rome, près des thermes de Dioclétien. Ancienne collection Borghèse, achat 1807. Ma231. Localisation : Paris, musée du Louvr. Photo © Musée du Louvre, Dist. RMN-Grand Palais / Thierry Ollivier.

Cover design by Peter Lang Ltd.

ISSN 1422-7320
ISBN 978-1-78997-648-9 (print)
ISBN 978-1-78997-649-6 (ebook)
ISBN 978-1-78997-650-2 (e-pub)

Published by Peter Lang Ltd, International Academic Publishers, Oxford, United Kingdom
oxford@peterlang.com, www.peterlang.com

This publication has been peer reviewed.

Géraldine Crahay

The Hermaphrodite, the Effete and the Butch

Sexual and Gender Ambiguities in Nineteenth-Century French Narratives

PETER LANG
Oxford · Bern · Berlin · Bruxelles · New York · Wien

FRENCH STUDIES
of the Eighteenth and Nineteenth Centuries

Edited by Fiona Cox, Associate Professor of French and Comparative Literature at the University of Exeter

PETER LANG
Oxford · Bern · Berlin · Bruxelles · New York · Wien

The Hermaphrodite, the Effete and the Butch

Acknowledgements

The writing of this book, based on my doctoral thesis, started in Durham and was completed on the Welsh island of Anglesey during the Covid-19 global pandemic. I would like to express my thanks to my PhD supervisors Nicola Frith and Manon Mathias, who have supported this project from the beginning; Christie Margrave, Stacie Allan and Catherine Dousteyssier-Khoze for their valuable advice; and Helena Chadderton for her careful proofreading of the manuscript. Finally, I would like to thank James Drysdale for his constant support and his interest in my work over the years.

– Durham, February 2022

INTRODUCTION

Masculinity and Gender Ambiguities during the July Monarchy (1830–1848)

In 1821, a young Honoré de Balzac writes a letter to his sister Laure Surville, to whom he is very close, to tell her about a contract signing party. However, the writer soon expresses his melancholy: he is indifferent to the social agitation in his family, he deplores the absence of love in his life, he recalls his father's coach accident and complains about his struggles to become independent through his work. In brief, he feels unhappy and dissatisfied. He concludes his letter as follows:

> À quoi bon la fortune et les jouissances quand ma jeunesse sera passée ? Qu'importent les habits d'acteur si l'on ne joue plus de rôle ? Le vieillard est un homme qui a dîné et qui regarde les autres manger; et moi, jeune, mon assiette est vide et j'ai faim ! Laure, Laure, mes deux seuls et immenses désirs, *être célèbre* et *être aimé*, seront-ils jamais satisfaits ? …[1]
>
> What is the point of wealth and pleasures when my youth is over ? What is the use of an actor's clothes if one is no longer playing a role ? The old man is a man who has finished his dinner and watches the others eating; as for me, I am still young, but my plate is empty and I am hungry ! Laure, Laure, my only two huge desires, *to be famous* and *to be loved*, will they ever be satisfied ? …

At age 22, Balzac expresses his discontent with his life, which seems to be both empty and slipping through his fingers. This discontent concerning the discrepancy between social expectations and reality to which Balzac refers has, since the nineteenth century, been labelled *le mal du siècle*.

1 Honoré de Balzac, letter XIV, to his sister Laure Surville (1821), *Correspondance de H. de Balzac*, 1819–1850, 2 vols, I (Paris: Calmann Lévy, 1876), 45–48 (48). All translations are my own unless otherwise noted.

About fifteen years later, in a letter to a woman named Louise, Balzac paradoxically attempts to minimize her admiration while aggrandizing himself: 'Sachez que tout ce que vous présumez chez moi de bon est meilleur encore; [...] que ma sensibilité est féminine et que je n'ai de l'homme que l'énergie; [...] tout est contraste en moi, parce que tout est contrarié' [Be aware that everything good that you presume in me is even better; [...] that my sensitivity is feminine and that I only have the energy of man; [...] everything is contrast in me because everything is thwarted].[2] Balzac's letter shows that he conceives of gender roles in opposition: feminine sensitivity versus masculine energy. However, the barrier between masculine and feminine identities does not seem to be insurmountable. Indeed, Balzac himself transcends this dichotomy, as his male identity is made up of feminine features. In highlighting his hybrid nature, he emphasizes his own alleged singularity.

Balzac's letters illustrate the fact that French society in the first half of the nineteenth century believed in a strict separation of gender roles. France in the 1830s and 1840s was indeed shaped by a resolutely virile model of masculinity encompassing values such as strength, bravery and honour. However, at the same time, there was awareness that a binary system of sexes and genders was not fully adequate to depict society. Like Balzac's letters, literary narratives of this period were populated with androgynous characters: epicene men, eunuchs, hermaphrodites and mannish women. Protagonists were often portrayed as suffering from the so-called *mal du siècle*. This expression designates young men's anxiety regarding the upheavals caused by the French Revolution, followed by the fall of the Empire, and, accordingly, their own loss of position in French society. While this representation of male youth in the early nineteenth century was largely ideological and therefore phantasmatic, it was nevertheless grounded in a genuine social inertia. Although Margaret Waller and Chantal Bertrand-Jennings have identified a concomitant female *mal du siècle*, this 'disease' was seen as affecting mainly men, and particularly young men from the middle and upper classes.[3] In *Balzac et le mal du siècle* [*Balzac and the*

2 Ibid., letter CLXXV, ii, to Louise (1836–37), 345–47 (346–47).

3 Margaret Waller, *The Male Malady: Fictions of Impotence in the French Romantic Novel* (New Brunswick: Rutgers University Press, 1993), Chapter Six: 'Toward a

Disease of the Century], Pierre Barbéris highlights the paradox that characterizes the early nineteenth century:

> avec la Révolution, s'était opérée une brusque accélération, avec une double conséquence: d'une part, s'étaient ouvertes d'extraordinaires possibilités d'emploi de soi. D'autre part, les barrières, un moment levées, s'étaient refermées. Tout était apparu possible, et tout était à nouveau mis en question.[4]
>
> [With the Revolution, a hasty acceleration had taken place which had a double consequence: on the one hand, extraordinary opportunities of employment were made available. On the other hand, the barriers, opened for a moment, had once again been closed. Everything had seemed possible and everything was once again brought into question.]

An emblematic literary text which discusses this indeterminate social phenomenon (although it is never called *le mal du siècle*) is the second chapter of Alfred de Musset's novel *La Confession d'un enfant du siècle* [*The Confession of a Child of the Century*], published separately in *La Revue des Deux Mondes* [*The Journal of the Two Worlds*] on 15 September 1835. In this chapter, Musset generalizes the ontological unease of young men of his generation, who see themselves as the main actors of History whilst being excluded from it.[5] The narrator, Octave, calls it a 'maladie morale abominable' [abominable moral disease], talks of a 'sentiment de malaise inexprimable' [indescribable feeling of malaise] and the 'maladie du siècle' [disease of the century].[6] According to Octave, this malaise stemmed from nostalgia for the grandeur attributed to Napoleonic times and the lack of opportunity to gain glory in post-imperial society. Living

---

Feminist Mal du Siècle: Sand's *Lélia*', 136–75; Chantal Bertrand-Jennings, *Un autre mal du siècle: Le romantisme des romancières, 1800–1846* (Toulouse: Presses Universitaires du Mirail, 2005).

4 Pierre Barbéris, *Balzac et le mal du siècle: Contribution à une physiologie du monde moderne*, 2 vols (Paris: Gallimard, 1970), I, 44.

5 See the special issue of the journal *Romantisme* on the Musset generation, notably José-Luis Diaz's introduction 'Génération Musset?', *Romantisme*, 147 (2010), 3–12, and his article '"Nous tous, enfants perdus de cet âge critique": La "génération de 1830" par elle-même' in the same issue (13–28).

6 Alfred de Musset, *La Confession d'un enfant du siècle*, Sylvain Ledda, ed. (Paris: Garnier-Flammarion, 2010), 59, 70 and 82.

in an era of uncertainty and nihilism, young men resigned themselves to a mediocre existence in a tedious, idle and lustful society. Importantly, many of Musset's contemporaries shared his gloomy reading of *la Monarchie de Juillet* [the July Monarchy; 1830–48] and men's role within French society in the light of a glorious past.

Focusing on this period of French history, the premises of this book are (1) that the sexual and gender yoke imposed on unconventional individuals such as individuals with intersex traits (or hermaphrodites, as they were called at the time), male homosexuals and masculine women also affected men who attempted to conform to traditional masculinity and (2) that the malaise generated by these sexual and gender constraints was at once illustrated, denounced and arguably remedied in the contemporaneous literature. Its aim, accordingly, is to investigate the complex relations between traditional definitions of sexual and gender roles and identities, especially those of men, as well as ambiguous genders and 'deviant' masculinities in the July Monarchy. This study focuses on the definition, depiction and critique of hermaphroditism, male effeminacy, homosexuality and female masculinity in medical, scientific and literary texts. I distance myself from other readings of this period, which claim that fictional works of the July Monarchy simply mirrored the contemporary masculine malaise in deploring the inability of young men to conform to an ideal model of masculinity.[7] Rather, I argue that literary narratives challenged traditional concepts of masculinity by suggesting alternative models arising from the very flaws that were supposed to characterize a masculine malaise. The hermaphrodite, the effete and the butch of the title play a central role in literary narratives and in scientific texts of the July Monarchy, I argue, because they reflect the destabilization of gender and sexual categories during this period. In spite of, or even thanks to, their 'abnormality' they in fact contributed to redefining masculine values.

7 One such study is Victoria Thompson, 'Creating Boundaries: Homosexuality and the Changing Social Order in France, 1830–1870', in Jeffrey Merrick and Bryant T. Ragan Jr, eds, *Homosexuality in Modern France* (New York and Oxford: Oxford University Press, 1996), 102–27.

Masculinities Studies and the 'Crisis' of Masculinity

Masculinity constitutes a relatively recent object of academic studies in comparison with women's studies.[8] As a branch of gender studies, masculinities studies were developed in reaction to the former, not in opposition to feminist principles, but rather as their extension.[9] If women's role in society and history needed to be reassessed in the light of feminist theories, men's role should meet a similar fate. Indeed, masculinity used to be conceived of as an 'empty', 'invisible', 'neutral' or 'abstract' category, as opposed to femininity.[10] In other words, it was deemed to represent the human norm and was therefore considered to be self-explanatory. The predominance of masculinity and maleness as the norms is illustrated,

8 Stephen M. Whitehead and Frank J. Barrett, 'The Sociology of Masculinity', in Stephen M. Whitehead and Frank J. Barrett, eds, *The Masculinities Reader* (Cambridge: Polity Press, 2001), 1–26 (1); Pierre Bourdieu, *La domination masculine* (Paris: Seuil, 'Points/Essais', 2002 [1998]), 22; Daniel Maira and Jean-Marie Roulin, 'Constructions littéraires de la masculinité entre Lumières et Romantisme', in Daniel Maira and Jean-Marie Roulin, eds, *Masculinités en révolution de Rousseau à Balzac* (Saint-Étienne: Publications de l'Université de Saint-Étienne, 2013), 9–28 (10).

9 See Michael Kimmel, *Angry White Men: American Masculinity at the End of an Era* (New York: Nation Books, 2017 [2013]), 128–29.

10 Élisabeth Badinter, *XY: De l'identité masculine* (Paris: Librairie Générale Française, 'Le Livre de Poche', 1994 [Odile Jacob, 1992]), 22–23; Antony Rowland, Emma Liggins and Eriks Uskalis, 'Introduction', in Antony Rowland, Emma Liggins and Eriks Uskalis, eds, *Signs of Masculinity: Men in Literature 1700 to the Present* (Amsterdam and Atlanta: Rodopi, 1998), 3–36 (3–8); Stefan Dudink, Karen Hagemann and Anna Clark, 'Editors' Preface: Historicizing Male Citizenship', in Stefan Dudink, Karen Hagemann and Anna Clark, eds, *Representing Masculinity: Male Citizenship in Modern Western Culture* (Houndmills, Basingstoke and New York: Palgrave Macmillan, 2007), ix–xv (xii). Michael Kimmel points out: '[...] we don't talk about men. [...] we don't name straight, white, middle-class, able-bodied, cisgender men. [...] If our political goal is to de-center that center, the first task is to make it visible': Michael Kimmel and Lisa Wade, 'Ask a Feminist: Michael Kimmel and Lisa Wade Discuss Toxic Masculinity', *Signs: Journal of Women in Culture and Society*, 44/1 (2018), 233–54 (235).

for example, in Thomas Laqueur's study of sixteenth-century anatomical books. The historian shows that a male body was more frequently depicted on the cover of these books than a female one and, more importantly, that women's organs (including genitalia) looked almost identical to male ones.[11]

Stephen Whitehead and Frank Barrett have cautiously defined masculinities – the use of the plural helps indicate the complexity of the concept – as 'those behaviours, languages and practices, existing in specific cultural and organizational locations, which are commonly associated with males and thus culturally defined as not feminine'.[12] Several ideas are contained in this definition. Firstly, masculinity is a cultural and social concept, not a biological one, and it is therefore variable. Secondly, despite being cultural, masculinity is generally linked to biological data, as it is associated with male individuals. Finally, it is defined in opposition to another concept: femininity, which is traditionally linked to female individuals. Masculinity is therefore perceived as the characteristic, the prerogative and even the duty of men, whereas women are and must be feminine.

The French equivalent term to *masculinity*, *masculinité*, is ambivalent because the semantic difference between *masculinité* and *virilité* (the latter derives from *vir* 'man', as opposed to both *mulier* 'woman' and *puer* 'child')[13] is far from simple. On the one hand, Alain Corbin and, to a lesser extent, Daniel Maira and Jean-Marie Roulin, consider that *masculinité* designates the biological characteristic of being a man – *maleness* in English – whereas *virilité* refers to the characteristics that are culturally associated with the male sex.[14] On the other hand, in *'Sois un homme!': La construction de la masculinité au XIX*e *siècle* [*'Be A Man!': The Construction of Masculinity in*

11 Thomas Laqueur, *Making Sex: Body and Gender from the Greeks to Freud* (Cambridge, MA, and London: Harvard University Press, 1990), 70–88. For more on Laqueur's hierarchical model of sexes, see Chapter 1 of this book.

12 Whitehead and Barrett, 'The Sociology of Masculinity', 15–16.

13 Alain Rey et al., Viril', in *Dictionnaire historique de la langue française*, 2 vols (Paris: Le Robert, 1992), II, 2263.

14 Alain Corbin, 'Introduction', in Alain Corbin, ed., *Histoire de la virilité*, II: *Le triomphe de la virilité: Le XIX*e *siècle* (Paris: Seuil, 2011), 7–11 (10); Maira and Roulin, 'Constructions littéraires de la masculinité', 15.

the Nineteenth Century] (2009), Anne-Marie Sohn points out the rarity of the concept of *masculinité* in the nineteenth century and considers *virilité* exclusively from a biological perspective.[15] However, a glance through French dictionaries of the nineteenth century reveals the interpenetration of the two concepts. The definitions of *masculin* and *viril* (the derivative substantives were not frequently used) in the sixth edition of the *Dictionnaire de l'Académie française* [*Dictionary of the French Academy*] (1835) and Émile Littré's *Dictionnaire de la langue française* [*Dictionary of the French Language*] (1873–77) do not clearly distinguish between biological and cultural connotations.[16] While primarily referring to the biological sex and to manhood, the adjectives also encompass physical and moral attributes that supposedly derive from the male attributes of an individual, such as strength, vigour, sexual potency and courage. To put it simply, maleness and manliness were so deeply intertwined in the nineteenth century – and still are nowadays – that questioning the latter would equate to questioning the former. Accordingly, one of the aims of this study is to understand what linked these two concepts in the July Monarchy, as well as the reasons for such a strong connection and the consequences of questioning them in literary texts.

A predominant component of masculinity was men's 'natural' superiority to women, because activity, culture and reason, the values associated with the former, were ascribed greater importance than so-called feminine values: passivity, nature and emotion.[17] Accordingly, men were encouraged to exercise their authority over women. This model of *hegemonic*

15 Anne-Marie Sohn, *'Sois un homme!': La construction de la masculinité au XIXe siècle* (Paris: Seuil, 2009), 11.

16 'Masculin', 'Masculinité', 'Viril' and 'Virilité', in *Dictionnaire de l'Académie française*, 6th edn, 2 vols (Paris: Firmin-Didot frères, 1835), II, 174 and 940–41, and in Émile Littré, *Dictionnaire de la langue française* <http://www.littre.org>, accessed 29 March 2017.

17 David S. Gutterman, 'Postmodernism and the Interrogation of Masculinity', in *The Masculinities Reader*, 56–71 (58). See also Arthur Brittan, *Masculinity and Power* (Oxford: Blackwell, 1989), 199.

masculinity, as R. W. Connell calls it,[18] has been characterized by its historical permanence and by the simplicity of its script.[19] This expected pattern of behaviour is summarized by David Gilmore as: 'impregnate women, protect dependents from danger, and provision [*sic*] kith and kin'.[20] Men's 'natural' supremacy was reinforced by laws such as the Napoleonic Code in nineteenth-century France. In 'Fonction paternelle et Code Napoléon' [Paternal Function and Napoleonic Code], Xavier Martin shows that, by punishing the petty infractions with which the government could not deal, paternal power complemented the legal and penal systems in post-revolutionary times.[21] As Maleville, one of the writers of the Civil Code, claims: 'La puissance paternelle est la providence des familles, comme le gouvernement est la providence de la société' [Paternal authority is families' Providence, just as the Government is society's Providence].[22]

However, in the early nineteenth century as it is nowadays, this apparently universal model of dominant masculinity is more a construct than a reality. In *Le mythe de la virilité* [*The Myth of Virility*] (2017), the French philosopher Olivia Gazalé analyses how the discourse of male superiority has not only led to women's submission but has also caused men's own oppression, as the latter are unable to cope with the multiple and sometimes contradictory constraints that are expected from them.[23] They display a

18 R. W. Connell, *Masculinities*, 2nd edn (Cambridge: Polity Press, 2005 [1995]), esp. 77.

19 Bourdieu, *La domination masculine*, 113–14; Alain Corbin, 'Préface', in Régis Revenin, ed., *Hommes et masculinités de 1789 à nos jours: Contributions à l'histoire du genre et de la sexualité en France* (Paris: Autrement, 2007), 7–11 (10).

20 David Gilmore, *Manhood in the Making: Cultural Concepts of Masculinity* (New Haven and London: Yale University Press, 1990), 223.

21 Xavier Martin, 'Fonction paternelle et Code Napoléon', *Annales historiques de la Révolution française*, 305 (1996), 465–75 (473–74). See also Robert A. Nye, *Masculinity and Male Codes of Honor in Modern France* (Berkeley and Los Angeles: University of California Press, 1998 [Oxford: Oxford University Press, 1993]), 55.

22 Maleville, quoted by Martin, 'Fonction paternelle et Code Napoléon', 473.

23 Olivia Gazalé, *Le mythe de la virilité: Un piège pour les deux sexes* (Paris: Laffont, 2017), esp. 9–23. Among the many contradictions of the myth of virility, Gazalé (303–12) highlights the simultaneous valorization of sexual proof (through the

persistent anxiety of not conforming to the ideal virile model and insecurity towards their social role. From this perspective, masculinity appears to be constantly in crisis. Originally, scholars have developed this notion of a masculine 'crisis' to refer to a post Second World War phenomenon in Western societies. This masculine malaise would stem from social changes that promoted the legal rights of individuals generally excluded from the masculine sphere (women and homosexuals), whilst seemingly undermining male dominance.[24] Men's insecurity would also originate in a re-assessment of traditional manly values increasingly perceived as flaws (strength now equates to violence and rationality to coldness, for instance).[25] If the traditional model of masculinity is stigmatized, men cannot refer to it; however, nor can they refer to alternative models, which are generally scarce, complex and contentious.[26] Whilst being invited to display vulnerability, the new 'sensitive' man is often mocked in mass media, generates scepticism and may be accused of lacking virility.[27]

Whilst allowing scholars to describe the recent changes in the representation of masculinity, the concept of a *crisis* is nonetheless problematic. Although seemingly seeking the opposite effect, it in fact essentializes masculinity by presenting it as a stable, universal and everlasting entity,

---

number of conquests, for instance) and of chastity (as a sign of control of one's sexual urges).

24 For more on the causes of a 'crisis' in masculinity, see Brittan, *Masculinity and Power*, 180–81; Whitehead and Barrett, 'The Sociology of Masculinity', 6–9; John Beynon, *Masculinities and Culture* (Buckingham and Philadelphia: Open University Press, 2002), 76–89; Connell, *Masculinities*, 85–86; Tim Edwards, *Cultures of Masculinity* (London and New York: Routledge, 2006), 8–16.

25 John MacInnes, *The End of Masculinity: The Confusion of Sexual Genesis and Sexual Difference in Modern Society* (Buckingham and Philadelphia: Open University Press, 1998), 47. See also Badinter, *XY*, 213–14; Gazalé, *Le mythe de la virilité*, 16–17; Kimmel and Wade, 'Ask a Feminist', 237.

26 Michael S. Kimmel, 'The Contemporary "Crisis" of Masculinity in Historical Perspective', in Harry Brod, ed., *The Making of Masculinities: The New Men's Studies* (Boston: Allen & Unwin, 1987), 121–53 (121).

27 Connell, *Masculinities*, 41, 139–40 and 222; Christopher E. Forth and Bertrand Taithe, 'Introduction', in Christopher E. Forth and Bertrand Taithe, eds, *French Masculinities: History, Culture and Politics* (Houndsmills, Basingstoke and New York: Palgrave Macmillan, 2007), 1–14 (11).

which, in some historical periods, has come under attack. It is therefore important to remember that, while the masculine crisis may not be 'real', what is at stake is its social construct. Historical studies have identified different periods in which the burden of masculinity has been voiced. In his study of Restoration England (1688–1714) and the turn of the twentieth century in the United States (1880–1914), the American sociologist Michael Kimmel has convincingly argued that some historical moments are particularly favourable for giving rise to gender issues, especially when they 'transform the institutions of personal life such as marriage and the family, which are sources of gender identity'.[28] My book argues that the July Monarchy is one such period, a breeding ground for gender issues.

Research on Masculinity in the Nineteenth Century

Masculinities studies have mostly focused on the twentieth and twenty-first centuries, but they have extended their field of analysis to previous periods and adopted historical and literary perspectives, rather than being confined to the realms of sociology. The extension to other fields of human sciences has nevertheless been a slow process. In 2007, Nigel Harkness argued: 'Whereas nineteenth-century studies have demonstrated the centrality of femininity to canonical fiction [...], masculinity has received little sustained treatment in either the literary or socio-historical fields.'[29] Literary scholars and historians have since started to view the French nineteenth century as a time of redefinition of masculine models, although they have predominantly focused on its extremities, that is, revolutionary and Napoleonic times and the *fin de siècle*.[30]

28 Kimmel, 'The Contemporary "Crisis" of Masculinity', esp. 123. See also Badinter, *XY*, 24–41.

29 Nigel Harkness, *Men of their Words: The Poetics of Masculinity in George Sand's Fiction* (London: Legenda, 2007), 17.

30 For instance: Annelise Maugue, *L'identité masculine en crise au tournant du siècle: 1871–1914* (Paris: Rivages, 1987).

Some historical studies of French masculinity in the nineteenth century tend to perceive this period as an era of continuity in the model of masculinity. This century is depicted by Anne-Marie Sohn as the paroxysm of virility: an era in which a manly ideal based on strength, courage and self-control emerged.[31] More strikingly, the second volume of *Histoire de la virilité* [*History of Virility*], dedicated to the nineteenth century and edited by Alain Corbin, is entitled *Le triomphe de la virilité* [*The Triumph of Virility*], whereas it is the volume dedicated to the twentieth century that is called *La virilité en crise?* [*Virility in Crisis?*].[32] While these studies analyse the construction of an ideal model of masculinity, they nevertheless overlook the struggles that men experienced when attempting to conform to such an unachievable benchmark.

Knowing how men 'really' experienced their masculine identities in the nineteenth century is nevertheless difficult because confiding feelings in diaries, for example, as women would be expected to, is frowned upon by the masculine script.[33] However, perceptions concerning the masculinity of this period can partially be accessed through contemporary literary and non-literary discourses. Nineteenth-century scientific discourses, as well as legal discourses, such as the Civil Code, reveal a propensity to a prescriptive definition of masculinity, enjoining men to conform to a strong model of masculinity.

In *Masculinity and Male Codes of Honor in Modern France* (1993), Robert Nye provides a thorough study of masculinity in France from the end of the Ancien Régime to the aftermath of the First World War. He examines the medical domain and the revival of the duel as a symbol of of bourgeois honour.[34] André Rauch's *Crise de l'identité masculine* [*Crisis of Masculine Identity*] (2000) offers a historical study of the changes in the role of men generated by the French Revolution. The passage from a hierarchical society to a relatively egalitarian one and the advent of the

31 Sohn, *'Sois un homme!'*.

32 See also Maira and Roulin, 'Constructions littéraires de la masculinité', 27.

33 Sohn, '*Sois un homme!*', 10.

34 Nye, *Masculinity and Male Codes of Honor*.

bourgeois ideal promoting family, professional success, social ascension and respectability contributed to altering the perception of masculine identity.[35]

Literary discourses, unlike medical or legal ones, could depict men who struggled with this virile norm and encourage a more positive reflection on gender roles in society through gender-bending protagonists. Several literary studies have examined changes in masculinity in the early nineteenth century, including Margaret Waller's outstanding monograph *The Male Malady* (1993). Her book takes as its subject post-revolutionary, Romantic fictions that focus on 'feminized' male protagonists in texts written by both male and female writers. Waller analyses the representation of male characters' antisocial and submissive behaviour, while also examining how female roles were redefined in this context.[36] Pierre Laforgue's *L'éros romantique* [*The Romantic Eros*] (1998) is another study that focuses on the end of the Restoration (1814–30) and the beginning of the July Monarchy, and points to the recurrence of narratives of (physically) impossible relationships during this period.[37] In an intuitive approach to gender identities, Laforgue argues that these texts denounced the materialism of a society dominated by money and lack of social cohesion.

The popularity of masculinities studies has increased at the beginning of the twenty-first century. Several edited collections have looked at French masculinity during the last two centuries, often from a historical perspective, as well as a literary one. *French Masculinities: History, Culture and Politics* (2007), edited by Christopher Forth and Bertrand Taithe, examines the connection between masculinity and national identity. *Hommes et masculinités de 1789 à nos jours* [*Men and Masculinities from 1789 to The Present Day*] (2007), edited by Régis Revenin, looks at the history of masculinities from the perspectives of sexuality, work, war, religion and race. *Masculinités en révolution* [*Masculinities in Revolution*] (2013), edited by Daniel Maira and Jean-Marie Roulin, synthetizes and expands French research on masculinity in the nineteenth century, as well as studying how

35 André Rauch, *Crise de l'identité masculine: 1789–1914* (Paris: Hachette Littératures, 'Pluriel', 2000).

36 Waller, *The Male Malady*.

37 Pierre Laforgue, *L'éros romantique: Représentations de l'amour en 1830* (Paris: Presses Universitaires de France, 1998).

various literary and non-literary works create categories of thinking and, thus, construct masculinity.[38]

Nevertheless, the period of the July Monarchy has thus far been somewhat overlooked despite being the theatre of important fractures in the biological and social model of sexual difference. A notable exception is Nigel Harkness's *Men of Their Words* (2007), which analyses the representation of masculinity in George Sand's novels and, in particular, the role of language in the construction of masculinity.[39] Harkness rightly argues that literary works do not simply and passively represent masculinity, but have an impact on societies, by advocating, deliberately or involuntarily, alternative models.

The July Monarchy deserves more sustained attention than it has received so far, owing to its numerous literary narratives that depict characters who do not conform to gender expectations, such as the hermaphrodite, the effeminate man and the mannish woman. For instance, Camille, in Latouche's novel *Fragoletta* (which is often poorly regarded by critics), is a hermaphrodite who experiences their intersex identity as a burden.[40] Madeleine de Maupin, the heroine of Gautier's novel *Mademoiselle de Maupin*, cross-dresses to observe how men behave in the absence of women and realizes that she cannot stand the restrictive conventions of women's lives that are imposed by men. Lucien de Rubempré, the main protagonist of Balzac's novels *Illusions perdues* [*Lost Illusions*] and *Splendeurs et misères des courtisanes* [*Splendours and Miseries of Courtesans*], is a young man almost universally perceived as weak and effeminate. Félicité des Touches, in Balzac's novel *Béatrix*, is a talented female writer who is perceived as a monster by the local villagers because of her 'masculine' temperament and activities. All these sexually ambiguous characters, and many others, mirror July Monarchy society's concern regarding gender conventions, while also reinventing new models of gender relations.

38 Forth and Taithe, eds, *French Masculinities*; Revenin, ed., *Hommes et masculinités*; Maira and Roulin, 'Constructions littéraires de la masculinité', 24.

39 Harkness, *Men of their Words*.

40 The non-binary pronouns and possessive adjectives *they*, *them*, *their*, etc. will be often used to refer to hermaphrodites and intersex people.

Thus far, studies on masculinity in nineteenth-century France have also often neglected a significant corpus of texts that both promote and disrupt the ideal model of masculinity: scientific texts on sexual difference, homosexuality and intersexuality. As research on nineteenth-century French medical texts has been mostly limited to the end of the century, this study fills a scholarly void in closely analysing earlier nineteenth-century texts. Analyses that include the first part of the nineteenth century often cover it quickly before giving more sustained attention to later periods. For instance, only one out of the nine chapters of Nye's *Masculinity and Male Codes of Honor in Modern France* includes detailed analyses of scientific texts that predate the 1850s. Likewise, Laure Murat's *La loi du genre* [*The Law of Gender*] (2006) explores the umbrella term *troisième sexe* [third sex], which comprises gay men, lesbians, male prostitutes, trans people and emancipated women. It analyses a variety of French, German and British discourses, whether judicial, medical, literary or journalistic, and covers a period from the 1830s to the 1930s, but in fact mostly focuses on the late nineteenth and the early twentieth centuries.[41]

Moreover, chapters in medical dictionaries and monographs on homosexuality, hermaphrodites and women have been little studied in relation to literary texts and have accordingly rarely been linked to masculinities studies.[42] However, as Vernon Rosario points out in relation to the late nineteenth century, there was a link between patients reading medical texts to understand their own condition, novelists reading these texts to find inspiration and doctors finding both their case studies in patients' tales and their narrative strategies to write their reports in novels.[43] Connecting scientific writings and literary texts highlights the extent to which the society of the 1830s was obsessed with sexual and gender categories. The hermaphrodite

41 Nye, *Masculinity and Male Codes of Honor*; Laure Murat, *La loi du genre: Une histoire culturelle du 'troisième sexe'* (Paris: Fayard, 2006).

42 There are a few exceptions: Vernon A. Rosario, *The Erotic Imagination: French Histories of Perversity* (Oxford and New York: Oxford University Press, 1997); Murat, *La loi du genre* and Anne E. Linton, *Prescribed Fictions: Literary and Medical Representations of Hermaphrodism in Nineteenth-Century France* (doctoral thesis, Yale University, 2011).

43 Rosario, *The Erotic Imagination*, 10–11 and 110–11.

and the androgyne were the objects of particular attention in fields as diverse as medicine, literature and sociology.[44] As will be shown, scientific texts and literary texts, however, took a different stance: the former considered the figure of the hermaphrodite to be a monster, whereas the latter deemed it to be an incarnation of ideal beauty. Critics have commented on the popularity of this figure in the 1830s.[45] Kari Weil, for instance, shows how the androgyne constitutes a figure of progress while embodying conservative stability in a context of historical malaise; Frédéric Monneyron attributes this success to the 'elasticity' of the novel, which adapts to the polymorphous constitution of the androgyne; Anne Linton states that the popularity of the literary hermaphrodite is linked to the growing importance of the hermaphrodite in nineteenth-century medicine.[46] As thought-provoking as these studies are, they struggle to explain convincingly the social implications of this phenomenon. Examining thoroughly the nineteenth-century scientific discourse on intersex people allows me to insert it in a wider reflection on gender categories and their subversion, notably in literary texts, and to study the connections between the latter and scientific and medical texts.

44 On the androgyne in the texts of socialist thinkers: Naomi J. Andrews, 'Utopian Androgyny: Romantic Socialists Confronts Individualism in July Monarchy France', *French Historical Studies*, 26/3 (Summer 2003), 437–57.

45 A. J. L. Busst, *The Image of the Androgyne in the Nineteenth Century* (London: Routledge & Kegan Paul, 1967); Valeria Ramacciotti, 'L'esilio dell'androgino', *Studi di letteratura francese*, 16 (1990), 188–201; François Kerlouégan, 'Une relecture de l'androgynie romantique: La quête du genre dans *Fragoletta*, *Sarrasine* et *Mademoiselle de Maupin*', *Insignis*, 1 (May 2010), 36–53.

46 Kari Weil, *Androgyny and the Denial of Difference* (Charlottesville and London: University Press of Virginia, 1992), 68–72; Frédéric Monneyron, *L'androgyne romantique: Du mythe au mythe littéraire* (Grenoble: Ellug, 1994), 8; Linton, *Prescribed Fictions*, 127.

Masculinity during the July Monarchy

Why could gender and sexual categories be perceived as being weakened during the July Monarchy? In what way did the historical and social context of the July Monarchy dampen young men's virility? Certainly, signs of a masculine malaise date back to at least the beginning of the nineteenth century. Some male characters in French novels written during the Napoleonic era and the Restoration, such as Chateaubriand's René (1802), Madame de Duras's Olivier (1822) and Stendhal's Octave in *Armance* (1827), express social and gender insecurity.[47] However, the political, economic and social upheavals that occurred during the July Monarchy increased the malaise regarding gender identities. The new monarchy was founded on the July Revolution, or *les Trois Glorieuses* [the Three Glorious (Days)], in reference to the three days during which the uprising occurred (27–29 July 1830). The July Revolution put an end to the power of the ageing dynasty of the Bourbons, who (except for the revolutionary period and the Empire) had been ruling France since 1589. The French ousted Charles X, the youngest brother of Louis XVI and Louis XVIII, after the publications of the Ordinances of Saint-Cloud. These decrees suspended the liberty of the press, dissolved the newly elected Chamber of Deputies and restricted the right to vote. In overthrowing the Bourbons and instituting a constitutional regime influenced by liberalism, the July Monarchy opted for rupture. The new king, Louis-Philippe I of Orléans, from the cadet branch of the Bourbons, was King of the French, rather than King of France, and developed the image of a *roi citoyen* [citizen king] promoting bourgeois values such as meritocracy and family interests. Nevertheless, the new regime also chose a certain element of continuity, as it conserved a monarchical system and installed a cousin of the Bourbons on the throne.[48]

47 On the Restoration's masculine malaise, see 'The Weak Man and the Aspiring Artist' in Chapter Three.

48 For more on the July Revolution, the establishment of the July Monarchy and Louis-Philippe I, see H. A. C. Collingham, *The July Monarchy: A Political History of France, 1830–1848* (London and New York: Longman, 1988), 6–33; Jo Burr Margadant, 'Gender, Vice, and the Political Imaginary in Postrevolutionary

The new bourgeois and liberal monarchy encountered a period of political, economic and social change. Economic growth accelerated during the July Monarchy, especially in the industrial sector (heavy metals, coal, textiles and transportation infrastructures), which set the scene for the full-scale industrial revolution that would take place during the Second Empire. Capital rose thanks to the development of financial institutions. Women's rights enjoyed a brief period of support, notably from Saint-Simonians and Fourierists. The bourgeoisie, a social group including individuals as diverse as bankers, lawyers, doctors, civil servants and shopkeepers, benefitted from these political and economic changes.[49] As Robert Gildea summarizes,

> French society [...] was clearly in motion, as individuals took advantage of the opportunities offered by economic change and the expansion of the state, to which was geared a developing system of education. Legally, too, careers were open to the talents and could not be confined to any particular caste or corporation.[50]

In brief, the political, economic and social contexts of the July Monarchy seemed favourable for the development of personal ambitions and achievements. The rise of market capitalism, Sharif Gemie argues, put money at the core of people's preoccupations. For this reason, Balzac (like many of his contemporaries) was concerned with its corrupting power and the resulting so-called demoralization of society and feared that the *notables* [elites] would be unable to contain this moral decline.[51]

France: Reinterpreting the Failure of the July Monarchy, 1830–1848', *The American Historical Review*, 104/5 (December 1999), 1461–96; Gabriel de Broglie, *La Monarchie de Juillet: 1830–1848* (Paris: Fayard, 2011), 15–64.

49 For more on the political, economic and social changes during the July Monarchy, see Roger Magraw, *France 1815–1914: The Bourgeois Century* (London: Fontana, 1983), 51–88; Collingham, *The July Monarchy*, 328–64; Robert Gildea, *Children of the Revolution: The French, 1799–1914* (London: Penguin Books, 2009 [Allen Lane, 2008]), 47–53 and 104–17; Broglie, *La Monarchie de Juillet*, 153–66, 181–92, 198–210 and 311–21.

50 Gildea, *Children of the Revolution*, 116.

51 Sharif Gemie, 'Balzac and the Moral Crisis of the July Monarchy', *European History Quarterly*, 19 (1989), 469–94. See also Pamela M. Pilbeam, *The 1830 Revolution in France* (London: Macmillan, 1991), 3; José-Luis Diaz, 'Balzac romancier de la

This mixture of booming economy and social change, as well as social inertia and alleged moral corruption led to the French citizens' perception that the new regime was unable to fulfil its promises. Although the bourgeoisie was on the rise, the social structure of the July Monarchy maintained former privileges, especially aristocratic ones, notably for those involved in administrative, diplomatic and military roles.[52] Accordingly, the paradox of continuity and rupture on which the July Monarchy seemed to be based is noticeable in the elites' treatment of younger generations coming from the upper bourgeoisie and the weakened aristocracy. On the one hand, existing elites sought progress and wanted to break with the society of the Ancien Régime. University studies and prestigious careers in the public service, for instance, became accessible to a larger number of people. On the other hand, those elites were incapable of taking steps in order to ensure the rupture with the past. As was the case during the Restoration period, the governing bodies rejected the involvement of intellectual middle-class youth in the ruling of the affairs of the state in order to consolidate their own power.[53] The educational system was deficient, since it did not prepare young men for the realities of the world and produced more members of the elite than were required. Numerous key posts within the July Monarchy administration and government were unavailable because they were occupied by former Bonapartists. Moreover, although the legal ages for voting and being elected were lowered, respectively, from 30 to 25 and from 40 to 30 during the July Monarchy, young men were still deprived of political influence. Only those who paid at least 200 francs of taxes could qualify as voters, that is, about five Frenchmen out of one thousand.[54] In

"grande maladie sociale" (1838–1847)', *L'Année balzacienne*, 3rd ser., 17 (2016), 221–40.

52 Pierre Barbéris, *Le monde de Balzac* (Paris: Arthaud, 1973), 488; Pilbeam, *The 1830 Revolution in France*, 3; Margaret Cohen, *The Sentimental Education of the Novel* (Princeton: Princeton University Press, 1999), 103–04; Judith Lyon-Caen, *La lecture et la vie: Les usages du roman au temps de Balzac*, pref. Alain Corbin (Paris: Tallandier, 2006), 19; Broglie, *La Monarchie de Juillet*, 184.

53 Peter Brooks, 'Balzac: Epistemophilia and the Collapse of the Restoration', *Yale French Studies*, 101 (2001), 119–31 (127–28).

54 Pilbeam, *The 1830 Revolution in France*, 88.

these conditions, young people were condemned to precarious living conditions, unsatisfactory careers and little power.[55]

Vautrin's depiction of the limited professional and financial opportunities of his reluctant protégé Eugène de Rastignac in Balzac's novel *Le Père Goriot* (1835) (unless he marries the wealthy but plain heiress Victorine Taillefer) reflects the failure of a system that aroused young men's ambitions but that was unable to fulfil them:

> Ayez des protections, vous serez procureur du roi à trente ans, avec mille écus d'appointements, et vous épouserez la fille du maire. Si vous faites quelques-unes de ces petites bassesses politiques [...], vous serez, à quarante ans, procureur général, et vous pourrez devenir député. [...] J'ai l'honneur de vous faire observer [...] qu'il n'y a que vingt procureurs généraux en France, et que vous êtes vingt mille aspirants au grade, parmi lesquels il se rencontre des farceurs qui vendraient leur famille pour monter d'un cran. Si le métier vous dégoûte, voyons autre chose. Le baron de Rastignac veut-il être avocat? [...] Si ce métier vous menait à bien, je ne dirais pas non; mais trouvez-moi dans Paris cinq avocats qui, à cinquante ans, gagnent plus de cinquante mille francs par an?[56]
>
> [If you have protections, you will be a Crown Prosecutor at thirty, with a salary of one thousand écus and you will marry the mayor's daughter. If you commit a few political base acts [...], you will be a General Prosecutor at forty and you will be able to become a Deputy. [...] I have the honour to point out to you [...] that there are only twenty General Prosecutors in France and that you are twenty thousand aspirants to this rank, among whom one can encounter some pranksters who would sell their families to climb one rung. If you are disgusted by this job, let's look at something else. Does Baron de Rastignac want to be a lawyer? [...] If this profession led

55 On young men's position and education in the July Monarchy: Richard Terdiman, 'Structures of Initiation: On Semiotics Education and Its Contradictions in Balzac', *Yale French Studies*, 63 (1982), 198–226 (204); Magraw, *France 1815–1914*, 83; Rauch, *Crise de l'identité masculine*, 109; Lyon-Caen, *La lecture et la vie*, 218–20; Broglie, *La Monarchie de Juillet*, 165–66.

56 Honoré de Balzac, *Le Père Goriot*, Stéphane Vachon, ed. (Paris: Librairie Générale Française, 'Le Livre de Poche Classique', 1995), 164–65. Gildea (*Children of the Revolution*, 105–06) also refers to this extract. Similarly, Magraw reports the words of an Orleanist bureaucrat who confirmed Balzac's observation by claiming that '[i]n Paris of very 1200 [students] reading for the law, 20 achieve fame, 80 fortune, 200 a modest living—the rest vegetate' (*France 1815–1914*, 83).

you to something good, I wouldn't say no, but find me five lawyers in Paris who, at fifty, earn more than fifty thousand francs a year?]

The estimation that Balzac puts in Vautrin's mouth is so accurate that the French economist Thomas Piketty dedicates several pages of his essay, *Le capital au XXIe siècle* [*Capital in the Twenty-First Century*], to analysing his speech in order to explain the difference between income from capital (such as inheritance, properties and lands) and income from work: 'dans la France du début du XIXe siècle [...], le travail et les études ne permettent pas d'atteindre la même aisance que l'héritage et les revenus du patrimoine' [in early nineteenth-century France [...] work and education do not lead to the same wealth as inheritance and property income].[57] As Piketty explains, the income of the 1 per cent of the population that received the highest inheritance was between twenty-five and thirty times higher than the average annual income (about 500 francs) during the period in which Balzac lived and wrote. In contrast, the salary of the 1 per cent of the best-paid employees (such as judges and prosecutors) was only ten times higher than the average income.[58] In other words, the society of the July Monarchy was aristocratic, not meritocratic, because it perpetuated the inequalities of the Ancien Régime. A talented, hard-working and experienced judge would never be as wealthy as the heir of a rich person, often an aristocrat.

Consequently, young men felt disillusioned about their future, suffered from an inability to satisfy what Laforgue calls their 'vouloir-vivre' [will to live] and doubted their role in the new society established by the July Monarchy.[59] Their insecurity undermined the traditional roles that they were, nonetheless, expected to fulfil as defenders of the nation, members of the body politic and fathers,[60] as well as the new roles of self-controlled and hard-working bourgeois. Since being a man implied fulfilling all these functions in the public sphere, expressing weakness undermined masculine

57 Thomas Piketty, *Le capital au XXIe siècle* (Paris: Seuil, 2013), 381.
58 Ibid., 647.
59 Laforgue, *L'éros romantique*, 130. See also Magraw, *France 1815–1914*, 83.
60 See Stacie Allan, *Writing the Self, Writing the Nation: Romantic Selfhood in the Works of Germaine de Staël and Claire de Duras* (Oxford: Peter Lang, 2019), 90.

virtues. Like their fictional counterparts that will be examined in this book, some men chose to renounce political and economic influence altogether, as well as to retire from the bourgeois lifestyle. This rejection sometimes took the form of bohemianism and artistic aspirations.[61] A text rooted in this *Zeitgeist* is Théophile Gautier's Préface to his novel *Mademoiselle de Maupin* (1835), published as an independent text in 1834. Gautier defended the disinterested nature of beauty as opposed to economic preoccupations, as well as the *jouissance* [(sexual) pleasure] that aesthetic experience conveys; by so doing, he 'articulated an aesthetic posture for the generation of artists and poets who defined modernity', as Marguerite Murphy claims.[62] But we shall look more closely at Gautier's novel in Chapter 2.

A Model of Masculinity: The Napoleonic Soldier

If the July Monarchy period was characterized by its malaise towards, and its bending of, gender conventions, which model of masculinity was challenged? Periods of war often provide countless virile models and promote manly values. It is not surprising that the so-called twentieth-century 'crisis' of masculinity followed the Second World War. The Nazi and Fascist ideologies, for instance, promoted a particularly virile model of manhood.[63] Likewise, men's insecurity during the Restoration and the July Monarchy succeeded the revolutionary and Napoleonic Wars. The problem that men might have encountered in the first half of the nineteenth century was the following: how could they prove their masculinity when allegedly virile perfection had already been reached? Indeed,

61 Magraw, *France 1815–1914*, 83–84.

62 Margueritte Murphy, 'Pure Art, Pure Desire: Changing Definitions of "l'art pour l'art" from Kant to Gautier', *Studies in Romanticism*, 47/2 (Summer 2008), 147–60 (147).

63 See George L. Mosse, Chapter Eight: 'The New Fascist Man', in *The Image of Man: The Creation of Modern Masculinity* (New York and Oxford: Oxford University Press, 1996), 155–80.

the model of a manly, brave and honourable masculinity embodied by Napoleon and his soldiers of the Great Army seemed impossible to surpass, or even equal.

Neither the end of Napoleon's Empire, nor the Emperor's deportation to Saint Helena in October 1815 along with the dismissal of his soldiers, who had to reintegrate into civilian life (with varying levels of success), brought an end to the Napoleonic legend. Jean Tulard and, more recently, Sudhir Hazareesingh have studied the legend surrounding Napoleon during the nineteenth century and beyond.[64] The fantasy of the fallen Emperor as a lawgiver, a defender of freedom and the values of the Revolution, and as a Romantic hero, still prevailed. Different generations and social groups have worshipped a legendary Napoleon for decades and in many guises (sometimes as unexpected as tobacco boxes, saucers and vases with his effigy). The Emperor was not revered solely for himself, but also because he was seen as a role model promoting the grandeur of France, along with the values of the Revolution and of virility.[65]

Napoleonic soldiers, or *grognards*, were also signifiers of the Napoleonic myth of French magnificence. Although mistreated and even persecuted by Restoration authorities, they still benefitted from extensive popularity. This popularity became more visible after 1830, when Louis-Philippe's regime attempted to make amends and to define an institutionalized framework for Napoleon's cult (with, for example, the repatriation of his remains in 1840).[66] Napoleonic soldiers played an 'active role in the propagation of the Napoleonic "legend" ', which 'earned them a privileged position in artistic

64 Jean Tulard, *Le mythe de Napoléon* (Paris: Armand Colin, 1971); Sudhir Hazareesingh, *The Legend of Napoleon* (London: Granta Books, 2004).

65 Barbéris, *Balzac et le mal du siècle*, II, 1641; Rauch, *Crise de l'identité masculine*, 81.

66 Jean Lucas-Dubreton, *Le culte de Napoléon: 1815–1848* (Paris: Albin Michel, 1960), 27–48 and 298–328; Barbéris, *Balzac et le mal du siècle*, II, 1641; Tulard, *Le mythe de Napoléon*, 63; Tiziana Goruppi, 'Gli eredi de "l'Enfant du siècle": Il mito di Napoleone nella *Confession* di Musset', *Studi Francesi*, 48/1 (January–April 2004), 101–07 (106); Hazareesingh, *The Legend of Napoleon*, 151, 154–55 and 236; Michael J. Hughes, 'Making Frenchmen into Warriors: Martial Masculinity in Napoleonic France', in *French Masculinities*, 51–66 (63); Brian Joseph Martin, *Napoleonic Friendship: Military Fraternity, Intimacy, and Sexuality in Nineteenth-Century France* (Durham, NH: University of New Hampshire Press, 2011), 152–57.

and literary representations of the Napoleonic cult'.[67] Not only did they contribute to this cult, but they were themselves objects of popular admiration, mainly outside the official networks of the regime, notably through the oral transmission of their successes.[68]

The generation of the soldiers of the Great Army was a military one and, *de facto*, a virile one. The Napoleonic soldier was seen as the incarnation of an idealized masculinity defending values inherited from the Revolution and promoting martial virtues, such as honour, glory and an eagerness to protect the fatherland.[69] National cohesion henceforth emanated from the fatherland rather than a kingly father.[70] Citizenship and military life were interconnected: military service was an essential means of becoming a citizen during the Empire, whilst the perfect citizen was depicted in speeches and artistic works (paintings and theatre plays) as endowed with the qualities of a warrior: strength, virility, patriotism and courage.[71] In other words, the warrior was established as a model of masculinity during revolutionary and imperial times.[72] This model was the yardstick by which men's masculine identity was measured and perpetuated during the July Monarchy despite a different political and sociocultural context. Individuals were accordingly compelled to define new gender models. The aim of this book is precisely to examine how fiction explored alternative models of masculinity.

67 Hazareesingh, *The Legend of Napoleon*, 237.

68 Ibid., 240

69 Hughes, 'Making Frenchmen into Warriors', 51–52. Honour became a true institution with Bonaparte's creation of the Légion d'honneur in 1802: Jean-Paul Bertaud, 'La virilité militaire', in *Le triomphe de la virilité*, 157–202 (171).

70 Allan, *Writing the Self*, 91.

71 Anna Clark, 'The Rhetoric of Masculine Citizenship: Concepts and Representations in Modern Western Political Culture', in *Representing Masculinity*, 3–22 (11); Alan Forrest, 'Citizenship and Masculinity: The Revolutionary Citizen-Soldier and His Legacy', in *Representing Masculinity*, 111–29 (112–17); Hughes, 'Making Frenchmen into Warriors', 63; Bertaud, 'La virilité militaire', 167–69 and 176–78.

72 Mosse, *The Image of Man*, 51; Rauch, *Crise de l'identité masculine*, 14 and 47; Deborah Gutermann, 'Le désir et l'entrave: L'impuissance dans la construction de l'identité masculine romantique: première moitié du XIXe siècle', in *Hommes et masculinités*, 55–74 (55).

Organization of This Book

The starting point of this book is the analysis of the definition of sexual and gender categories in early nineteenth-century scientific and medical texts. It then investigates how literary narratives from the 1830s and 1840s, written by Honoré de Balzac (1799–1850), Henri de Latouche (1785–1851), George Sand (1804–76), Alfred de Musset (1810–57) and Théophile Gautier (1811–72), interpreted the weakening of strong gender categories and the development of alternative masculinities through three literary figures: the hermaphrodite, the effeminate man and the masculine woman. Finally, this study analyses the ambiguity of their treatment of a masculine malaise. On the one hand, the writers often acknowledge the necessity of sexual difference between men and women through the negative depiction of weak young men. On the other hand, however, they challenge the organization of July Monarchy bourgeois society and the restrictive effect that traditional gender categories had on those men.

Chapter 1, 'Sexual Dichotomy and Its Contradictions', provides contextual and cultural background by studying French scientific writings of the first half of the nineteenth century. The examination of these texts determines how French society was defined by the separation of the sexes during this period in which the concept of gender was not yet acknowledged, let alone theorized. According to this principle, men and women exercised different social duties that were 'naturally' determined by their physical and psychological differences. In the first instance, this chapter analyses scientists' often unconscious slippage from a bodily definition of sex to what is deemed nowadays as a gendered definition. Secondly, it examines what was seen at the time as the problems raised by individuals who did not conform to physical and behavioural norms and therefore challenged the legitimacy of the sexual dichotomy: masculine women and lesbians, intersex people and male homosexuals. Finally, it analyses attempts within scientific discourse either to assimilate intersex individuals and homosexuals to one sexual category, or to marginalize them.

The framework in which men and women's characteristics and social roles were considered within nineteenth-century scientific thought having been determined, Chapter 2, 'The Hermaphrodite', focuses on the depiction of hermaphroditic characters in literary narratives. It reveals how this representation in Latouche's *Fragoletta* (1829), Balzac's *Sarrasine* (1831) and Gautier's *Mademoiselle de Maupin* (1835) allows writers to challenge the norm of the separation of the sexes, whilst paradoxically showing their reluctance to transgress this norm. Although the hermaphrodite is portrayed as a physical and ontological monster, this figure is also perceived as the embodiment of a neoclassical aesthetic model that promoted the beauty of effeminate men. The transformation of hermaphroditic figures into real or fantasized works of art enables writers to discuss homosexuality allusively, to glorify the concomitance of masculine and feminine qualities in an individual, and, thus, to deny sexual difference.

Chapter 3, 'The Effete', looks at literary narratives that apparently criticize gender ambiguities and deplore the loss of stable masculine models in order to reassess masculine norms and expectations in July Monarchy society. Balzac's Lucien de Rubempré in *Illusions perdues* (1837–43) and *Splendeurs et misères des courtisanes* (1838–47), and Musset's Octave in *La Confession d'un enfant du siècle* (1836) contrast with the unattainable virile model embodied by the Napoleonic soldier. They are characterized by so-called feminine and puerile flaws, such as weakness, carelessness and passivity. This chapter also shows how implicit homosexuality is employed as a means of discrediting young men in their leadership capacities, notably through similar rhetorical strategies as those used in scientific writings. Finally, it argues that a homosexual, like Balzac's criminal protagonist Vautrin, can nevertheless embody a positive alternative to masculine insecurity through his combination of masculine and feminine qualities.

Finally, claiming that masculinity is not limited to male individuals, Chapter 4, 'The Butch', examines the construction of female masculinity in Sand's first edition of *Lélia* (1833), as well as Balzac's novels *Béatrix* (1839–45) and *La Cousine Bette* (1846). It reassesses the fictional depiction of masculine women. They display physical and behavioural characteristics, while occupying social functions that were believed to be typical of men: artists, breadwinners and leaders. This chapter highlights the ambiguous position of writers about these women. On the one hand, the writers present them as superior beings, due to a mixture of masculine and maternal qualities. On the other hand, they appear as menaces to the social order, owing to non-conventional rejection of heteronormative sexuality and marriage. Accordingly, this chapter reveals that the power conveyed by women's masculinity is contained through their disappearance, seclusion and/or death, whilst exemplifying and advocating women's possible agency and equality.

In brief, I argue that French narratives of the July Monarchy period renegotiated sexual and gender categories. The representation of 'deviant' genders enables narratives to construct a new model of gender relations that challenges hegemonic masculinity. The images of hermaphrodites, effeminate men and mannish women are rarely exhibited as frightening counter-models in these texts. Rather, they are employed to suggest the extent to which masculinity and society in general may be transformed through alternative models and through underestimated agents of masculinity, promoting values traditionally perceived and despised as feminine. In their exploration of ambiguous genders, the July Monarchy literary narratives paved the way for the construction of a sexual and gender model that distanced itself from a strict separation of the sexes.

CHAPTER 1

Sexual Dichotomy and Its Contradictions[1]

This chapter analyses scientific discourse on gender and sexual ambiguities in the first half of the nineteenth century (from 1802 to 1857). It looks at articles in medical dictionaries, scientific monographs, forensic essays and works that combine scientific, philosophical and moral considerations from that period. The results of this study of more than fifty years of scientific texts sustain the argument that the predominant scientific thought firmly put forward the separation of the sexes and identified the moral, social and ideological implications of such a separation in the foundation of sexual norms, particularly for men. For early nineteenth-century French society – and Western societies more generally – sex was a concept that was, or should be, devoid of ambiguity. The sixth edition of the *Dictionnaire de l'Académie française* (1835), for example, defined *sexe* as 'différence physique et constitutive du mâle et de la femelle' [constitutive physical difference between the male and the female].[2] The author of the *sexe* article in the *Dictionnaire des sciences médicales* [*Dictionary of Medical Sciences*] (1821) published by *une société de médecins et de chirurgiens* [a society of physicians and surgeons], Julien-Joseph Virey – an author whom we will encounter frequently in this chapter – focused on the etymology of the word: '*sexus*,

1 Earlier versions of some parts of the material found in this chapter have been published in two separate places: Géraldine Crahay, 'Valorising Genitalia: Homosexuality and Hermaphroditism in French Scientific Writings in the Nineteenth Century', *Harts & Minds*, 1/2 (October 2013) and '"Voyez les femmes les plus hommasses, ces *viragines* audacieuses": La domestication de la femme masculine dans les traités savants de la première moitié du dix-neuvième siècle', in *Medicine and Maladies: Representing Affliction in Nineteenth-Century France* (Leiden and Boston: Brill-Rodopi, 2018), 19–40.

2 *Dictionnaire de l'Académie française*, II, 740.

de *secdre*, diviser, parce qu'il distingue ou sépare le mâle de la femelle' [*sexus*, from *secdre*, to divide, because it distinguishes between or separates the male from the female].[3] This etymological reference emphasizes the clear separation between two groups of humans (and other living things). The social concept of the separation of the sexes, which I also call *sexual dichotomy*, was often claimed to be the result of 'natural', biological differences between men and women.

For French doctors and scientists in Western societies, there was often little doubt that only two sexes existed. Other cultures, in contrast, recognized the existence of a third sex (which could include homosexuals, intersex individuals, trans people and cross-dressers). They designated this third sex with a specific name and sometimes – but not always – gave it a dignified status. For instance, the *Mahus* in Polynesia and the *Hijras* in India, who were either born intersex or chose a genderfluid lifestyle, were assigned a specific, non-binary identity.[4] These examples suggest that the separation between a male sex and a female sex is not as simple as it was believed to be. The biologist Anne Fausto-Sterling claims that it does not correspond to the complexity of the human body for which '[t]here is no either/or'; rather, she argues that instead of two polar opposites 'there are shades of difference'.[5] The corpus of this chapter shows that, for at least two centuries, physical, but also psychological and social oppositions between a male sex and a female sex were largely accepted. However, the existence of individuals who did not fit within this sexual dichotomy revealed the limits of the traditional concepts of sex.

3 Julien-Joseph Virey, 'Sexe', in *Dictionnaire des sciences médicales*, 60 vols (Paris: Panckoucke, 1812–22), LI (1821), 218–44 (218). The spelling of our corpus has been modernized and corrected when necessary.

4 Rommel Mendès-Leite, 'Inconstance des sexes et des genres dans les sociétés non occidentales', in Jean-Yves Tamet, ed., *Différenciation sexuelle et identités: Clinique, art et littérature*, pref. Pierre Chatelain (Paris: In Press, 2012), 111–32 (114–17). For more on third sexes in non-Western cultures, see also David Gilmore, *Manhood in the Making*, 222; Anne Fausto-Sterling, *Sexing the Body: Gender Politics and the Construction of Sexuality* (New York: Basic Books, 2000), 108–09.

5 Fausto-Sterling, *Sexing the Body*, 3.

This chapter demonstrates that the ideal Frenchman was deemed to be a figure of opposition to different sexually 'unconventional' individuals. Firstly, I will look at how women were considered to be biologically complementary to men and how masculine women infringed this expected complementarity. Secondly, we will see how men were contrasted to intersex individuals, or hermaphrodites, as they were called at the time, who were seen as 'monstrous' beings, combining physical characteristics from the male and the female sexes. Finally, I will show how men were contrasted to homosexuals.[6] The latter were regarded as 'abnormal' because of their sexual attraction to the same sex and, as such, could be seen as 'psychosexual hermaphrodites', to borrow the phrasing of German and English sexologists (Albert Moll, Johann Ludwig Casper, Richard von Krafft-Ebing and Henry Havelock Ellis).[7] This chapter highlights scientific attempts to deny the existence of individuals with sexual and gender ambiguities (while paradoxically studying their cases), by assimilating them to the norm that the science of the time had defined. It argues that the study of these individuals itself challenged the essentialist and normative separation of the sexes that scientists intended to uphold. Indeed, by excluding these persons from the sexual dichotomy, they revealed its instability.

This chapter is inscribed within the field of medical humanities. Nineteenth-century medical discourse has attracted sustained attention since the 1970s. The first volume of Michel Foucault's *Histoire de la sexualité* [*History of Sexuality*] is a pioneering example of this interest. While some of its claims have been justifiably challenged by later researchers, its key features remain ground-breaking. In his book, Foucault opposes the widely

6 And to trans people. However, transidentity was hardly acknowledged at that time.

7 Pierre Hahn, *Nos ancêtres les pervers: La vie des homosexuels sous le second empire*, pref. Michael Sibalis (Béziers: H&O, 2006 [Paris: Orban, 1979]), 91; Alice Domurat Dreger, 'Hermaphrodites in Love: The Truth of the Gonads', in Vernon A. Rosario, ed., *Science and Homosexualities* (New York: Routledge, 1997), 46–66 (59). See also 'hermaphrodites de l'esprit': Régis Revenin, 'Conceptions et théories savantes de l'homosexualité masculine en France, de la monarchie de Juillet à la Première Guerre mondiale', *Revue d'Histoire des Sciences Humaines*, 17/2 (2007), 23–45 (28).

shared precept according to which the nineteenth century was a sexually repressed era. On the contrary, the French historian argues that, since the seventeenth century, and especially during the nineteenth century, discourses about sex had proliferated in European societies. They often emanated from the institutional power itself, in the fields of science, medicine and, later, psychiatry. However, whilst sex was a common topic in scientific and medical discourses, he claims, it took on the form of a secret: 'Ce qui est propre aux sociétés modernes, ce n'est pas qu'elles aient voué le sexe à rester dans l'ombre, c'est qu'elles se soient vouées à en parler toujours, en le faisant valoir comme *le* secret' [What is characteristic of modern societies is not that they relegated sex to the shadows, but that they devoted themselves to always speaking of it, while asserting it as *the* secret].[8] Likewise, the physicians and scientists in this corpus certainly considered it their duty and prerogative to interpret the sexual identity of a person.

This chapter brings together three areas of medical studies that are often treated separately: women, intersexuality and homosexuality. Whereas the French medical gaze on homosexuals in the – late – nineteenth century has engendered increasing scholarly interest,[9] intersex people have been relatively overlooked. Alice Dreger's pioneering historical monograph on hermaphroditism, *Hermaphrodites and the Medical Invention of Sex* (1998), focuses on the second half of the nineteenth century and the beginning of the twentieth century.[10] It analyses how the French and British medical communities perceived and treated intersexuality and argues that doctors considered the gonads (testes and ovaries) to be the only reliable markers of the 'real' sex of an individual. This radical position, as will be shown,

8 Michel Foucault, *Histoire de la sexualité*, I: *La volonté de savoir* (Paris: Gallimard, 1976), 49 (emphasis in original).

9 Vernon A. Rosario, 'Pointy Penises, Fashion Crimes, and Hysterical Mollies: The Pederasts' Inversions', in *Homosexuality in Modern France*, 146–76; Rosario, *The Erotic Imagination*; Graham Robb, *Strangers: Homosexual Love in the Nineteenth Century* (London: Picador, 2003); William A. Peniston, *Pederasts and Others: Urban Culture and Sexual Identity in Nineteenth-Century Paris* (New York: Harrington Park Press, 2004); Revenin, 'Conceptions'.

10 Alice Domurat Dreger, *Hermaphrodites and the Medical Invention of Sex* (Cambridge, MA, and London: Harvard University Press, 1998).

was far from shared by their predecessors, who employed contrived reasoning to acknowledge and deny at once the existence of hermaphrodites. Another impressive research on intersexuality in the nineteenth century and the beginning of the twentieth century in France and Germany is Geertje Mak's *Doubting Sex* (2012). The gender historian analyses the evolution of hermaphrodites' and physicians' experiences of doubting and assigning one's sex throughout the century. She highlights how, in the early nineteenth century, the sex of intersex people was strongly connected to their place and roles within a community. From the 1860s on, she argues, sex progressively became an integral part of the medical examination of the body, notably due to progress in surgery; at the turn of the century, a person's sex was increasingly connected to their own sense of self.[11]

In order to investigate how doctors perceived the sexual identity of their patients, we must use the key concept of *gender* 'to refer to the social (as opposed to the so-called biological or "sexed") aspects of femaleness or maleness'.[12] In the nineteenth century, this concept did not exist. The French word *genre* was restricted to the fields of natural history, literature, fine arts, music, fashion and grammar.[13] The sexologist John Money may have coined the current acceptation of the word *gender* in 1955.[14] Although his methods were controversial to say the least, he was a pioneer of the idea of gender as an entity which was devoid of biological influence, but which was strongly affected by education and environmental factors.[15] This

11 Geertje Mak, *Doubting Sex: Inscriptions, Bodies and Selves in Nineteenth-Century Hermaphrodite Case Histories* (Manchester and New York: Manchester University Press, 2012).

12 Nikki Sullivan, 'The Matter of Gender', in Lisa Downing, Iain Morland and Nikki Sullivan, eds, *Fuckology: Critical Essays on John Money's Diagnostic Concepts* (Chicago and London: University of Chicago Press, 2015), 19–40 (21).

13 'Genre', in *Dictionnaire de l'Académie française*, I, 832.

14 Sullivan, 'The Matter of Gender', 21.

15 For instance, the highly mediatized John/Joan case. John Money encouraged the parents of a little boy whose penis had been accidentally severed during an operation to raise him as a girl. Money monitored the 'girl''s adaptation over years and claimed the process to be successful. However, it turned out that 'Joan' was very unhappy as a girl and that Money had changed the results of his sociological experiment. For more on John Money and the John/Joan case, see Fausto-Sterling, *Sexing the Body*, 66–71; Judith Butler, 'Doing Justice to Someone: Sex Reassignment

conception distinguishes between the biological reality of sex and the cultural construct of gender. The American philosopher Judith Butler states:

> Originally intended to dispute the biology-is-destiny formulation, the distinction between sex and gender serves the argument that whatever biological intractability sex appears to have, gender is culturally constructed: hence, gender is neither the causal result of sex nor as seemingly fixed as sex.[16]

Butler claims, nonetheless, that the meaning of *construction* is problematic. In *Gender Trouble* and *Bodies That Matter*, she argues that the idea of a natural sex is itself constructed and she eradicates the sex/gender distinction. Although popular, Butler's theories have attracted criticism. Toril Moi, for instance, has written an insightful critique of Butler's rejection of biological facts when discussing the distinction between sex and gender. She points out that poststructuralists (including Butler) consider that anything that is not discursively constructed is natural and stable, thus ignoring that natural processes can also be variable.[17] Likewise, Mak points out that a body cannot be reduced to a discourse ('If a body cannot ejaculate sperm, it cannot ejaculate sperm') but that a discourse can nevertheless give it meaning.[18] It remains true that, thanks to Butler's work and to gender studies more broadly, the sexual dichotomy and the correlations between male and masculine, and female and feminine have been entirely reassessed in contemporary academic discourses.

Nineteenth-century doctors and scientists did not take the distinction between sex and gender into consideration, given that they had little understanding of the latter concept. While they were aware of the existence of

---

and Allegories of Transsexuality', in Susan Stryker and Stephen Whittle, eds, *The Transgender Studies Reader* (New York and London: Routledge, 2006), 183–93; Georgiann Davis, *Contesting Intersex: The Dubious Diagnosis* (New York and London: New York University Press, 2015), 58–64; Downing, Morland and Sullivan, eds, *Fuckology*.

16 Judith Butler, *Gender Trouble: Feminism and the Subversion of Identity* (New York: Routledge, 1990), 6.

17 Toril Moi, *What Is a Woman? And Other Essays* (Oxford: Oxford University Press, 1999), 45–54 (esp. 51).

18 Geertje Mak, *Doubting Sex*, 8.

individuals with ambiguous genitalia, they could not, or would not, draw conclusions about the fluidity of sexual identities. The concept of gender would even have seemed superfluous to many scientists of that time because any gendered trait was believed to be the consequence of a person's biological sex. In other words, gender was regarded as being as 'natural' as genitalia. However, I argue in this chapter that, in practice, scientists often thwarted this rigid conceptual framework and, unconsciously, hinted at a gendered representation of sexual norms.

Nineteenth-century French scientific discourse on sex and gender identities was shaped by *extrascientific* considerations, to borrow this term from Nikki Sullivan, who describes them as 'preconceived ideas and values'.[19] This means that scientific texts were influenced by, and therefore often mirrored, the traditional, non-scientific perception of the sexes as being separated. They set out arguments based both on scientific examinations and on common preconceptions. By examining genitalia, scientists and physicians emphasized the bodily definition of the sexes and, in turn, they helped to establish a sexual dichotomy that was omnipresent throughout the nineteenth century. They were also concerned with the implications of scientific observations about this dichotomy in daily life, for instance in matters relating to birth certificates, marriages and procreation. The 'Prospectus' introducing the first issue of the journal *Annales d'hygiène publique et de médecine légale* [*Annals of Public Hygiene and Forensic Medicine*] claimed: 'La médecine n'a pas seulement pour objet d'étudier et de guérir les maladies, elle a des rapports avec l'organisation sociale' [Medicine does not only aim to study and cure diseases; it also has connections with social organization].[20] We will see that one of medicine's main aims, for many doctors and scientists, was actually to cure society of its 'abnormal' members.

This study gives voice to early nineteenth-century scientists and physicians whose almost intuitive conceptualization of the notion of gender deserves more attention than they are usually granted. Importantly, connecting homosexuality, intersexuality and femaleness demonstrates that these concepts belong to the same medical discourse on a sexual and

19 Sullivan, 'The Matter of Gender', 28.

20 'Prospectus', *Annales d'hygiène publique et de médecine légale*, 1 (1829), v–viii (v).

gendered heteronormative identity. I argue that discourses on sexual and gender ambiguities, which were *a priori* confined to scientific niches, inscribed the medical discourse on homosexuality within a wider sociocultural context. They echoed, but also influenced, a certain construction of French society based on a separation of the sexes that rigorously defined the roles assigned to each sex. The sexual and gendered yoke imposed on 'abnormal' individuals such as masculine women, intersex individuals and homosexuals affected all men, including the so-called 'normal' ones – with consequences that the literary texts discussed in later chapters illustrate.

The Separation of the Sexes

Where sexual dichotomy was concerned, at the beginning of the nineteenth century, French physicians generally argued that men and women were different, not simply because of their genitalia, but also because of their physiology, which influenced their ideas and their mindsets. This principle of fundamental difference between the sexes, or sexual dichotomy, however, did not emanate exclusively from scientific discourse. In his highly influential treatise *Émile ou de l'éducation* [*Émile, or On Education*] (1762), Jean-Jacques Rousseau states: 'En tout ce qui tient au sexe la femme et l'homme ont partout des rapports et partout des différences' [Where sex is concerned, woman and man have connections everywhere and differences everywhere].[21] The philosopher claims that the sexual difference between men and women determines moral differences that influence the duties assigned to each sex. He argues that the main function of a woman is to please men, and especially her husband, by her virtue, modesty, common sense and docility. Scientific writings of the beginning of the nineteenth century echoed and expanded, through methodical argumentation, Rousseau's educational doctrine. Nineteenth-century French doctors and scientists such

21 Jean-Jacques Rousseau, *Émile ou de l'éducation*, André Charrak, ed. (Paris: Garnier-Flammarion, 2009), 515.

as Pierre-Jean-Georges Cabanis (1757–1808), Jacques Louis Moreau de la Sarthe (1771–1826), Julien-Joseph Virey (1775–1846) and Charles François Menville de Ponsan were all categorical in their judgement that men and women's different genitalia (penis, scrotum and testes versus vagina, uterus and ovaries) were the foundation upon which to differentiate between the male and female sexes.

However, according to Thomas Laqueur, the concept of two fundamentally opposite sexes was a relatively recent one. Prior to the nineteenth century, and since the time of Aristotle and Galen, Laqueur argues, a one-sex model prevailed. According to this model, female genitalia were seen as identical to male ones, but simply internal, and they were judged inferior to those of men. The connection between male and female genitals was that of hierarchy. From 1800, on the contrary, scientists employed a two-sex model, which promoted a relation of incommensurability, or opposition, between the sexes.[22] The perspective of two opposite sexes that was sustained by countless nineteenth-century French scientists (including those whose texts are analysed in this chapter) is challenged by modern endocrinologists and paediatricians. When assigning a sex to an individual, they also consider factors that were unknown in the nineteenth century, such as chromosomes, genes and hormones. These 'new' elements may have in turn altered their conception of the sexes. In contrast, the anatomy of external and internal genital organs was the only tangible evidence of the sex of an individual in the nineteenth century.

In practice, genital difference was nevertheless neither sufficient nor absolute, because, as the case of hermaphrodites will show, genitalia were not always devoid of ambiguity. It was therefore necessary to determine other bodily signs of maleness or femaleness, arising from the genitals themselves. The latter were not only important because of their reproductive functions, but also because of their influence on other organs. This 'incontestable' influence, as Cabanis argues, was different in men and

22 Thomas Laqueur, 'Orgasm, Generation, and the Politics of Reproductive Biology', in Catherine Gallagher and Thomas Laqueur, eds, *The Making of the Modern Body: Sexuality and Society in the Nineteenth Century* (Berkeley, Los Angeles and London: University of California Press, 1987), 1–41 (esp. 2–3); Laqueur, *Making Sex*, esp. 4–6 and 23.

women.[23] Virey claims that women were more influenced by their genitalia (especially their uterus) than men, and even goes as far as to argue that their reproductive organs were the principle of their entire biological structure: 'Tout individu femelle est uniquement créé pour la propagation; ses organes sexuels sont la racine et la base de toute sa structure' [Every female individual is solely created for reproduction; her sexual organs are the root and the foundation of her whole structure].[24] Moreau de la Sarthe also states: 'Le sexe le plus faible et le plus sensible doit plus à la nature, le sexe le plus fort doit davantage à la civilisation' [The weaker and more sensitive sex owes more to nature, the stronger sex owes more to civilization].[25] Both scientists illustrate what Sherry Ortner has questioned in her article 'Is Female to Male What Nature Is to Culture?' (1974): the virtually universal conviction that women are inferior to men. Ortner's analysis shows that women's inferiority is generally believed to be due to their reputed closeness to nature, which is determined by their procreative functions, and their psyche is allegedly modelled by their mothering role.[26] The importance of the body in a woman's identity is supposedly an obstacle to the development of her cultural capacities, notably her intelligence. Many texts in this corpus confirm Ortner's analysis.[27]

23 Pierre-Jean-Georges Cabanis, *Rapports du physique et du moral de l'homme*, 2 vols (Paris: Crapart, Caille et Ravier, 1802), I, 310–97 (320).

24 Julien-Joseph Virey, *De la femme, sous ses rapports physiologique, moral et littéraire*, 2nd edn (Paris: Crochard Libraire, 1825 [1823]), 2. Rousseau similarly observes: 'Le mâle n'est mâle qu'en certains instants, la femelle est femelle toute sa vie, ou du moins toute sa jeunesse; tout la rappelle sans cesse à son sexe' [The man is male only at certain moments; the woman is female all her life, or at least all her youth; everything constantly reminds her of her sex] (*Émile*, 520).

25 Jacques Louis Moreau de la Sarthe, *Histoire naturelle de la femme suivie d'un traité d'hygiène appliquée à son régime physique et moral aux différentes époques de la vie*, 2 vols (Paris: Duprat, 1803), I, 698. See also 187–88, 203–06 and 680.

26 Sherry B. Ortner, 'Is Female to Male What Nature Is to Culture?', in *A Cultural Studies Reader: History, Theory, Practice* (London and New York: Longman, 1995), 492–508.

27 Moreau de la Sarthe, *Histoire naturelle*, I, 119–21 and II, 403; Virey, *De la femme*, 171 and 173.

The influence of genitalia on other organs was considered to produce physical characteristics, or secondary sexual characteristics, that differentiated each sex and contributed to the sexual identity of an individual. Cabanis, Moreau de la Sarthe, Virey and Menville de Ponsan established portraits of men and women that were based on the principle of the opposition between these characteristics. For instance, women were smooth, men were hairy; women were cold, men were warm; women were moist, men were dry; women were pale, men were tanned and so forth.[28] Moreover, body and physiology were essential because they were seen as influencing the personality and behaviour of an individual. A keyword in these scientific texts is *sensibilité* [sensitivity], which covered physical and psychological characteristics. Women were considered to be more sensitive than men.[29] This sensitivity was physically caused by their weaker constitution and nervous system, and was believed to affect their culinary, musical, olfactory and tactile tastes. Women, for instance, would have a more developed sense of smell and would therefore be more affected by the smell of newly painted apartments.[30] In relation to personality and taste, sensitivity determined women's emotions and instincts,[31] whereas men were seen as less emotional and more rational. Women's emotional sensitivity and nervous weakness were held responsible for their perceived passivity and changeability.[32]

The difference between male and female, physical and psychological characteristics was attributed to the social functions that each sex had to exercise, as Cabanis summarizes: 'la nature n'a pas simplement distingué les sexes par les seuls organes, instruments directs de la génération: entre l'homme et la femme il existe d'autres différences de structure, qui se

28 Cabanis, *Rapports*, I, 316; Moreau de la Sarthe, *Histoire naturelle*, I, 88–210; Virey, *De la femme*, 172–88; Charles François Menville de Ponsan, *Histoire médicale et philosophique de la femme considérée dans toutes les époques principales de sa vie avec tous les changements qui surviennent dans son physique et son moral avec l'hygiène applicable à son sexe et toutes les maladies qui peuvent l'atteindre aux différents âges*, 3 vols (Paris: Amyot-Labé, 1845), I, 17–21.

29 Cabanis, *Rapports*, I, 334–35 and 350; Moreau de la Sarthe, *Histoire naturelle*, I, 112–13 and 118; Virey, *De la femme*, vii and 103–04; Menville de Ponsan, *Histoire médicale*, I, 55–61.

30 Moreau de la Sarthe, *Histoire naturelle*, I, 114–16.

31 Cabanis, *Rapports*, I, 361.

32 Moreau de la Sarthe, *Histoire naturelle*, I, 693–95; Virey, *De la femme*, 220–23.

rapportent plutôt au rôle qui leur est assigné' [nature has not simply distinguished between the sexes on the basis of organs alone, as direct tools of reproduction: there are other structural differences between men and women, which rather relate to the role that is assigned to them].[33] A woman's 'natural' and desired mission was to be a mother.[34] The alleged smoothness of her body and the gentleness of her character enabled her to exercise her maternal duties, from the point of conception to the care of her children.[35] Conversely, a man was judged too clumsy, too impatient and too rough to take care of an infant.[36] It was therefore 'natural' that he performed physical and intellectual work, which required his strength and intelligence, in the public sphere.[37] All these differences were judged necessary. Cabanis summarizes the portrait of man and woman: 'Il faut que l'homme soit fort, audacieux, entreprenant; que la femme soit faible, timide, rusée' [Man must be strong, audacious and resourceful; woman must be weak, timid and sly], before solemnly adding: 'Telle est la loi de la nature' [Such is the law of nature].[38]

We might well question such a statement. The necessity of an opposition between male and female physiologies and behaviours defies logic. It is unlikely that a woman would physiologically benefit from being weak in one of the most defining events of her life, namely during childbirth. It was still a life-threatening procedure in Cabanis's day.[39] Moreau de la Sarthe and Virey, for instance, mention that the female death rate was the highest between 20 and 25 years old, when women were more likely to bear children.[40] Likewise, the justification for man's need for constant strength can certainly be found outside the realms of biology. Cabanis's categorical declaration reveals anxiety regarding women's role in society. His excessive

33 Cabanis, *Rapports*, I, 315–16. See also Menville de Ponsan, *Histoire médicale*, I, 48–49.

34 Moreau de la Sarthe, *Histoire naturelle*, II, 146 and 186.

35 Cabanis, *Rapports*, I, 349–52 and 357; Virey, *De la femme*, 215.

36 Cabanis, *Rapports*, I, 355–56.

37 Ibid., I, 349; Virey, *De la femme*, 174.

38 Cabanis, *Rapports*, I, 348.

39 Ibid., I, 349.

40 Moreau de la Sarthe, *Histoire naturelle*, II, 341; Virey, *De la femme*, 140.

reliance on nature can be interpreted as a secular attempt to maintain a social order based on a sexual dichotomy. If the physical and psychological differences between men and women were revealed to be insignificant, the whole organization of society would collapse.

As a result, Virey justifies these sexual differences by referring to the only functions for which these differences are in fact necessary: reproduction. However, psychological difference is added to genital difference during procreation. Virey argues that reproduction is possible only if man and woman complete each other, that is, if each of them bring to the relationship qualities of which the other is deprived.[41] He emphasizes the importance of complementarity between man and woman, a concept that is analogous with Laqueur's *incommensurability*: 'L'homme sans la femme n'est point un être complet' [Man without woman is not a complete being].[42] Thus, man and woman become complementary precisely because of their physiological and emotional opposition. In the following excerpt, the anthropologist explains almost mathematically the mechanisms of the complementarity of sexes:

> Comment donc s'établit l'amour le plus pénétrant, le plus parfait entre les sexes? C'est lorsque la femme est le plus femelle, et que l'homme est le plus viril; c'est quand un mâle brun, velu, sec, chaud et impétueux, trouve l'autre sexe délicat, humide, lisse et blanc, timide et pudique. L'un doit donner, et l'autre est constituée pour recevoir; le premier, par cette raison, doit avoir un principe de surabondance, de force, de générosité, de libéralité, qui aspire à s'épancher; la seconde, au contraire, étant constituée *en moins*, doit, par sa timidité, tendre à recueillir, à absorber, avec une sorte de besoin et d'économie, le *trop* de l'autre, pour établir l'égalité, le niveau complet. Ainsi le résultat de l'union conjugale, ou le but de la procréation d'un nouvel être, ne peut être rempli que par cette unité physique et morale dont parlent Pythagore

41 A similar precept can be found even in the work of so-called social constructionists as John Money: 'The irreducible requirement for the survival of humanity is that men and women cooperate *as* men and women at least well enough to survive, reproduce, and rear a new generation. […] Gender stereotypes, with all their many more or less sex arbitrary distinctions, provide the framework for that cooperation': John Money and Patricia Tucker, *Sexual Signatures*, quoted by Sullivan, 'The Matter of Gender', 28 (emphasis in original).

42 Virey, *De la femme*, 2.

> et Platon, au moyen de laquelle les deux sexes s'égalent, se saturent pour ainsi dire réciproquement.[43]
>
> [How is the most penetrating, the most perfect love established between the sexes? It is when the woman is the most female and when the man is the most virile; it is when a brown, hairy, dry, warm and impetuous man finds the other sex delicate, wet, soft, white, shy and modest. The former must give and the latter is constructed to receive. For this reason, the former must have a principle of overabundance, strength, generosity and liberality that aspires to pour out; the latter, on the contrary, being constructed as *minus* and, out of timidity, must tend to receive, absorb, with some kind of need and economy, the *excess* of the other, to establish equality and complete level. The result of conjugal union, or the purpose of the procreation of a new being, can indeed be fulfilled only through this physical and moral unity that Pythagoras and Plato discussed, thanks to which the two sexes equal one another, reciprocally saturate each other, so to speak.]

According to Virey, men are characterized by overabundance, force and generosity, which are sustained by their sperm; they are *trop* [too much]. Women, conversely, are timorous beings, characterized by the mathematical sign *minus*. Commenting on this extract, Nye points out that women are not defined by any fluid, but, *in absentia*, by their lack of semen.[44] They must absorb men's excess in order to find a balance. The influence of man and woman's respective inherent mathematical signs is such that it affects their exterior. As this excerpt illustrates, a man's physiological manhood was believed to be reflected in his virile appearance. This shows how close the nineteenth century thought the link between secondary sexual characteristics and sex to be.

The physical and psychological differences between men and women were regarded as necessary for them to fulfil their procreative functions and to engender ideally virile offspring. This implied that there should be more male than female births. Virey puts the ratio at 96 girls for 100 boys in France. This surplus of male births was compensated for by the higher mortality rate of male adults who died from their masculine activities, such as war.[45] Furthermore, a virile progeny consisted of manly boys. A woman

43 Ibid., 195 (emphasis in original).
44 Nye, *Masculinity and Male Codes of Honor*, 60.
45 Virey, *De la femme*, 138–40. For more on the sex ratio, Nye, *Masculinity and Male Codes of Honor*, 86–87.

who dominated her husband would not only have more girls, but she would also mother effeminate sons, according to Virey. In his opinion, virility was the sign of superior nations, whereas effeminacy led to the degeneration of the 'race', a great concern in nineteenth-century France.[46]

Linked to this perspective, the Orient was viewed as feminized and thus judged to be inferior to the West.[47] Consequently, it should not be sought as a model for sexual and gender relations between men and women. Yet, the oriental harem generated considerable fantasy among Western men (and women) in the nineteenth century.[48] Since 'fantasy and desire, as unconscious processes, play a fundamental role in the colonial relation that is established with the colonized',[49] Virey must fight against the prejudices of his male reader and belittle the fantasy of polygamy as a virile lifestyle. In order to do so, he does not replace this fantasy with accurate, anthropological accounts of the harem, but rather creates a new, negatively connoted image of polygamy that emphasizes its degenerative consequences. Polygamy is judged to be non-virile because the authority of the polygamous husband is divided between his wives and so is his semen, which produces feminized offspring.[50] Virey inserts his arguments within a nationalist discourse. He affirms the superiority of European and, unsurprisingly, French culture, precisely because it supposedly followed the example of nature in opting for monogamous marriages.[51] Virey justifies his

46 Virey, *De la femme*, 147–48. On the French scientific community's increasing concern for degeneration throughout the nineteenth century, see Nye, *Masculinity and Male Codes of Honor*, 72–95.

47 See Edward W. Said, *Orientalism: Western Conceptions of the Orient* (London: Penguin Books, 1995 [Routledge & Kegan Paul, 1978]).

48 Mary Roberts, *Intimate Outsiders: The Harem in Ottoman and Orientalist Art and Travel Literature* (Durham, NC, and London: Duke University Press, 2007), esp. 9.

49 Meyda Yeğenoğlu, *Colonial Fantasies: Towards a Feminist Reading of Orientalism* (Cambridge: Cambridge University Press, 1998), 2.

50 Virey, *De la femme*, 144 and 147–48. Joan DelPlato's study shows that the belief that the multiple pleasures provided by multiple wives had a weakening effect on men (but not on their progeny) was shared by other Westerners: *Multiple Wives, Multiple Pleasures: Representing the Harem, 1800–1875* (London: Associated University Presses, 2002), 19–20.

51 Virey, *De la femme*, 132.

claim with carefully chosen examples, such as the monogamy of apes, the fairly equal number of men and women throughout the world, the social happiness resulting from monogamy and the necessity to be a couple in order to raise children.[52] As can be seen, his reasoning shifts from biology to sociology, thus moving away from the 'natural' argument. In contrast, Eastern cultures, as fashioned by European fantasies of exoticism and eroticism, were judged to be imperfect. Yet, Virey's text reveals the scientist's anxiety that the superiority of the French nation may be threatened by degeneration due to the lack of complementarity, namely biological and psychological opposition, between male and female partners. He consequently prescribes stereotypical sexual and psychological behaviour in order to establish national and racial supremacy.

The physiological and psychological differences between the sexes, which would lead to the prevalence of virile offspring, were thus justified by the greater good: to ensure the survival of the human species and, additionally, the superiority of Western cultures. For Virey, these differences did not mean that women were inferior to men. A woman should not attempt to abolish such differences and behave like a man because her body had not been created to assume such a role. Virey argued that there was a natural order: social distinctions between men and women were due to physiological differences. Fighting against nature was pointless and even dangerous to women, as it would lead to diseases such as cancer.[53]

The scientist's chain of reasoning originates in the physiology of male and female sexes and leads to objectivizing the traits and social roles attributed to each sex. The social differences based on sexual and physiological variations determined both male hegemony and the contempt in which scientists held women at the beginning of the nineteenth century.[54] Early

52 The debate about the naturalness of monogamy is old. Whilst some species (most birds, for instance) are monogamous, polygyny dominates among animals, notably mammals. Unlike what Virey claims, most apes are not monogamous: Frank Cézilly, 'La monogamie est-elle naturelle?', *Sciences humaines*, 188 (December 2007): <https://www.scienceshumaines.com/la-monogamie-est-elle-naturelle_fr_21607.html>, accessed 05 June 2018.

53 Virey, *De la femme*, 87–88 and 237–39.

54 This disdain characterized virtually all the depictions of women made by men in Western societies. See Nancy Tuana, *The Less Noble Sex: Scientific,*

nineteenth-century French scientists often used a contrived approach to express their disdain for women. Although they defined the relations between men and women as complementary rather than superior–inferior, they did not eradicate, but merely altered, the previous hierarchical conception.[55] The sexual equality between man and woman that Virey asserted when claiming that woman was, by her nature, as perfect as man was by his nature[56] is belittled in practice by the fact that woman's perfection was defined by her capacity to accept her *de facto* inferior status in the organization of French society. For instance, Virey declares that women are happier to live in 'une douce sujétion' [a gentle submission] than longing for 'une domination pour laquelle elles ne sont pas nées' [a domination for which they were not born].[57] Mentioning female happiness enables him to justify male hegemony and to convey the patriarchal and patronizing idea that men know best what is good for women, while actually reasserting women's inferiority.

The world depicted by Cabanis, Moreau de la Sarthe, Virey and Menville de Ponsan was that of French society in the early nineteenth century. Men and women, who are portrayed as models of mankind and womankind, were White, wealthy and from the upper classes. Certainly, Virey dedicates a chapter to women around the world ('Variétés du sexe féminin selon les divers climats et les différentes races d'hommes' [Varieties of the female sex according to the diverse climates and the different human races]), but this chapter is mostly a collection of 'exotic' anecdotes that apparently have little impact on his physiological and psychological theories. However, by rejecting polygamy as barbaric and harmful to men and women alike, by presenting polygamous societies as counter-models, rather than alternative models, which readers should avoid (despite the temptations of the fantasized harem), the exotic anecdotes highlight the Eurocentrism and thus the subjectivity of the norm.

Religious, and Philosophical Conceptions of Woman's Nature (Bloomington and Indianapolis: Indiana University Press, 1993).

55 Ibid., 82; Laqueur, 'Orgasm', 35.

56 Virey, *De la femme*, 10.

57 Ibid., 88. See also 254; Menville de Ponsan, *Histoire médicale*, I, 241.

According to these scientific writings, French society was fully organized through sexual dichotomy. Men and women were physically, emotionally, intellectually and even spatially separated. A woman's place was in the home and that of a man was on the battlefield, in political assemblies or at university. Individuals who did not belong to either side of the sexual dichotomy, such as masculine women, individuals with intersex traits and homosexuals, did not belong to this organized society either. They were deemed to be the exceptions that proved the rule rather than real alternatives. 'Other' beings nonetheless existed, and their existence was problematic because it destabilized these sexual and gender norms. The following sections analyse the increasing interest in masculine women, hermaphrodites and homosexuals in the nineteenth century and in so doing reveal the inherent fragility of this sexual dichotomy.

Characteristics of Masculine Women

In his chapter about elderly women, Moreau de la Sarthe points out: 'la constitution de la femme se rapproche de celle d'un homme, comme le prouvent la rigidité des organes, la pousse de la barbe dans plusieurs circonstances et la nature des maladies' [Woman's constitution becomes closer to that of a man; this is shown by the rigidity of the organs, the growing beard in some circumstances and the nature of diseases].[58] The ageing woman is, accordingly, a masculine woman, that is, a woman characterized by physical and behavioural traits generally attributed to men. Her female masculinity infringes the sexual dichotomy. *Female masculinity* is the title of Jack Halberstam's 1998 book on the same topic. Intending to 'conceptualiz[e] masculinity without men', Halberstam argues that masculinity is – still – incorrectly associated with maleness despite the rupture that has been established between sex and gender owing to the

58 Moreau de la Sarthe, *Histoire naturelle*, I, 404.

rise of gender studies.[59] Since masculinity is too often equated to White, middle-class maleness, he examines a usually neglected form of masculinity – female masculinity – which does not simply imitate maleness, but helps us appreciate by contrast what is considered to be normative masculinity.[60] Likewise, studying female masculinity in nineteenth-century scientific texts helps us understand how science conceived of male masculinity and feminine duties.[61]

In early nineteenth-century France, the masculine woman appears as a pseudo-scientific category that has been concocted to respond to scientists' concerns stemming from sociological observations: some women do not conform to gender stereotypes because of their activities or looks. According to nineteenth-century scientists, one was not born a masculine woman, but one became a masculine woman. In accordance with the principle of complementarity, a woman who absorbed the surplus of sperm that was produced by her partner during sexual intercourse acquired some masculine qualities, such as energy, strength and robustness.[62] Masculinization was therefore the fate of a married woman: 'Elle est maintenant, par rapport à la jeune vierge, ce que l'homme est à l'égard de la femme' [She is now, compared with a young virgin, what a man is, compared with a woman].[63] This partial virilization might be a way for scientists to reconcile women's so-called innate fragility and the strength that they may require to perform their duties, for instance the act of childbirth. However, the masculinizing process, doctors and scientists were careful to add, occurred in small proportions and did not have too great an impact on her overall femininity. Accordingly, it did not challenge the sexual dichotomy. In contrast, scientists claimed that an excess of male semen in a woman was deemed to be dangerous because she would develop excessively manly characteristics. In this case, science fits morals all too conveniently. The prostitute, or at least the promiscuous woman, was referred to as a masculine woman, or

59 Jack Halberstam, *Female Masculinity* (Durham, NC, and London: Duke University Press, 1998), 2 and 45.

60 Ibid., 2.

61 See also Chapter 4 of this book about masculine women in literary narratives.

62 Menville de Ponsan, *Histoire médicale*, I, 242.

63 Ibid. See also Virey, *De la femme*, 181.

'hommasse' [mannish], owing to her numerous relations with men and the large quantity of sperm that she had received.[64] The theory of sperm absorption allowed scientists to condemn her behaviour on medical grounds rather than moral ones.

Scientists identified different categories of masculine women, such as the ageing woman and the sterile woman.[65] Another typically masculine woman was the *vivandière* [sutler]. This unflattering portrait illustrates her female masculinity:

> Voyez les femmes les plus hommasses, ces *viragines* audacieuses, dont tous les organes sont très développés, tels que les parties sexuelles dilatées, les mamelons du sein ouverts, dont la parole est haute, criarde, arrogante, dont les gestes sont provocants, dont la démarche est délibérée, l'air hardi, la trogne masculine, même le ton grenadier. En effet, les courtisanes, les vivandières se présentent avec ce maintien et ces qualités demi-viriles, comme si elles étaient déjà transformées à moitié en l'autre sexe à force de cohabiter avec les hommes, et il en est plusieurs auxquelles pousse un peu de barbe au menton, surtout en avançant en âge.[66]
>
> [Watch the manliest women, these audacious *viragoes*, whose organs are very developed, such as dilated sexual parts and open teats; their voice is high-pitched, shrill and arrogant; their gestures are provocative; their gait is decisive, their air is bold, their face is masculine and their tone is martial. Indeed, courtesans and sutlers present themselves with this bearing and these semi-virile qualities, as if they were already half turned into the opposite sex through cohabiting with men. Some of them have a bit of beard growing on their chin as they get old.]

The association between courtesans and *vivandières* may be due to women who sold provisions to the army and laundresses being often regarded as prostitutes.[67] Following this line of reasoning, sutlers might receive a great quantity of sperm during encounters with soldiers. However, Virey's text paradoxically highlights the fact that masculine women were characterized by the development of some female secondary sexual characteristics

64 Virey, *De la femme*, 86, 104 and 181.

65 Moreau de la Sarthe, *Histoire naturelle*, I, 404; Cabanis, *Rapports*, I, 320–21; Virey, *De la femme*, 170–71.

66 Virey, *De la femme*, 86 (emphasis in original).

67 Jean-Paul Bertaud, 'L'armée et le brevet de virilité', in *Le triomphe de la virilité*, 63–79 (75).

(the large nipples and the piercing voice), whereas their gestures, gait, expression and tone were judged to be masculine. This implies that their masculinization was not simply the biological result of the absorption of sperm but was at least partly caused by a principle of imitation, according to which women living very close to men copied their attitudes. Accordingly, there is a contradiction in Virey's text, as he implies – rather than states – that masculinization is not solely a biological process. This suggests that a scientist like Virey acknowledged, at least unconsciously, that female masculinity was a socially constructed phenomenon rather than a purely physiological one. In other words, he showed an intuitive approach to gender.

Elsewhere, Virey states that working-class women are equally mannish:

> Aguerries par les travaux, habituées aux fortes émotions, les paysannes, par exemple, les femmes laborieuses du peuple, sont peu maladives; les passions exercent moins de ravages sur elles; enfin, dans la vieillesse, plusieurs de ces *viragines* deviennent hommasses, acquièrent de la barbe, des traits masculins, une voix rauque, une allure presque martiale.[68]
>
> [Seasoned by work, used to strong emotions, peasant women, for instance, working-class women, are not sickly. Passions are less damaging to them. Finally, when ageing, a few of these viragoes become mannish, get some beard, masculine traits, a husky voice and an almost martial look.]

These women are endowed with strength, masculine features, facial hair, a hoarse voice and a martial gait, as a result of their gruelling activities, allegedly reserved for men. Virey admits that their so-called virile qualities of strength and robustness are acquired and developed throughout their tough life. Working-class women worked hard in the fields or in the factory to earn a living and, for this reason, they were not considered as womanly as idle middle-class women, confined to the domestic sphere. This indicates that only the latter was regarded as the 'genuine' woman. Virey does not acknowledge the paradoxes that are present in his text, which oscillates between the idea of masculinity that is inherent in man's

68 Virey, *De la femme*, 104.

condition and parsimoniously transmitted to women, and that of masculinity that is developed through interaction with, or imitation of, the harshness of the male world.

In both excerpts, Virey describes mannish women as *viragines*. The sixth edition of the *Dictionnaire de l'Académie française* (1835) defines a virago as a 'fille ou femme de grande taille, qui a l'air d'un homme' [tall girl or woman, who looks like a man], adding: 'ne se dit que par dérision' [used only out of derision].[69] The masculine woman was blamed for her masculine allure, as her behaviour defied the modesty that was supposed to be typical of women. A woman who had been virilized by numerous encounters with men, because of alleged frivolity, paradoxically became unable to attract men. She was indeed judged repellent to the senses.[70] Virey states: 'Jamais une femme masculine ne sera bien chérie d'un homme; il croirait pécher avec elle comme avec son semblable, et il éprouve presque le même genre de répugnance' [A masculine woman will never be cherished by a man. He would think that he is committing a sin with her, as with a fellow man, and he feels almost the same kind of repugnance].[71] He claims that masculine women call the virility of their lovers into question because these men are attracted to the characteristics of their own gender. He thus implies that the latter are suspected of homosexuality.

As is becoming clear, masculine women could also be perceived as homosexual. References to lesbians were rare in scientific writings of the early nineteenth century, a phenomenon that Éric Bordas calls 'l'invisibilité lesbienne' [lesbian invisibility].[72] The word *tribade*, more common than *lesbienne* at the time (although the latter was employed by Moreau de la

69 'Virago', in *Dictionnaire de l'Académie française*, II, 940.

70 Virey, *De la femme*, 176–77.

71 Ibid., 194. See also Menville de Ponsan, *Histoire médicale*, I, 39.

72 Éric Bordas, 'Introduction: Comment en parlait-on?', *Romantisme*, 159 (2013), 3–17 (15). Likewise, Christopher Rivers points to the inability to speak explicitly about female homosexuality: '*Inintelligibles pour une femme honnête*: Sexuality, Textuality and Knowledge in Diderot's *La Religieuse* and Gautier's *Mademoiselle de Maupin*', *Romance Review*, 86/1 (January 1995), 1–29 (16–22). On late nineteenth-century doctors' indifference to lesbianism, see also Rosario, *The Erotic Imagination*, 108–09.

Sarthe), was defined in the *Dictionnaire de l'Académie française* as a 'femme qui *abuse* de son sexe avec une autre femme' [woman who *abuses* her sex with another woman].[73] Significantly, having defined it, the lexicographer immediately states that '[o]n évite d'employer ce mot' [one avoids using this word]. Moreover, lesbianism did not raise the same scientific interest as male homosexuality because, allegedly, it was not characterized by the same level of genital involvement as anal intercourse between men; as a result, it could not be regarded as a collection of acts involving genitalia.[74] On another level, the homophobia of physicians was not as strong in the case of lesbianism because same-sex relationships between women 'had a voyeuristic appeal to straight men',[75] as Leslie Choquette puts it, even though it defied the patriarchal model.

Lesbianism was, nonetheless, perceived as an 'aberration de l'amour' [aberration of love] according to Moreau de la Sarthe, who worried about its 'suites funestes' [dire consequences], but described none.[76] He thus conformed to this policy of discretion that was characteristic of the treatment of lesbianism in scientific writings throughout most of the nineteenth century. Moreau de la Sarthe recommended to mothers and governesses to watch closely over little girls in order to avoid such unacceptable behaviour. Girls appear in his text as perverse little beings who are prone to licentious excesses (homoerotic as well autoerotic ones). It is therefore manifest in Moreau de la Sarthe's text that observations about lesbians are sociological rather than medical. In the investigation of Alexandre Parent-Duchâtelet

73 Moreau de la Sarthe, *Histoire naturelle*, I, 53; 'Tribade', in *Dictionnaire de l'Académie française*, II, 882 (my emphasis).

74 Alain Corbin, 'La rencontre des corps', in Alain Corbin, ed., *Histoire du corps*, II: *De la Révolution à la Grande Guerre* (Paris: Seuil, 'Points', 2005), 151–219 (207–08); Louis Crompton, *Homosexuality and Civilization* (Cambridge, MA: Harvard University Press, 2003), 472–73.

75 Leslie Choquette, 'Homosexuals in the City: Representations of Lesbian and Gay Space in Nineteenth-Century Paris', in Jeffrey Merrick and Michael Sibalis, eds, *Homosexuality in French History and Culture* (New York: Harrington Park Press, 2001), 149–67 (149). See also Rosario, 'Pointy Penises', 166–67; Rosario, *the Erotic Imagination*, 108–09; Murat, *La loi du genre*, 74. For instance, Virey (*De la femme*, 34) connected two fantasies of the Western man: sapphism and the Turkish harem.

76 Moreau de la Sarthe, *Histoire naturelle*, II, 270.

(1790–1836) of prostitution in Paris, lesbianism is likewise presented in a sociological light. It is seen as a common practice among prostitutes and prisoners, but the physical characteristics of lesbian prostitutes are not described.[77] As prostitutes were seen as mannish, it can be inferred that lesbians were similarly judged to be mannish.

In Virey's text, lesbianism appears to be the consequence of a woman's virility. He points out that 'la femme hommasse, ayant trop de qualités masculines dans sa constitution, tend à se rejeter sur son sexe, comme pour s'efféminer, et afin de retrouver ses qualités naturelles' [The mannish woman, having too many masculine features in her constitution, tends to push herself towards her own sex, as if to feminize herself, and in order to get her natural qualities back].[78] His comment raises three issues. Firstly, the procedure through which the mannish woman re-acquired her 'natural', feminine qualities: Was there any transmission of fluids? If yes, why was the man having intercourse with a woman not feminized, according to the principle of sexual complementarity? Secondly, the consequences of re-acquiring her feminine qualities on her sexuality: Did the woman become heterosexual? Finally, the reasons why the mannish woman's partner, supposedly feminine, was attracted to the masculine woman: Would she in turn become masculine? Virey seems to imply that lesbianism could constitute a cure to female masculinity, a position that radically contradicts heteronormativity. The scientist does not give further thought to sapphism because such thought would compel him to defend principles at variance with the sexual dichotomy and the patriarchal model in which he believed. Accordingly, he hurries to conclude that these 'abominables et outrageux' [revolting and shocking] customs mostly occurred in hot climates.[79] The honour of France is safe!

Paradoxically, a masculine woman was not solely a woman who received too much sperm. Indeed, she could also be a woman who never

77 Alexandre-Jean-Baptiste Parent-Duchâtelet, *De la prostitution dans la ville de Paris, considérée sous le rapport de l'hygiène publique, de la morale et de l'administration: Ouvrage appuyé de documents statistiques puisés dans les archives de la Préfecture de police*, 2 vols (Paris: Baillière, 1836), I, 161–65.

78 Virey, *De la femme*, 198.

79 Ibid., 199.

received any. A spinster was criticized for her celibacy, because '[l]a continence [...] est un état contre nature' [abstinence [...] is a state against nature].[80] Like elderly women, the old maid was condemned for the virilization of her features. Whilst recommending abstinence before, and moderation during, marriage, nineteenth-century doctors considered that, after a certain age, excessive abstinence could generate diseases, such as anaemia, absence of menstruation, hysteria and cancers.[81] Such ailments could only be cured by marriage. Adopting an openly secular perspective, doctors thus criticized the celibacy that the Catholic Church imposed on nuns.[82]

The biological principle of woman's masculinization owing to excessive quantity of sperm yielded to psychological and behavioural explanations. The pedantic woman and the intellectual woman, in other words the *bas-bleus* [bluestockings], were also considered to be masculine women. While a woman was encouraged to be witty and perceptive, cerebral activities were not recommended because they were deemed incompatible with her delicate constitution. Intellectual pursuits damaged her beauty, caused sterility and a malfunctioning uterus.[83] However, Cabanis took pains to stress that few women were of exceptional intelligence.[84] Their ability to focus was limited, their knowledge was superficial, and their reasoning was flawed. As for women endowed with true intellectual disposition, Moreau de la Sarthe relegated them to the status of 'cas extraordinaires' [extraordinary cases] and 'phénomènes, desquels on ne peut rien conclure, sinon que la nature réalise tous les possibles' [phenomena, about which nothing can be concluded, except that in nature anything is possible].[85] In short, the intellectual woman was tolerated as long as she was the exception that proved the rule of the mediocrity of women's brains. Due to her status as 'natural' accident, she was not blamed, but rather pitied because she did not belong

80 Menville de Ponsan, *Histoire médicale*, III, 165. Male celibacy was also criticized, see for example, Martin, 'Fonction paternelle et Code Napoléon', 466–68.

81 Virey, *De la femme*, 83 and 169–70.

82 Moreau de la Sarthe, *Histoire naturelle*, II, 274; Virey, *De la femme*, 82–83.

83 Moreau de la Sarthe, *Histoire naturelle*, I, 119–21 and II, 403; Virey, *De la femme*, 171 and 173.

84 Cabanis, *Rapports*, I, 362–63.

85 Moreau de la Sarthe, *Histoire naturelle*, I, 119.

to any category: 'Dans la jeunesse, dans l'âge mûr, dans la vieillesse, quelle sera la place de ces êtres incertains, qui ne sont, à proprement parler, d'aucun sexe?' [In youth, in maturity and in old age, what will be the position of these uncertain beings who belong to no sex, strictly speaking?], Cabanis worries.[86] This rhetorical question reveals the scientist's doubts regarding his capacity to control social and sexual categories.[87] The idea that individuals of ambiguous gender do not belong to French society suggests that, in the end, what is at stake in scientific discourse is actually sociological rather than merely biological or medical. Masculine women and other ambiguous individuals are social anomalies because their ambiguity infringes gender role distribution in society: too masculine to be caring feminine figures, but not male enough to be allowed to fulfil masculine roles.

The unease of Cabanis and his colleagues is perceptible. While a woman was already judged mysterious, the masculine woman was even more enigmatic because she escaped the sexual and gender categories defined by science.[88] Menville de Ponsan refused this woman 'cette hardiesse, cette fermeté de caractère qui doit être l'attribut de notre sexe' [this boldness, this rigour that must be the attribute of our sex] – women are clearly excluded here as readers, although the doctor claimed to be writing for them as well.[89] As for women's qualities, Menville de Ponsan only highlights commonly held ones such as intuition, generosity, sensitivity and emotional ardour, all these traits being enhanced by feminine charm.[90] He summarizes the principle of a fusion between sex and gender that guided

86 Cabanis, *Rapports*, I, 363.

87 Virey expresses similar anxiety regarding gender categories and boundaries: 'Après avoir outrepassé les bornes de l'honneur, quelles seront les limites du vice chez elle [la fille qui a perdu la pudicité]? Avec son caractère extrême en tout, impétueux dans ses amours comme dans ses haines, où s'arrêtera-t-elle?' [After having overstepped the line of honour, what will the limits of vice be in the woman who has lost her modesty? With her extreme character in everything, impetuous in her love as well as in her hatred, where will she stop?] (*De la femme*, 86–87).

88 Described as 'abîmes impénétrables' [impenetrable abysses] and 'inextricable labyrinthe' [inextricable maze] (Ibid., 213).

89 Menville de Ponsan, *Histoire médicale*, I, xiv–xv and xxxii.

90 Ibid., I, xv.

the scientists of his time: 'tout doit être féminin dans la femme' [everything must be feminine in the woman].[91]

Characteristics of Hermaphrodites

An individual that disrupted the nineteenth-century sexual categories even more than the masculine woman was the hermaphrodite. If social functions were dictated by male or female physical characteristics, how could one determine the role of an individual whose anatomic characteristics were ambiguous? The challenge that nineteenth-century scientists faced was as follows: to observe bodies and acknowledge the existence of intersex individuals who contradicted the separation of the sexes, whilst integrating them into the system of the sexual dichotomy and, in order to do so, denying their intersexuality. Scientists had been fascinated by hermaphrodites for centuries; nineteenth-century doctors were no exception to this rule.[92] The main French text dedicated to intersexuality during the July Monarchy period was included in Isidore Geoffroy Saint-Hilaire's *Traité de tératologie* [*Treatise of Teratology*] (1832–37).[93] Geoffroy Saint-Hilaire (1805–61; son of the naturalist Étienne Geoffroy Saint-Hilaire) was the inventor of the modern science of teratology (as well as of the word *tératologie*), which consists of the scientific study of

91 Ibid., I, xiv.

92 An example of a Renaissance treatise on hermaphrodites is that of Ambroise Paré, *Des monstres et des prodiges* (1573). For more on hermaphrodites before the nineteenth century, see Lorraine Daston and Katharine Park, 'The Hermaphrodite and the Orders of Nature: Sexual Ambiguity in Early Modern France', *GLQ*, 1/4 (1995), 419–38. On the medical perspective on hermaphrodites in France and Britain in the late nineteenth and early twentieth centuries, see Dreger, *Hermaphrodites*.

93 Isidore Geoffroy Saint-Hilaire, *Histoire générale et particulière des anomalies de l'organisation chez l'homme et les animaux, ouvrage comprenant des recherches sur les caractères, la classification, l'influence physiologique et pathologique, les rapports généraux, les lois et les causes des monstruosités, variétés et vices de conformation, ou Traité de tératologie*, 4 vols (Paris: Baillière, 1832–37).

physical monstrosities. He considered that monstrosity, including hermaphroditism, was not an inexplicable supernatural phenomenon, but was rather determined by the same laws that governed the organization of 'normal' beings. His classification of different types of hermaphroditism was a reference for most French physicians during the nineteenth century.[94] By distancing himself from popular superstitions, Geoffroy Saint-Hilaire showed remarkable insight regarding individuals such as hermaphrodites who did not conform to a biological and physiological type. He nevertheless perpetuated existing ideas about the sexual norm.

Before analysing Geoffroy Saint-Hilaire and his nineteenth-century colleagues' discourse on hermaphrodites, I will examine the twenty-first-century definition of intersexuality in order to help us understand what is at stake here. Firstly, it is important to note that the term *hermaphroditism* is not employed by modern scientists. The terms *intersex* and *intersexuality* were widely used instead until 2005, when scientists recommended the designation DSD (*Disorder of Sexual Development*, translated in French as *Anomalies de la Différenciation Sexuelle* [ASD] or *Développement Sexuel Déficitaire*).[95] The frequent changes of name nowadays, along with the controversies that these names raise, attest to the difficulty in conceptualizing sexual ambiguities. This difficulty had already been highlighted by the author of the article 'Hermaphrodite' in the *Dictionnaire des sciences médicales* (1817), a Parisian doctor named Marc, who may be the same Marc (1771–1840) as the co-founder of the *Annales d'hygiène publique et de médecine*

94 Dreger, *Hermaphrodites*, 140. See also Fausto-Sterling, *Sexing the Body*, 36–37.

95 Jean-Yves Tamet, Patricia Bretonès and Claire-Lise Gay, 'Se représenter puis se nommer: Une courte histoire de la nomination de sujets atteints d'anomalie de la différenciation sexuelle', in *Différenciation sexuelle et identités*, 19–27 (25). This new designation is also controversial due to the negative connotations of *disorder/anomalies/déficitaires*. On the history of the designation DSD and the divergent receptions of this term, see Davis, *Contesting Intersex*, 43–46, 97–113 and 137–42; Cary Gabriel Costello, 'Intersex and Trans* Communities: Commonalities and Tensions', in Stefan Horlacher, ed., *Transgender and Intersex: Theoretical, Practical, and Artistic Perspectives* (New York: Palgrave Macmillan, 2016), 83–113 (esp. 87–88).

légale. He states that the expression, although incorrect, was 'conservée faute d'une meilleure' [kept for lack of a better one].[96]

Current biological knowledge regarding sexual differentiation – which is constantly evolving – shows that the sex of an individual is partly determined by the sexual chromosomes (XX or XY) and partly by other sexual genes that are not present on the sexual chromosomes. Moreover, each embryo possesses a double set of ducts: the Wolffian ducts (rudimentary male genitalia) and the Müllerian ducts (rudimentary female genitalia). The differentiation of the gonads (testes and ovaries) starts around the sixth or seventh week for males and about three to four weeks later for females. In a male individual, the Anti-Müllerian Factor prevents the Müllerian ducts from developing, whereas testosterone influences the development of Wolffian ducts. Conversely, Müllerian ducts will develop and become female genitals under the influence of oestrogen in a female individual. The complexity of human sexual development facilitates irregularities. Genetic anomalies, mutations and endocrine disruptors, such as viruses and pollutants, can interrupt the differentiation of the sexes and generate cases of undetermined sex.[97] Causes of intersexuality notably include an extra X chromosome in male individuals, the lack of a second X chromosome in females, the malfunction of the enzymes involved in the production of steroid hormones in females, the insensitivity of male bodies to androgens, the lack of an enzyme called 5-alpha-reductase that is able to convert testosterone into dihydrotestosterone and the opening of the urethra at the base, rather than at the tip, of the penis (hypospadias).[98]

The current explanations of sexual development involve elements that were unknown in the nineteenth century, such as genes, chromosomes

96 Charles-Chrétien-Henri Marc, 'Hermaphrodite', in *Dictionnaire des sciences médicales*, XXI (1817), 86–121 (89). On Marc, see Rosario, *The Erotic Imagination*, 52–53 and 57.

97 For more on human sexual differentiation, see Pierre Chatelain, Yves Morel and Marc Nicolino, 'Le développement et la différenciation sexuels', in *Différenciation sexuelle et identités*, 31–44; Anne Fausto-Sterling, *Sex/Gender: Biology in a Social World* (New York and London: Routledge, 2012), 16–26 and 119–20.

98 Dreger, *Hermaphrodites*, 35–41; Fausto-Sterling, *Sexing the Body*, 52; Intersex Society of North America website <https://isna.org/faq/conditions/>, accessed 02 April 2020.

and hormones. These elements constitute what the twenty-first-century French endocrinologist Pierre Mouriquand calls the internal sex.[99] However, nineteenth-century doctors were aware only of the external sex of their patients: the genitalia. As a result, they mostly based their sexual classifications on this external sex. Definitions of hermaphroditism varied from one author to the next, but all considered that it designated the assemblage of some sexual characteristics that partook of both sexes:

> On entend par hermaphrodite un individu qui réunit les deux sexes, ou les parties sexuelles ou naturelles du mâle et de la femelle.[100]
>
> [The hermaphrodite is defined as an individual who combines both sexes, or the sexual or natural parts of the male and the female.]
>
> HERMAPHRODITE, s. m. et adj., *hermaphroditus*, individu qui réunit les deux sexes[101]
>
> [HERMAPHRODITE, m. s. and adj., *hermaphroditus*, individual who combines both sexes.]
>
> […] l'hermaphrodisme consiste en un mélange, soit apparent, soit plus ou moins réel et distinct des attributs de la génération départis aux deux sexes[102]
>
> [Hermaphroditism consists of an either apparent or more or less real and distinct combination of the reproductive attributes allotted to both sexes.]
>
> L'hermaphrodisme est la réunion chez le même individu des deux sexes ou de quelques-uns de leurs caractères.[103]
>
> [Hermaphroditism is the union in the same individual of both sexes or of some of their characteristics.]

These definitions support the following belief. For nineteenth-century scientists, hermaphroditism was a phenomenon affecting plants and 'inferior' animals, such as molluscs and shellfishes. The 'real', perfect or

99 Pierre Mouriquand, 'S'asseoir dehors', in *Différenciation sexuelle et identités*, 53–64 (60–61).

100 Paul-Augustin-Olivier Mahon, 'Hermaphrodites', in *Médecine légale et police médicale*, 3 vols (Paris: Bertrand, 1807), I, 90–112 (90).

101 Marc, 'Hermaphrodite', 86.

102 Ibid., 88.

103 Geoffroy Saint-Hilaire, *Traité de tératologie*, II [1836], 31.

absolute hermaphrodite, namely the one that had male and female reproductive systems, and who was therefore able both to impregnate and to be impregnated did not exist in human beings.[104] Scientists accordingly set out to prove that even the individual closest to this ideal did not have both complete sexes. For instance, Hugues Maret (1726–86), a physician who is extensively quoted by Paul-Augustin-Olivier Mahon (1752–1801), professor at the School of Medicine in Paris, denied a hermaphrodite named Jean Pierre a clear position in society:

> Mais, quoique la nature ait paru en quelque sorte prodigue en sa faveur, les dons qu'elle lui avait faits ne devaient pas exciter sa reconnaissance, puisque, par cette prodigalité, il avait été rendu inhabile aux fonctions auxquelles l'un et l'autre [sexes] sont destinés. [...] Jean Pierre, qui était sensiblement homme et femme, n'était cependant, dans le fait, ni l'un ni l'autre; et son état, qui augmente le nombre de cette espèce de monstres, rend l'existence des *hermaphrodites* parfaits bien peu vraisemblable.[105]
>
> [But, although nature seemed somehow to have been generous towards them, the gifts that were bestowed on them would not arouse their gratitude, since, because of this generosity, they had been made incompetent in the functions to which both sexes are destined. [...] Jean Pierre, who was apparently both man and woman, was however factually neither one nor the other. And their state, which increases the number of this kind of monsters, makes the existence of perfect *hermaphrodites* very unlikely.]

Maret's text presents intersexuality as a poisonous gift from nature. Moreau de la Sarthe, who was as eager as Mahon to demonstrate the impossibility of true hermaphrodites, condemned his patients to a 'triste neutralité' [sad neutrality].[106] Far from the Ovidian myth of Hermaphroditus, hermaphrodites were regarded as monsters, notably because of their ineffectiveness in the process of perpetuating the human species, an essential function for human beings.[107]

104 Moreau de la Sarthe, *Histoire naturelle*, I, 213; Mahon, 'Hermaphrodites', I, 90; Marc, 'Hermaphrodite', 86–87; Geoffroy Saint-Hilaire, *Traité de tératologie*, II, 30.

105 Mahon, 'Hermaphrodites', I, 110–11.

106 Moreau de la Sarthe, *Histoire naturelle*, I, 229.

107 Chatelain, Morel and Nicolino, 'Le développement et la différenciation sexuels', 31. For more on Hermaphroditus, see Chapter 2.

The term *monster* was not always pejorative. Geoffroy Saint-Hilaire considered monstrosity to be determined by the same laws that influenced the *specific type*. Any deviation from this specific type was called an *anomaly*, whereas *monstrosities* designated the most severe or most apparent anomalies.[108] In other words, the norm solely depended on quantity. A monster was monstrous because it was rare. What about the hermaphrodite, a separate category in Geoffroy Saint-Hilaire's classification of monsters? Scientists were interested in intersexuality because of its rarity, but this rarity is relative. While no estimation of the number of individuals with intersex traits has been made for nineteenth-century France, current figures exist and vary. Alice Dreger estimates that 0.05 to 0.15 per cent births in the United States are intersex; Anne Fausto-Sterling and Gabrielle Houbre go as far as to claim respectively, that they represent 1.7 and 2 per cent of all births.[109] These disparities can be partly explained by the fact that intersexuality is a broad category that consists of many different conditions; some are very rare and some are fairly common. The variety depends on which conditions scholars have included in their data. These figures nevertheless show that intersexuality is not a very rare condition. Fausto-Sterling points out that it is more common than albinism, given that the latter affects only 0.005 per cent births.[110] Moreover, Mak shows that many intersex individuals and their relatives were reluctant to consult a doctor regarding their conditions and, consequently, known cases of intersexuality published in medical texts were only 'the tip of the iceberg'.[111]

The reader of early nineteenth-century reports on hermaphrodites was nonetheless under the impression that intersex individuals were rare. This is notably due to scientists constantly quoting previous sources rather than presenting original observations. Although he suggests that intersex cases were numerous, Geoffroy Saint-Hilaire, for instance, mentions only one

108 Geoffroy Saint-Hilaire, *Traité de tératologie*, I [1832], 30.

109 Dreger, *Hermaphrodites*, 40–43; Fausto-Sterling, *Sexing the Body*, 51; Gabrielle Houbre, 'Dans l'ombre de l'*hermaphrodite*: Hommes et femmes en famille dans la France du XIXe siècle', *CLIO: Histoire, femmes et sociétés*, 34 (2011), 85–104 (85, footnote 2). See also Davis, *Contesting Intersex*, 3.

110 Fausto-Sterling, *Sexing the Body*, 51–53.

111 Mak, *Doubting Sex*, 22.

observation that he himself made on a human being – but quotes the report of a colleague.[112] This practice of intertextuality often goes beyond a simple reference to the work of a colleague and can consist of long quotations displayed over many pages.[113] Marc and Mahon, for instance, quote Maret's report on the hermaphrodite Jean Pierre in its entirety. These citations confuse the identity of the *je* and suggest that authors are interchangeable, since the quoter seems to embrace fully the opinion of the quoted. The vagueness of the grammatical subject supports the idea that opinions about hermaphrodites are objective and misleadingly hints at the fact that scientists meet no contradiction among the scientific community. Furthermore, the habit of employing long quotations reveals a lack of critical distance from the original text. Marc, for example, quotes Béclard who describes the case of the female hermaphrodite Marie-Madeleine Lefort (an individual with intersex traits that had been assigned a female sex), who had a twenty-seven-centimetre long clitoris (referred to as a 'corps conoïde' [cone-shaped body]). Geoffroy Saint-Hilaire cites the same report but modifies one element (the change is explained in a footnote): 'centimètres' became 'millimètres'.[114] Marc copied Béclard's mistake without realizing the incongruity of the figures.

The determination of the intersexuality of an individual was based on the examination of their body, especially their genitalia, because the body was deemed to hide priceless information. Consequently, it emerges from the scientists' comments that, for the purposes of science, a good hermaphrodite is a dead hermaphrodite. For instance, in less than two pages, Moreau de la Sarthe twice points out that any certainty regarding a case can be acquired only after the hermaphrodite is dead.[115] Death allowed

112 Geoffroy Saint-Hilaire, *Traité de tératologie*, II, 41–42 and 102.

113 By *intertextuality*, I am referring to Gérard Genette's definition as 'une relation de coprésence entre deux ou plusieurs textes' [a relation of co-presence between two or more texts], that is, 'la présence effective d'un texte dans un autre' [the effective presence of a text within another one], which focuses on the text and its relation with other texts: *Palimpsestes: La littérature au second degré* (Paris: Seuil, 1982), 8.

114 Marc, 'Hermaphrodite', 98; Geoffroy Saint-Hilaire, *Traité de tératologie*, II, 103, footnote 1.

115 Moreau de la Sarthe, *Histoire naturelle*, I, 226–27. Such a perspective is also that of later experts. See Dreger, *Hermaphrodites*, 52, 86 and 149; Linton, *Prescribed*

physicians to carry out the dissection of the body and to collect information about the 'real' genitalia. Marc recommends that physicians probe all the openings of the living intersex body in order to determine their extent and direction, 'sans blesser ni exciter une vive douleur' [without causing injury or severe pain].[116] Indeed, the body of a living hermaphrodite could be injured and, consequently, the scrutiny of the organs was limited in practice by their degree of pain tolerance. The physician Béclard declared that he could not perform a deeper examination because Marie Lefort refused further examination after a tube was inserted at least three times in her bladder and her anus to collect urine and, generally, to probe her urinary tracts and her genitals.[117] Medical examinations could indeed be invasive. Geoffroy Saint-Hilaire, for instance, indicates that two fingers were introduced into what was believed to be the vagina of an Italian hermaphrodite during their examination.[118]

Moreover, living hermaphrodites were not considered reliable sources for commenting on their own sex because they could have been misled by having received the 'wrong' education or because they had lied for their own interest. Marc's text suggests that some female hermaphrodites were aware of their 'real' sex but wanted to benefit from the same privileges as men and therefore feared being reassigned as women.[119] In so doing, they deliberately challenged the boundaries of the sexual dichotomy.

The examination of intersex bodies allowed Marc and Geoffroy Saint-Hilaire to classify sexual ambiguities based on genitalia. Although different, their respective classifications share many characteristics. Marc's classification (Figure 1), just like that of Geoffroy Saint-Hilaire, identifies two types that are not considered to be proper hermaphroditisms, that is,

---

Fictions, 59. Mahon ('Hermaphrodites', I, 95) mocks those who believe in the existence of the perfect hermaphrodite, but never check their 'conjecture' by dissecting bodies. The importance that Mahon and Marc give to the case of Jean Pierre is probably due to the fact that he was dead when his body was examined by Maret.

116 Marc, 'Hermaphrodite', 116.

117 Ibid., 99; Geoffroy Saint-Hilaire, *Traité de tératologie*, II, 103–04.

118 Geoffroy Saint-Hilaire, *Traité de tératologie*, II, 77.

119 Marc, 'Hermaphrodite', 116.

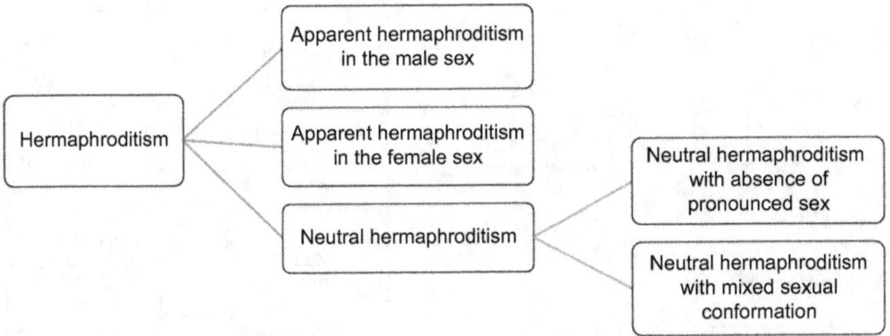

Figure 1. Marc's classification of hermaphrodites (based on data provided by his article 'Hermaphrodite' in the *Dictionnaire des sciences médicales*).

cases in which genitalia are deemed to be deceptive because they resemble the genitalia of the other sex. The first is male hermaphroditism despite feminized looks, caused by atrophied penises and undescended testes, for instance. The second is female hermaphroditism in spite of masculinized looks, due to hypertrophied clitorises and external ovaries. These scientists also determine one group in which it is almost impossible to identify a predominating sex – I will shortly discuss this notion – which is called neutral hermaphroditism, including one sub-group, named mixed hermaphroditism, in which both sexes are considered to be equally important. In order to exemplify their classifications, the doctor and the scientist used evidence from cases provided by their predecessors.[120]

Geoffroy Saint-Hilaire's classification (Figure 2) is more complex, because he establishes two macro-categories. The first macro-category, called hermaphroditism without excess, is analogous to the categories previously mentioned. Individuals have only one reproductive system, partly male and partly female. In the other macro-category, called hermaphroditism with excess, complex hermaphrodites have two more or less

120 Generally, the categories to which cases belong do not vary from one classification to the other, with the exception of a soldier from Plymouth, observed by Everard Home, who is listed by Geoffroy Saint-Hilaire (*Traité de tératologie*, II, 66–67) as a male hermaphrodite, and by Marc ('Hermaphrodite', 102) as a neutral hermaphrodite.

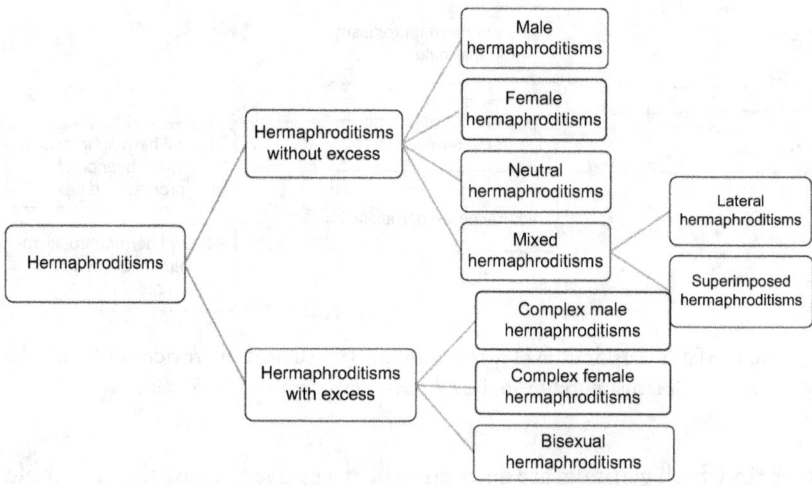

Figure 2. Geoffroy Saint-Hilaire's classification of hermaphrodites (based on data provided by the chapter 'Des hermaphrodites' in his *Traité de tératologie*).

complete reproductive systems. Such intersex cases seem to have been rarer than other types of hermaphroditism. Geoffroy Saint-Hilaire gives only two examples of human cases, which Marc listed as mixed hermaphrodites.[121] This suggests that Geoffroy Saint-Hilaire was under the impression that previous categorizations were imprecise and he created the macro-category of hermaphroditism with excess to include any hermaphrodite that had been too complex to be identified.

Marc's, and above all Geoffroy Saint-Hilaire's, obsession with intersex taxonomy appears as a means to control individuals that eluded their understanding. However, this subjugation is not fully successful. Differences in classifying cases highlight the ambivalence of the dominant sex discourse since scientists did not agree on establishing a clear border between male and female categories based on genitalia. The divergences could also be caused by the fact that they were based on second-hand observations. Scientists complained about the lack of information provided by colleagues

121 Geoffroy Saint-Hilaire, *Traité de tératologie*, II, 158–59; Marc, 'Hermaphrodite', 110–11.

and considered that the best examination was the one that they performed themselves.[122]

As can be noted from their remarks, Marc and Geoffroy Saint-Hilaire attached great importance to genitalia in their classification of hermaphrodites, just as Cabanis, Moreau de la Sarthe, Virey and Menville de Ponsan did when studying women. Their approach contrasts with that of several, but not all, twenty-first century scientists, who avoid such classifications. It is true that Fausto-Sterling has jokingly described a five-sex system similar to Marc's classification: to male and female categories, she adds *herms* (true hermaphrodites), *merms* (male hermaphrodites) and *ferms* (female hermaphrodites).[123] She nevertheless actually classifies hermaphroditisms according to the causes of incomplete sexual differentiation.

If nineteenth-century scientists seemed to neglect the causes of intersexuality, it is because they found them difficult to determine.[124] Geoffroy Saint-Hilaire nonetheless used the theory of unity of composition to explain the origins of intersexuality. According to this theory, male and female genitalia were established according to the same anatomic type, although with opposite physiological functions. In other words, he rightly believed that genitalia were originally undifferentiated in the embryo[125] and highlighted the similarity between the sexes rather than their difference:

> En effet, si chaque partie de l'appareil mâle est essentiellement analogue par sa composition élémentaire à une partie de l'appareil femelle, si leur diversité apparente résulte seulement de quelques différences dans le mode ou dans le degré de leur développement, rien de plus facile à concevoir que l'existence d'états intermédiaires entre ces deux états extrêmes, entre ces deux formes opposées qui constituent l'état normal de l'un et de l'autre sexe.[126]

> [Indeed, if each part of the male apparatus is essentially analogous, by its elementary composition, to a part of the female apparatus, if their apparent diversity only comes from a few differences in the mode or degree of their development, nothing easier to

122 Geoffroy Saint-Hilaire, *Traité de tératologie*, II, 72 and 116.

123 Fausto-Sterling, *Sexing the Body*, 78.

124 Marc, 'Hermaphrodite', 111–12; Geoffroy Saint-Hilaire, *Traité de tératologie*, II, 42 and 53.

125 Geoffroy Saint-Hilaire, *Traité de tératologie*, II, 43–44.

126 Ibid., II, 44–45.

conceive than the existence of intermediate states between these two extreme states, between these two opposite forms that constitute the normal state of both sexes.]

This overall likeness between genitalia explains the reason why a penis can resemble a clitoris and testes can resemble ovaries. Geoffroy Saint-Hilaire's affirmation contradicts Moreau de la Sarthe's claim that men and women have a different structure, but continues Marc's suggestion that testes and ovaries were at first formed in the same area of the body.[127] Let us note that uncertainties regarding sexual development remain nowadays. The development of the ovaries, for instance, is as little-known today as during Marc's and Geoffroy Saint-Hilaire's times.[128] The latter admitted that it was impossible to identify the causes of hermaphroditism as long as scientists ignored how sex was 'normally' constituted.[129] Similarly, Mouriquand observes: 'qu'est-ce qui fait qu'un garçon est un garçon et une fille une fille? […] à l'aube du troisième millénaire on est incapable de répondre à cette question dans sa globalité' [What makes a boy a boy and a girl a girl? […] At the beginning of the third millennium, we are still unable to completely answer this question].[130]

Nineteenth-century scientists were more interested in the social consequences of intersexuality and envisaged the medico-legal applications of scientific observations. For Marc, the role of the physician was twofold: to identify the sex of a hermaphrodite and to determine whether a hermaphrodite was able to procreate. Geoffroy Saint-Hilaire, however, relativized the second function, since sterility was no grounds for marriage annulment.[131] The task of doctors and scientists was not for the sake

127 Geoffroy Saint-Hilaire, *Traité de tératologie*, I, 185–86; Marc, 'Hermaphrodite', 112. The former's statement was made prior to Laqueur's timeline of the theorization of an androgynous embryo: Laqueur, 'Orgasm', 3; Laqueur, *Making Sex*, 10.

128 Marc, 'Hermaphrodite', 113–14; Geoffroy Saint-Hilaire, *Traité de tératologie*, II, 56; Chatelain, Morel and Nicolino, 'Le développement et la différenciation sexuels', 43; Fausto-Sterling, *Sex/Gender*, 17.

129 Geoffroy Saint-Hilaire, *Traité de tératologie*, II, 53.

130 Mouriquand, 'S'asseoir dehors', 58.

131 Marc, 'Hermaphrodite', 116–20; Geoffroy Saint-Hilaire, *Traité de tératologie*, III [1836], 572. For more on marriage and hermaphroditic impotence, see Houbre, 'Dans l'ombre de l'*hermaphrodite*', 94.

of science, but, allegedly, for the sake of society. Geoffroy Saint-Hilaire, unable to foresee that same-sex marriage would one day become a reality in France, mentioned what he considered to be the moral danger of uniting two people of the same sex, especially if their marriage could not be annulled.[132] As for Marc, he showed the importance that parents, doctors, the government and the Church (that is, society) gave to sex, by the simple question: 'Is it a boy or a girl?'

> Les premiers mots d'une femme délivrée du travail de l'accouchement sont: qu'est-ce que j'ai fait? Est-ce une fille, est-ce un garçon, s'écrient à la fois le père, la famille et les assistants? De quel sexe est l'enfant qui vient de naître, demande l'officier de l'état civil? Quel enfant présentez-vous à l'église, dit le ministre des autels?[133]
>
> [The first words of a woman freed from childbirth are: 'What have I done?' 'Is it a girl? Is it a boy?', the father, the family and the assistants cry together. 'What is the sex of the child who has just been born?', the civil registrar asks. 'Which child are you introducing to the Church?', the priest asks.]

Marc believed that every hermaphrodite had a predominant sex that the physician should be able to identify if he was cautious enough when examining their body.[134] For the rare cases of neutral or mixed hermaphrodites, he should highlight the difficulty of his task and 'rechercher, s'il est possible, dans les goûts et les habitudes de l'hermaphrodite, le sexe qu'il convient de déclarer prédominant' [to seek, if this is possible,

132 Geoffroy Saint-Hilaire, *Traité de tératologie*, III, 573–74.

133 Marc, 'Hermaphrodite', 90–91. Judith Butler (*Bodies That Matter*, 7) made the same observation almost two centuries later.

134 Interestingly, Suzanne Kessler shows that the endocrinologists and urologists whom she interviewed in 1985 used a similar discourse: 'The message in these examples is that the trouble lies in the doctor's ability to determine the gender, not in the baby's gender per se. The real gender will presumably be determined/proven by testing, and the "bad" genitals (which are confusing the situation for everyone) will be "repaired"': *Lessons from the Intersexed* (New Brunswick and New York: Routledge, 1998), 23. Whilst nineteenth-century scientists and doctors considered that they merely *identified* the 'true' sex of a hermaphrodite, twenty-first ones are more likely to *assign* a gender to an individual with intersex traits.

in the hermaphrodite's tastes and habits, the sex that must be declared predominant].[135]

Marc's statements are instructive for two reasons. Firstly, they highlight the contradictions within scientific discourse. Although they had observed ambiguous bodies, scientists could not admit the existence of individuals with undetermined sex. Here, my position contrasts with that of Linton, who claims that French scientists acknowledged the existence of 'true' hermaphroditism.[136] As we have seen, admitting openly that individuals with male and female genitalia were real was too significant an infringement of the sexual dichotomy to be tolerated. Scientists had therefore to tolerate ambiguous and contradictory reasoning. On the one hand, Marc and Geoffroy Saint-Hilaire declared that some of the cases observed were neutral, mixed or even bisexual, which reveals their inability to assign a prevalent sex – Geoffroy Saint-Hilaire even confessed that the determination of the sex could be impossible.[137] On the other hand, they claimed on a regular basis that just such an absence of a sex was impossible and adamantly denied the existence of the 'perfect' hermaphrodite. They rejected the idea that one could genuinely be deprived of a specific sex. Certainly, failing to assign a sex could be due to the lack of information that was available to them, as they established their categories by following the reports of their colleagues. However, their certainty that such sex existed highlights the pressure exercised by society's reliance on a strict separation of the sexes to function harmoniously. The only bending of the sexual dichotomy that Moreau de la Sarthe, Marc and Geoffroy Saint-Hilaire allowed themselves to make was the indication of the *predominance* of a sex instead of its complete existence.[138]

135 Marc, 'Hermaphrodite', 115.

136 Linton (*Prescribed Fictions*, 40–72) thus disagrees with Dreger (*Hermaphrodites*, 139–66), who claims that the gonads were increasingly seen as the only reliable markers of the 'real' sex of an individual, a definition that put an end to the possibility of 'true' hermaphroditism. This exclusive focus on the gonads does not appear in the texts of this corpus.

137 Geoffroy Saint-Hilaire, *Traité de tératologie*, III, 574–75.

138 Moreau de la Sarthe, *Histoire naturelle*, I, 221; Marc, 'Hermaphrodite', 115; Geoffroy Saint-Hilaire, *Traité de tératologie*, III, 577.

Secondly, Marc's commentaries take psychological factors into consideration when assigning a sex to a hermaphrodite. Many cases show that taste and behaviour were as essential as the anatomical and physiological description of the hermaphrodites, given that genitalia were judged to be deceptive. Each sex, they thought, was predisposed to specific gendered behaviour, produced by the physiological sex. While education could hide the 'real' sex of a hermaphrodite, their activities revealed it. Consequently, biographical details about hermaphrodites were frequent in scientific writings in order to assign the sex of a hermaphrodite. For example, Marc explained that the tastes of the male hermaphrodite Marie-Marguerite, who was considered to be of female sex at birth, changed along with 'her' body at puberty. 'She' preferred working in the fields to taking care of the home and the poultry yard; once 'her' sex had been officially changed, 'she' became a good ploughman.[139] The evolution of Marie-Marguerite's tastes, or rather the tastes imposed by education being replaced by 'natural' tastes, displays 'her' inability to be a 'real' woman.

Scientists of the first half of the nineteenth century collectively denied the existence of true hermaphrodites because their intersexuality questioned the so-called 'natural' separation of the sexes. This increasing rejection of hermaphrodites was coupled with the claim for an absolute sexual difference and was mirrored by their equally important interest in, and rejection of, homosexuals.

Characteristics of Homosexuals

Whereas intersexuality had long been part of medical discourse, scientific considerations on homosexuality were still in their infancy during the July Monarchy. Homosexuality, however, was well-known through its representation in religious, moral and legal texts.[140] In the first volume

139 Worbe, quoted by Marc, 'Hermaphrodite', 91 and 93. See also Mak, *Doubting Sex*, 51–52.

140 On the legal discourse of homosexuality, see Robb, *Strangers*, 17–39; Peniston, *Pederasts and Others*, 13–22.

of his *Histoire de la sexualité*, Michel Foucault suggests that the figure of the homosexual, and more generally that of the so-called pervert, were creations of the second half of the nineteenth century. Previously, only the existence of the sodomite had been acknowledged. The difference between the pervert and the sodomite was that the latter was defined by his actions. He was deemed to be a criminal who acted against the law by having anal intercourse (sodomy being a crime in France until 1791). Conversely, the pervert's labelling related to his identity. He was considered mentally ill and was scrutinized by physicians because his mindset, not his actions, was contentious.[141] The term *homosexuality*, almost universally used nowadays to talk about same-sex relationships, did not exist in the first half of the nineteenth century.[142] Coined in German by the Hungarian Karl-Maria Kertbeny in the 1860s, it first appeared in French scientific discourse in the 1890s and became common in the French language only after the Second World War.[143] The terms *inversion* and *inverti*, which are now outdated, were popularized in French by Jean-Martin Charcot and Valentin Magnan during the Third Republic and were inspired by the works of German scientists, notably Karl Westphal.[144] The most frequently used terms in scientific writings at that time were *sodomie* and *pédérastie*. Their usage was, in various contexts, both broader and narrower than the current one because they focused on sexual intercourse between men almost exclusively. The nineteenth-century use of the term *pederasty* did not necessarily imply the relationship between an adult and an adolescent. Additionally, academic articles about sodomy usually disregarded anal intercourse with women.

During the first half of the nineteenth century, homosexuality was never the object of a scientific treatise. *Sodomie* and *pédérastie* appeared in sentences within treatises relating to other topics (such as prostitution,

141 Foucault, *Histoire de la sexualité*, I, 50–67. For more on the difference between the sodomite and the homosexual, see Rosario, 'Pointy Penises', 167–68; Revenin, 'Conceptions', 26.

142 I use it nonetheless for practical reasons.

143 Régis Revenin, 'Homosexualité et virilité', in *Le triomphe de la virilité*, 369–401 (369).

144 Rosario, 'Pointy Penises', 146–47 and 156.

prison and venereal diseases) or in short articles in medical dictionaries. What strikes the modern reader of these texts is the predominance of an embedded anxiety. Physicians and scientists, afraid of being mistaken for the objects of their studies, were led by homophobia. Accordingly, they immediately informed their audience that they disapproved of the customs and conducts that they described. Their vocabulary was both deprecating and moralistic: *vice*, *perversity*, *immorality*, *abominable*, *vile*, *depraved* and *shameful* are frequently used in the articles 'Pédérastie' and 'Sodomie' in the *Dictionnaire des sciences médicales*.[145] Their authors stressed the fact that they were discussing homosexuality begrudgingly, for the sake of science and social order. Nigel Smith has called this strategy the *reluctance-discourse*.[146] This is illustrated by the opening sentence of François-Emmanuel Fodéré (1764–1835)'s article on sodomy: 'Que ne puis-je éviter de salir ma plume de l'infâme turpitude des pédérastes!' [How can I avoid not dirtying my pen with the repugnant turpitude of pederasts?]. The French forensic scientist Ambroise Tardieu (1818–79) quoted this sentence at the beginning of the section of his *Étude médico-légale sur les attentats aux mœurs* [*Forensic Study on Indecent Assault*] dedicated to pederasty.[147] What is not mentioned by Smith is the so-called heroic attitude adopted by these scientists. Tardieu claimed his bravery in dealing with such an 'appalling' topic: 'J'ai dit que

145 Reydellet, 'Pédérastie', in *Dictionnaire des sciences médicales*, XL (1819), 37–45; François Fournier Pescay, 'Sodomie', in *Dictionnaire des sciences médicales*, LI (1821), 441–48. See also Claude Courouve's inventory of substantives and adjectives associated with male homosexuality from the Renaissance until the twentieth century: *Vocabulaire de l'homosexualité masculine* (Paris: Payot, 1985), 26–28; Jacques-Philippe Saint-Gérand, 'Homosexualité des *alphadécédets*: Remarques sur un innommable des dictionnaires conformes, et recours aux excentriques', *Romantisme*, 159 (2013), 19–34.

146 Nigel E. Smith, 'Silence, Secrecy, and Scientific Discourse in the Nineteenth Century', in Dominique D. Fisher and Lawrence R. Schehr, eds, *Articulations of Difference: Gender Studies and Writing in French* (Stanford: Stanford University Press, 1997), 83–99 (87). See also Hahn, *Nos ancêtres les pervers*, 49–50.

147 François-Emmanuel Fodéré, 'Sodomie', in *Traité de médecine légale et d'hygiène publique ou de police de santé*, 6 vols (Paris: Mame, 1813), IV, 374–76 (374); Ambroise Tardieu, *Étude médico-légale sur les attentats aux mœurs*, 2nd edn (Paris: Baillière, 1858 [1857]), 112.

je ne reculerais pas devant l'ignominie du tableau; c'est ici qu'il faut en retracer les traits les plus hideux, et emprunter jusqu'au langage des êtres dégradés, dont je veux essayer d'ébaucher la repoussante image' [I said that I would not retreat from the ignominy of the picture; it is here that one must draw the most hideous traits and even borrow from the language of these degraded beings, whose repulsive image I want to try to sketch].[148] However, Tardieu's zeal in examining pederastic bodies and the numerous revisions of his *Étude médico-légale* call his declared repulsion into question and show the increasing interest of the scientific community in, and the popularity of, such a topic.[149]

Parts of articles from the *Dictionnaire des sciences médicales* that are solely dedicated to scientific considerations are short. For instance, in François Fournier Pescay's 'Sodomie' entry, only one fifteenth of the article relates to medical concerns, such as the description of the symptoms of sodomites. The rest of the article mostly develops a history of sodomy, which highlights the opposition between the Hebrews, who condemned it, and other ancient and modern peoples who practised it freely. These references stress the idea that sodomy was always seen as a foreign vice.[150] Likewise, almost all Reydellet's articles on pederasty attempted to demonstrate the inferiority of the Ancients where morals were concerned. Reydellet elusively defined *pédérastie* as a 'vice infâme que la morale, la nature et la raison réprouvent également' [infamous vice that morals, nature and reason equally condemn].[151] This constitutes an example of what Smith calls the *avoidance-discourse*, according to which doctors admitted the existence of a 'vice' in their writings but were reluctant to be more specific regarding its nature.[152] Dwelling on historical and moral considerations rather than scientific facts contrasts with the *Dictionnaire*'s editorial promise to promote impartiality, experimental method and empirical observations.[153]

148 Tardieu, *Étude médico-légale*, 118. See also 7.

149 Smith, 'Silence', 89.

150 Courouve, *Vocabulaire de l'homosexualité masculine*, 28; Robb, *Strangers*, 7; Murat, *La loi du genre*, 241–42.

151 Reydellet, 'Pédérastie', 37.

152 Smith, 'Silence', 86–87.

153 L. J. Renauldin, 'Introduction', in *Dictionnaire des sciences médicales*, I (1812), xix–clxix (clxix).

Nevertheless, although the relevance of Smith's analysis of scientific texts on homosexuality is unquestionable, his labelling of these texts as 'pseudo-scientific' is an anachronistic judgement.[154] They indisputably appeared as scientific writings to nineteenth-century readers due to the prominence of their authors and their publication in medical dictionaries.

A common belief that characterized the discourse on homosexuality during the first third of the nineteenth century is that the sexual identity of homosexuals was located in their sexual organs. Physicians distinguished between the roles played during anal intercourse, that is, between the sodomizer (the active partner) and the sodomized (the passive partner), and identified specific symptoms accordingly. During the first part of the nineteenth century, physicians focused mostly on men who took the passive role in sodomy and pointed to numerous alleged deformations of the anal area and the buttocks, such as contusions, excrescences, ulcerations, inflammation, excessively large buttocks, relaxation of the sphincter, anal tearing and swollen anal rolls of fat that are similar to labia, among others. Being sodomized was also believed to generate diseases, including haemorrhoids, cancerous tumours called *squirres* [scirrhi] and syphilis.[155] These long lists of obscure medical jargon demonstrate the willingness of physicians to identify any peculiarity of the anal area as a medical symptom of homosexuality. Thus, not only did they represent homosexuality as a collection of sexual acts, but as a collection of acts which had a pathological effect on the body. As Régis Revenin states: 'la médecine légale insiste [...] sur les stigmates physiques supposés de l'homosexualité, bien plus que sur les causes physiologiques ou psychologiques de cette nouvelle perversion sexuelle' [forensic medicine stresses [...] the alleged physical stigmata of

154 Smith, 'Silence', 83 and 86.

155 Mahon, 'Sodomie', in *Médecine légale et police médicale*, I, 138–39 (138); Fodéré, 'Sodomie', 375; Reydellet, 'Pédérastie', 45; Fournier Pescay, 'Sodomie', 447; Eusèbe-Fr. de Salles, 'Médecine légale', in Jean-Louis Alibert et al., eds, *Encyclopédie des sciences médicales ou Traité général, méthodique et complet des diverses branches de l'art de guérir, 2nd division: Médecine* (Paris: Bureau de l'Encyclopédie, 1835), 1–291 (224–25); J.-G. Human, *Nosographie des maladies vénériennes ou Étude comparée des divers agents thérapeutiques qui ont été mis en usage pour combattre ce genre d'affections* (Paris: Béchet Jeune, 1838), 84–85.

homosexuality, much more than the physiological or psychological causes of this new sexual perversion].[156] These detailed descriptions were seldom the result of personal observations. As was the case for experts in intersexuality, scientists copied each other's articles because they judged symptoms to be commonplace. For example, Fodéré enumerated the same symptoms of sodomy in the same order as Mahon.[157]

The most spectacular of the passive homosexual's symptoms was the so-called *anus infundibuliforme* [funnel-shaped anus].[158] Reydellet, Parent-Duchâtelet and Tardieu attributed the discovery of the funnel-shaped anus to a doctor called Cullerier, probably Michel Cullerier, a physician and surgeon who studied syphilis.[159] However, the source of this information is not clearly referenced by any of the three physicians. The invocation of Cullerier as the 'father' of the *anus infundibuliforme* seems to be used to reinforce the assertion that such a 'monstrous' anus was genuine. However, the objectivity suggested by the mention of a medical authority is called into question by the absence of bibliographical references that would be expected from scientific works. The uncertainty concerning the specificity of Cullerier's *anus infundibuliforme* theory, his method of observation and his sample therefore undermine his authority. Parent-Duchâtelet is the only one to question the existence of this funnel-shaped anus on the grounds that he did not notice similar symptoms in female prostitutes. He concludes that this demonstrates a 'nouvelle preuve de la réserve extrême qu'il faut mettre en médecine légale dans toutes les décisions, et du danger de se fier d'une manière trop exclusive à ces assertions qu'on rencontre malheureusement trop souvent dans les livres de médecine' [new proof of the utmost caution which must be adopted with regard to all decisions in forensic medicine,

156 Revenin, 'Homosexualité et virilité', 377. See also Peniston, *Pederasts and Others*, 53.

157 Fodéré, 'Sodomie', 375; Mahon, 'Sodomie', I, 138. Admittedly, Fodéré cited Mahon, but it is in relation to another observation (his colleague's warning against mistaking symptoms of venereal diseases for sodomy).

158 Reydellet, 'Pédérastie', 45; Hubert Lauvergne, *Les forçats considérés sous le rapport physiologique, moral et intellectuel, observés au bagne de Toulon* (Paris: Baillière, 1841), 289; Tardieu, *Étude médico-légale*, 135–37.

159 Reydellet, 'Pédérastie', 45; Parent-Duchâtelet, *De la prostitution*, I, 226; Tardieu, *Étude médico-légale*, 135.

and of the danger of relying too heavily on these assertions which are unfortunately all too frequently encountered in medical books].[160] Tardieu, however, we will go on to see, did not share the same reservations.

Tardieu's *Étude médico-légale* is an important source for those interested in French scientific discourse on homosexuality. However, its interest is not rooted in its novelty, as many of Tardieu's observations had already been made by his colleagues. Rather, it resides in the systematization of his discourse. Indeed, for the first time, homosexuality was not confined to a short dictionary article focusing mostly on cultural and historical aspects or briefly mentioned in an essay that focused on another topic; it was now a fundamental object of study, since a whole section was dedicated to what Tardieu calls pederasty.[161] In *Étude médico-légale*, homosexuals are divided into categories; descriptions are detailed; statistics are provided … although they are not devoid of mathematical mistakes, such as: 71 + 18 = 88![162] With the publication of the first edition of Tardieu's study, homosexuality entered the field of science and the era of medicalization. As Laure Murat points out, this study perfectly combined 'un réquisitoire moral et [...] une science *vérifiable*' [moral indictment and [...] *verifiable* science].[163] Homosexuality became pathologized; it was deemed to be a disease that must be examined, treated and cured.

In his study, Tardieu emphasizes the difference between an active partner and a passive one, and states that signs of pederasty vary according to the role that the pederast was likely to play during anal intercourse, although his sample contradicted his theory. Indeed, among the 206 bodies that Tardieu claimed to have observed, almost half of the cases were exclusively passive (ninety-nine) and only eighteen were exclusively active,

160 Parent-Duchâtelet, *De la prostitution*, I, 227.

161 The previous sections are devoted to public immorality and rape; the order chosen suggests that these offences are listed from apparently less to more serious types of crime. On the popularity of Tardieu's treatise, see Jean-Paul Aron and Roger Kempf, *Le pénis et la démoralisation de l'Occident* (Paris: Grasset, 1978), 85; Rosario, 'Pointy Penises', 148; *The Erotic Imagination*, 72; Thompson, 'Creating Boundaries', 114; Murat, *La loi du genre*, 112 and 116–19.

162 Tardieu, *Étude médico-légale*, 144.

163 Murat, *La loi du genre*, 117 (emphasis in original).

whereas more than one third were both active and passive (seventy-two) and seventeen had 'habitudes non caractérisées' [uncharacterized habits] – whatever Tardieu meant by this phrase.[164] These figures make the distinction between different categories of homosexual unconvincing.

Despite his small sample, Tardieu, unlike his colleagues who focused on passive homosexuals, also examined the signs of active homosexuality. This constituted his 'tour de force', to borrow Murat's words.[165] He thoroughly analysed the shape of homosexuals' penises.[166] According to Tardieu, such a penis was either long, thinner at the end and resembled the penis of a dog; or it was very big, with a thin head that looked like the muzzle of an animal. Perhaps unsurprisingly, active homosexuals were dehumanized and compared to animals in his study.[167] Tardieu also argued that the penis of a pederast was twisted like a corkscrew. This shape would be due to the torsion that it operated when penetrating the anus. Furthermore, he observed that some homosexuals had a twisted mouth, short teeth and thick, deformed lips due to their 'plus abjectes complaisances' [most abject indulgences] and 'usage infâme' [repugnant use].[168] These words designated, one can assume, fellatio.

Tardieu's representations of the indelible marks that sodomy leaves on homosexual bodies highlight his eagerness to reduce homosexuals to their physical identity and to anal intercourse. This perspective is manifest in his biased interpretation of Parent-Duchâtelet's lack of evidence of funnel-shaped anuses in female prostitutes having anal intercourse. Rather than calling Cullerier's theory into question, as Parent-Duchâtelet implicitly invited the reader to do, Tardieu concludes that sodomy does not affect the female body in the same way as it affects the male, homosexual body.[169] Tardieu does not attempt to explain why the same actions do not have the same consequences, but this absence of explanation attests to his willingness to distort the facts in order to comply with his interpretation of sodomy.

164 Tardieu, *Étude médico-légale*, 129.
165 Murat, *La loi du genre*, 115.
166 Tardieu, *Étude médico-légale*, 144–47.
167 Smith, 'Silence', 91.
168 Tardieu, *Étude médico-légale*, 143.
169 Ibid., 135.

Vernon Rosario considers that the notion of pederasty in Tardieu's writings 'undermines the oft-recited Foucauldian tenet that an epistemic break exists between pederasty and homosexuality'.[170] Certainly, Tardieu's *Étude médico-légale* cannot be unequivocally read in the light of sodomitic or homosexual relations. However, rather than dismissing Foucault's theory, I view it as an illustration of a period of transition from the conception of same-sex relationships as a collection of sodomitic acts to homosexuality as an identity. On the one hand, Tardieu focused on *acts* 'against nature', mostly between men, but did not exclude anal intercourse with women. For instance, the first of his *Observations* (a collection of medical cases) is that of a woman who had anal sex with her husband.[171] On the other hand, other sections recognize homosexuality as an identity. Indeed, Tardieu dedicates many pages to pederastic prostitution in which young male prostitutes, called *tantes* [aunts], or their pimps, blackmailed their clients.[172] He observes that 'ce sont surtout ceux que l'on appelle *tantes*, c'est-à-dire ceux qui se prostituent aux véritables pédérastes, qui recherchent parfois à leur tour les rapports avec les femmes' [It is above all those so-called *aunts*, that is, those who prostitute themselves to the true pederasts, who in turn sometimes seek relations with women].[173] Here, the physician clearly distinguishes between acts and identity. The *tantes* are not 'true' pederasts because, although they practise sodomy with other men, they do it for financial reasons and, allegedly, do not feel pleasure in such acts, unlike their clients. Likewise, Eusèbe de Salles (1796–1873) differentiates 'real' sodomites from those who practised anal sex because they were deprived of women on ships, in prisons, boarding schools and seminaries.[174]

Tardieu's study epitomizes the trend, starting from the end of the 1830s and the 1840s, of longer essays that focused, although indirectly, on

170 Rosario, 'Pointy Penises', 150. See also *The Erotic Imagination*, n. 12, 192.

171 Tardieu, *Étude médico-légale*, 162–63.

172 *Tante* stems from prison slang from the beginning of the nineteenth century and originally designated the male homosexual prisoner, the male homosexual by analogy, then a male prostitute: Courouve, 'Tante', in *Vocabulaire de l'homosexualité masculine*, 207–09.

173 Tardieu, *Étude médico-légale*, 122 (emphasis in original).

174 Salles, 'Médecine légale', 224.

homosexuality. These texts gave a definition of homosexuality that began to compete with that grounded strictly on bodies and sexual acts.[175] This definition was based on the behaviour and psychological mindset of homosexuals. Certainly, homosexuality was only defined as a mental illness in France after the publication of Charcot and Magnan's article 'Inversion du sens génital et autres perversions sexuelles' [Inversion of the Genital Sense and Other Sexual Perversions] (1882). Pathologizing homosexuality occurred considerably later in France than it did in Germany, where the works of psychiatrists, such as Johann Casper, Karl Heinrich Ulrichs and Karl Westphal, defined homosexuality in psychiatric terms in the 1850s and 1860s.[176] Yet, earlier, French scientists had acknowledged the idea that homosexuality influenced the personality and the tastes of pederasts, and had claimed that the homosexual was not only defined by his participation in anal sex, but also by the effect that his attraction to other men supposedly had on his personality. However, in the first half of the nineteenth century, these personality traits were only noted in passing or were only seen as the result of the effeminate physiology of a male homosexual. Progressively, the 'real' sex of the homosexual came into increasing conflict with his gendered traits. Whilst his body was male, he allegedly manifested behaviour socially defined as feminine, and thus defied the arguments defending a 'natural' separation of the sexes. In the remainder of this section, I will explore how physiological interpretations of homosexuality progressively yielded ground to behavioural ones and how nineteenth-century scientists conflated sexuality and gender.

Like eunuchs and impotent men whose physiologies and personalities were associated with those of women because they lacked sperm,[177]

175 For Philippe Lejeune, the change is even later (the 1860s), although the primary texts analysed in the following paragraphs challenge this claim: 'Autobiographie et homosexualité en France au XIXe siècle', *Romantisme*, 56 (1987), 79–94 (80).

176 Robert A. Nye, 'Sex Difference and Male Homosexuality in French Medical Discourse, 1830–1930', *Bulletin of the History of Medicine*, 63 (1989), 32–51 (41); Rosario, 'Pointy Penises', 155–56; Daniel Grojnowski, ed., *Confessions d'un inverti-né suivies de Confidences et aveux d'un Parisien* (Paris: Corti, 2007), 29; Revenin, 'Conceptions', 30–31; Julie Mazaleigue-Labaste, 'De l'amour socratique à l'homosexualité grecque', *Romantisme*, 159 (2013), 35–46 (41).

177 Virey, *De la femme*, 180.

homosexuals were considered to be deprived of the beneficial influence that sperm was supposed to have on men. Homosexuals, like eunuchs, were thought to display female physiological characteristics. Their skin was smooth and soft, their muscles were weak and their voices were high-pitched.[178] Hubert Lauvergne (1797–1859), chief medical officer at the hospital of the *bagne* [labour camp] of Toulon, pointed out the similarity that was believed to exist between a homosexual and a castrato: 'La voix de cet homme est faible, tourne à celle du *castrat*' [The voice of this man is weak, similar to that of the *castrato*].[179] For Cabanis, their walk was also a symptom of their physiological difference, since it was supposed to be the result of the width of their pelvis, similar to that of a woman. When they walked, they allegedly drew a large curve around their centre of gravity.[180]

The psychiatrist Étienne Esquirol (1772–1840) also observed the particularity of the homosexual gait but was unusual in attributing it to imitation rather than to body shape, noting that one of his patients 'imitait leur démarche [des femmes] en se promenant' [imitated women's gait when walking].[181] Indeed, Esquirol was one of the first doctors to adopt a psychiatric perspective on homosexuality, or, what is in fact, transidentity. Esquirol dedicated the last three pages of his chapter on demonomania, or religious melancholia, to the 'maladie des Scythes' [Scythian disease], following Hippocrates's terminology.[182] As he gave little information on this disease, adopting the *avoidance-discourse* strategy mentioned earlier, it is difficult to assess whether he distinguished between what is considered nowadays to be two separate concepts: homosexuality and transidentity. The psychiatrist provided no definition, symptoms or cure, and described

178 Cabanis, *Rapports*, I, 385 and 388–89; Virey, *De la femme*, 180.

179 Lauvergne, *Les Forçats*, 288 (emphasis in original).

180 Cabanis, *Rapports*, I, 388–89.

181 Étienne Esquirol, *Des maladies mentales: Considérées sous les rapports médical, hygiénique et médico-légal*, 2 vols (Paris: Baillière, 1838), I, 525.

182 Ibid., I, 523. The ancient Greeks thought that the Scythians, a people from Central Asia, suffered from a disease which made them either sterile or inclined to adopt feminine behaviour: Marie Delcourt, *Hermaphrodite: Mythes et rites de la Bisexualité dans l'Antiquité classique* (Paris: Presses Universitaires de France, 1958), 60–61.

only two cases: that of Madame M…, a widow who believed themselves to be a man since their husband had died, and a 26-year-old man (the one with the feminine gait) who 'se persuada qu'il était femme et chercha à en convaincre tout le monde, même les membres de sa famille' [persuaded himself that he was a woman and tried to convince everyone, even the members of his family, that he was].[183] Esquirol did not believe that his patients truly belonged to the other sex and likened them to those who think that they are wolves, dogs or cats. Although he showed empathy towards his patients and provided them with a safe space in which they could express themselves, he nonetheless encouraged them to accept their biological sex.[184] However, more importantly for this analysis is the fact that Esquirol highlighted the psychological dimension of gender and sexual identity construction.

During this period, the physiological definition of homosexuality became increasingly stretched as scientists tried to use it to explain a wide variety of behaviours socially gendered as feminine. It was not only the deviations from the 'normal' male body that constituted homosexuals' identity (their effete physical constitution, their canine penis or their funnel-shaped anus), but also their general look, choice of clothes, allure and tastes. Salles summarizes the opinion of most authors by saying: 'Les individus portés à la sodomie ont en général un caractère efféminé' [Individuals disposed to sodomy generally have an effeminate character].[185] According to William Peniston, it was a commonly held opinion in the nineteenth century that the appearance, mannerisms and behaviour of homosexuals were effeminate (and indeed this opinion is still widely shared nowadays), although it was hardly supported by facts.[186]

183 Esquirol, *Des maladies mentales*, I, 525. Madame M …'s case is described in I, 523–24.

184 Murat, *La loi du genre*, 197–98.

185 Salles, 'Médecine légale', 224.

186 Peniston's study of the police ledger of arrests for public offences against decency in the 1870s challenges this notion, given that only four cases of overt effeminacy are reported out of 328 cases: two of cross-dressing and two of men acting like women (*Pederasts and Others*, 102–04). See also Michael Sibalis, 'The Palais-Royal and the Homosexual Subculture of Nineteenth-Century Paris', in *Homosexuality in French History and Culture*, 117–29 (127).

Lauvergne and Tardieu were the most explicit regarding the effeminacy of homosexuals. The physician of the *bagne* describes convicts walking around Toulon who smelled of musk and amber, and wore silk and batiste, elegant clothes, sophisticated shoes, white stockings and small chains around their neck.[187] Likewise, Tardieu enumerates general signs that are relative to the external appearance of the pederast:

> Les cheveux frisés, le teint fardé, le col découvert, la taille serrée de manière à faire saillir les formes, les doigts, les oreilles, la poitrine chargés de bijoux, toute la personne exhalant l'odeur des parfums les plus pénétrants, et dans la main, un mouchoir, des fleurs, ou quelque travail d'aiguille, telle est la physionomie étrange, repoussante, et à bon droit suspecte, qui trahit les pédérastes.[188]
>
> [Curly hair, painted face, open collar, waist tightly fitted to enhance the figure, fingers, ears and chest adorned with jewels, the whole body exuding the smell of the most penetrating perfumes, and in hand a handkerchief, flowers or some embroidery work: such is the strange, repulsive and rightfully suspicious physiognomy which betrays pederasts.]

Although there is no explicit comparison with women, these descriptions highlight the allegedly feminine nature of homosexuals.[189] Virey and Menville de Ponsan point out similar female tastes and concerns with appearance when they declare, for instance, that young girls liked shiny jewellery.[190] Similarly, Esquirol, Lauvergne and Tardieu describe men who spent a considerable amount of time on their *toilette*, curled their hair, wore feminine outfits and admired themselves in the mirror.[191] Jenyjoy La Belle's analysis of looking-glasses in literature shows that looking at oneself in the mirror is usually associated with women's behaviour and is considered an act of vanity.[192] Nineteenth-century scientists shared this

187 Lauvergne, *Les Forçats*, 290.

188 Tardieu, *Étude médico-légale*, 130.

189 On the connections between misogyny and homophobia, see Gazalé, *Le mythe de la virilité*, 285.

190 Virey, *De la femme*, 70; Menville de Ponsan, *Histoire médicale*, I, 50.

191 Esquirol, *Des maladies mentales*, I, 525; Lauvergne, *Les Forçats*, 292–93; Tardieu, *Étude médico-légale*, 130–31.

192 Jenijoy La Belle, *Herself Beheld: The Literature of the Looking Glass* (Ithaca and London: Cornell University Press, 1988), 14–20.

opinion and built their reasoning upon it. The term *coquetterie* [vanity] is often used, notably by Lauvergne, to qualify the behaviour of homosexuals.[193] Women's vanity, in reasonable proportions, was deemed a quality which responded to a biological impulse for the perpetuation of the species. It was the natural way for a woman to attract a man, secure his protection and start a family.[194] In contrast, male vanity was 'abject' for Tardieu not only because biology supposedly dictated that men must seduce women by their strength, but also because this elegance was thought to hide the 'malpropreté' [uncleanliness] of homosexuals.[195] Nineteenth-century definitions of masculinity dictated that men were not naturally vain, and thus a foppish man was not a 'real' man. Smith rightly argues that the stereotypical depiction of homosexuals as effeminate in Tardieu's essay is designed to inspire scorn for them.[196] His reflection can be extended to Tardieu's colleagues, especially Lauvergne. These physicians suggest that the hybrid gender of homosexuals lacks the positive attributes of both genders, as Cabanis claimed for eunuchs, whose humanity was allegedly altered by their physical mutilation.[197] Homosexuals are depicted as lacking masculine courage and honour, without gaining feminine tenderness. In short, the gendered characteristics of homosexuals contravened society's strict definitions used in the division of the sexes, and thus made it impossible to categorize them as either male or female.

In the early nineteenth century, homosexuals were identified as such by many physicians based on the shape of their penises and, even more commonly, of their anuses. The concept of homosexuality grounded exclusively in men's physical constitution, however, faced obstacles, as it did not enable doctors to explain the so-called feminine gendered traits, such

193 Lauvergne, *Les Forçats*, 289 and 292–94.

194 Moreau de la Sarthe, *Histoire naturelle*, I, 712–14 and II, 401; Virey, *De la femme*, 191, 200 and 242; Menville de Ponsan, *Histoire médicale*, I, 44–48 and 103–04.

195 Tardieu, *Étude médico-légale*, 130. See also Salles, 'Médecine légale', 224.

196 Smith ('Silence', 89) notes that Tardieu adopts the dominant role and 'penetrates his "patient" in the name of science'.

197 Cabanis, *Rapports*, I, 386.

as *coquetterie*. Gradually, as the century advanced, this identification would give way to more psychological interpretations of same-sex relationships.

From Genitalia to Behaviour: The Assignment of the Sex to an Individual

All the texts reviewed in this chapter were written by scientists from diverse backgrounds who regarded the body as the key to assigning a sexual identity to an individual. Accordingly, medical observations and descriptions of human bodies, especially of the genitalia, constituted an essential part of nineteenth-century science.[198] Marc and Tardieu advised on the methods of practising medical examinations on hermaphrodites and homosexuals. Tardieu even indicated 'tricks' in order to overcome prisoners' ways of hiding their alleged homosexual condition (for instance, make them kneel on the edge of a chair or draw out the examination).[199] Examinations of intersex and homosexual bodies implied the use of exploratory techniques, such as the introduction of a finger into the vagina or the anus, a gesture that was not only highly invasive, but had sexual and even homoerotic connotations in the latter case.[200] This medical behaviour entailed a paradox: scientists' dedication to demonstrating the harmful effects of homosexuality led them to adopt what could be regarded as homosexual behaviour. It would nevertheless be simplistic to interpret these acts and writings as being produced by men repressing their homosexuality. Rather than expressing 'a closeted desire to engage

198 For instance, Mahon, 'Hermaphrodites', I, 102–04; Marc, 'Hermaphrodite', 93, 97–100, 102–03 and 106–07; Geoffroy Saint-Hilaire, *Traité de tératologie*, II, 71–72, 75, 97–99, 102–04, 117–18, 136–39, 154–55, 159 and 164–65; Tardieu, *Étude médico-légale*, 133–40 and 145–47.

199 Tardieu, *Étude médico-légale*, 150. On how to practise medical examinations: Marc, 'Hermaphrodite', 116; Tardieu, *Étude médico-légale*, 149–51.

200 Geoffroy Saint-Hilaire, *Traité de tératologie*, II, 77 and 81; Tardieu, *Étude médico-légale*, 150. See also Smith, 'Silence', 92.

in homosexual acts',[201] these texts displayed a scientific bias. Scientists judged the context of sodomitic acts to be as relevant as the acts themselves, although they never described the nature of these contexts in detail. Tardieu's text, for instance, shows that the introduction of fingers and objects into the anus is considered to be acceptable and even recommended in a medical examination. However, the introduction of foreign objects (bottles and pieces of wood) is condemned in other contexts, such as during relations between prostitutes and clients.[202] Contexts, such as anal intercourse with a prostitute, rather than solely acts, are fundamental to determine what is considered acceptable behaviour. They reveal that external and internal sexual organs, secondary sexual characteristics (hair, voice and breasts) and participation in sodomitic acts alone were not regarded as convincing evidence to determine the homosexuality of an individual.

Certainly, genitalia and physiological characteristics were still deemed of primary importance in assigning the sex of intersex individuals and defining the sexual identity of homosexuals because they were often immediately visible and did not always necessitate in-depth medical examination. They were nonetheless sometimes judged unreliable. Accordingly, behaviour, activities, tastes and other psychological characteristics of homosexuals and intersex individuals were regarded as relevant because they were thought to be caused by the sexual organization of the body. The uterus, for example, was believed to influence the sensitivity of women, and their sensitive constitution allegedly affected their tastes and personality. Likewise, a correlation was established between the supposedly womanly physiology of homosexuals (such as their wide hips), and their alleged effeminate tastes and allure. Experts also considered that the habits and tastes of a hermaphrodite were relevant when assigning their predominant sex. Geoffroy Saint-Hilaire concludes about Marie-Jeanne, a male hermaphrodite identified as a woman: 'ses goûts, ses plaisirs étaient […] ceux d'un homme, son caractère viril se trahissait en lui par des traits […] évidents' [his tastes and pleasures were […] those of a man, his virile personality betrayed itself by

201 Smith, 'Silence', ibid.
202 Tardieu, *Étude médico-légale*, 141–43.

[…] obvious traits].[203] In other words, an individual's 'true' sex was seen as going beyond their genitalia, including their behaviour for instance.

In another extract, Geoffroy Saint-Hilaire notes that tastes, instincts and habits 'révèle[nt]' [reveal] and 'dévoile[nt]' [unveil] the male sex of a hermaphrodite.[204] The use of the verbs *révéler* and *dévoiler* is not incidental. As Linton has shown with regard to late-nineteenth-century doctors, physicians liked to pose in their articles as detectives solving a mystery.[205] Accordingly, these verbs indicate that an individual cannot hide their 'real' sexual identity, whether or not they try to, in spite of 'trompeuses apparences' [deceptive looks]; the shrewd physician-sleuth will find out the truth… or at least this is what he would have the reader believe.[206]

Likewise, Lauvergne and Tardieu choose the lexical fields of betrayal and denunciation to discuss homosexuals' identity. Homosexuals are *betrayed* by their physiognomy according to Tardieu, and Lauvergne pointed out that '[l]es sexes douteux se dénoncent d'eux-mêmes aux regards exercés d'un médecin' [Dubious sexes denounce themselves to the expert gaze of a doctor].[207] The pejorative connotation of the verbs *trahir* [to betray] and *se dénoncer* [to denounce oneself] is clear. Homosexuals' effete tendencies do not belong to their biological sex. Consequently, they 'naturally' emerge in order to reveal their 'deviant' nature to scientists. Geoffroy Saint-Hilaire also used *se trahir* [to betray oneself] in the extract about the hermaphrodite Marie-Jeanne, but with a positive purpose in mind.[208] According to the teratologist, Marie-Jeanne is right to behave like a man and to express masculine tastes, as *he* is a man. Conversely, homosexuals should not behave like women, as they are not biological women. The different connotation that scientists gave to a verb such as *trahir* may reveal that they conceived of, and treated, homosexuals and hermaphrodites differently. Whereas the ambiguity of intersex bodies did not hinder their re-categorization if their gendered behaviour belied their apparent physical sex, homosexuals could

203 Geoffroy Saint-Hilaire, *Traité de tératologie*, II, 73.
204 Ibid.
205 Linton, *Prescribed Fictions*, 38–40.
206 Geoffroy Saint-Hilaire, *Traité de tératologie*, II, 65.
207 Tardieu, *Étude médico-légale*, 130; Lauvergne, *Les Forçats*, 288.
208 Geoffroy Saint-Hilaire, *Traité de tératologie*, II, 73.

not be re-categorized as women, owing to their undeniably male sexual organs. They thus defied the gender dichotomy expected by society. This challenge to patriarchal categories was therefore seen as a betrayal of the very idea of manliness.

Comparing nineteenth-century scientific writings with those of the twenty-first century shows that the importance of gendered habits and tastes when assigning the sex of an individual has not disappeared nowadays, but that the perspective has shifted. Habits do not reveal or betray the 'real' sex – that is, the genetic and anatomical sex – they indicate what the urologist Mouriquand calls 'l'identité sexuelle individuelle' [individual sexual identity], in other words, how individuals define themselves irrespective of their biological sex.[209] Mouriquand reports the twenty-first-century case of a patient who was identified as a girl at birth owing to a biological anomaly and raised accordingly, even though she had a pair of XY chromosomes. She had testes, a long clitoris and facial hair, and neither a vagina nor breasts. However, despite her male characteristics, she identified as a girl, was operated upon and treated with female hormones to feminize her body.[210] Conversely, the reports of nineteenth-century scientists focused exclusively on the 'real' biological sex and did not take individual sexual identity into account. They mention hermaphrodites who truly believed themselves male or female, but not what they felt when they were told that they were mistaken, as if their perception of their own body was irrelevant.[211] Doctors presented their change of sexual identity as unproblematic, because it was supposed to restore the natural order.

Some evidence, nevertheless, contradicts the implied smoothness of this process. For example, the official request of the male hermaphrodite Marie-Marguerite to change their birth certificate makes them appear keen to be a man. However, according to Worbe's report, they shed many tears

209 Mouriquand, 'S'asseoir dehors', 59.

210 Ibid., 53–54.

211 Marc, 'Hermaphrodite', 100; Geoffroy Saint-Hilaire, *Traité de tératologie*, II, 72. On the absence of hermaphrodites' accounts about their own bodies and state of mind, see Daston and Park, 'The Hermaphrodite', 426; Dreger, *Hermaphrodites*, 167; Houbre, 'Dans l'ombre de l'*hermaphrodite*', 92; Mak, *Doubting Sex*, 10–13 and Chapter Two: 'Early Sex Reassignments and the Absence of a Sex of Self', 43–65.

and had to become accustomed to the idea of renouncing their psychological female identity.[212] Mak rightly describes Worbe's report of Marie-Marguerite's reaction as 'one of the most explicit and moving physician's descriptions from the first half of the nineteenth century of the shock of disruption of the self with regard to sex'.[213] The later case of Adélaïde Herculine Barbin also demonstrates that the change of sexual identity can be traumatic. Barbin was identified as a girl at birth, but reassigned as a man in 1860, at the age of 21 and had their birth certificate altered accordingly. Their account, a unique testimony of the living conditions of a nineteenth-century individual with intersex traits, reveals that, once they 'became' a man named Abel, they lost their friends, their job as a teacher in a school for girls and even their social position. More and more isolated, Abel committed suicide.[214] Tardieu presented the case of Barbin as proof of the dangers caused by an erroneous sex assignment at birth but neglected to mention that the hermaphrodite's misfortunes mostly commenced only after this mistake was 'corrected'.[215]

The definition of sexual habits in nineteenth-century writings did not take individual sexual identity into account but did include sexuality. Geoffroy Saint-Hilaire, for instance, reported the case of Joséphine Badré, a male hermaphrodite mistaken for a woman who was sexually attracted to women.[216] Badré's erotic attraction to women was not judged to be lesbianism, since it revealed *his* 'real' sex. In the doctors' biased accounts, there were no such individuals as homosexual hermaphrodites. They considered that hermaphrodites' sexuality had to be compatible with their predominant

212 Quoted by Marc, 'Hermaphrodite', 92.

213 Mak, *Doubting Sex*, 52.

214 The 'Souvenirs' of Herculine Barbin were published by Michel Foucault in *Herculine Barbin dite Alexina B.* (Paris: Gallimard, 1978). For more on Barbin's social position, see Nathaniel Wing, *Between Genders: Narrating Difference in Early French Modernism* (Newark, DE: University of Delaware Press, 2010), 122–24; Claude Burgelin, 'Le cas Herculine B.', in *Différenciation sexuelle et identités*, 93–100 (99–100); Mak, *Doubting Sex*, 66–89.

215 Ambroise Tardieu, 'Question médico-légale de l'identité dans les rapports avec les vices de conformation des organes sexuels' (1874), in Foucault, ed., *Herculine Barbin*, 135–36.

216 Geoffroy Saint-Hilaire, *Traité de tératologie*, II, 85.

sex in a heteronormative context. In their opinion, if intersex individuals were attracted to the 'wrong' sex, it was due to the education that they had received and it certainly did not correspond to their psychological mindset. Even for the *hermaphrodite bisexuel* (the hermaphrodite who is closest to the perfect hermaphrodite), Geoffroy Saint-Hilaire assumed that 'il est fort douteux que ses penchants le portassent à la fois vers les deux sexes' [It is very unlikely that their tendencies should incline them towards both sexes].[217] In the present day, the correlation between predominant physiological sex and sexuality has not changed. Fausto-Sterling notes that, for many doctors, 'a healthy intersexual is a straight intersexual'.[218] In short, then and now, a hermaphrodite who was not attracted to the sex opposite to their predominant sex was viewed as 'abnormal'.

Nineteenth-century scientists' rejection of intersex homosexuality tallies with the need to classify hermaphrodites into categories that comply with the dominant discourse of sexual dichotomy and 'compulsory heterosexuality'.[219] The goal of these categories is to offer a stable identity framework for individuals and the suggestion of not conforming to one of these categories would appear to deprive them of their own identity. Butler argues:

> [A]s 'identity' is assured through the stabilizing concepts of sex, gender, and sexuality, the very notion of 'the person' is called into question by the cultural emergence of those 'incoherent' or 'discontinuous' gendered beings who appear to be persons but who fail to conform to the gendered norms of cultural intelligibility by which persons are defined.[220]

Combinations of categories including 'incoherent' or 'discontinuous' gendered beings, to borrow Butler's terminology, not only call 'the very notion of "the person"' into question, but also threaten the sexual dichotomy. What if, in a French society prior to the legalization of marriage between same-sex individuals, a male hermaphrodite, previously mistaken for a woman, had married a man? The need to identify the predominant

217 Ibid., 171.

218 Fausto-Sterling, *Sexing the Body*, 71. See also Davis, *Contesting Intersex*, 93 and 95.

219 Butler, *Gender Trouble*, ix.

220 Ibid., 17.

sex of a hermaphrodite with invasive examinations not only fulfilled scientific curiosity, it was also in line with a legal approach intended to validate patriarchal discourse, since medical conclusions regarding the sex of an individual could lead to the dissolution of their marriage.

Normalization and Moralization of Sex

In examining and describing individuals who did not conform to social norms, scientists displayed two goals: firstly, and above all, to assert the norm, and secondly, to marginalize individuals who infringed this norm. The latter goal is illustrated by Tardieu and other scientists who wrote on homosexuality. They exhibited a strong aversion to homosexuality and 'a tone of moral outrage [that] took precedence over one of clinical detachment'.[221] Their texts confirm Foucault's interpretation of the nineteenth-century science of sex: 'C'était [...] une science subordonnée pour l'essentiel aux impératifs d'une morale dont elle a, sous les espèces de la norme médicale, réitéré les partages' [This was [...] a science mostly subordinated to the imperatives of morals, whose divisions were repeated in the guise of the medical norm].[222] The scientific heteronormativity promoted by scientists is above all a moral one. Tardieu, for instance, stressed his horror towards these acts which were 'against nature' and played the role of the martyr for science who only agreed to explore such an 'ignominious' topic in order to advance science and progress.[223] With the appalling descriptions of dog-like or corkscrew-shaped penises, enormous buttocks and funnel-shaped anuses, Tardieu managed to make homosexuals look 'abnormal', even hybrid: partly human and partly animal or object, partly male and partly female. The figure of the homosexual created by scientists was a fantastical monster whose mission was to keep 'respectable' male citizens steadfast against the temptation of homosexuality.

221 Nye, 'Sex Difference', 33.

222 Foucault, *Histoire de la sexualité*, I, 72.

223 Tardieu, *Étude médico-légale*, 118.

The hegemonic discourse of scientists regarding homosexuality confirms Anna Marie Smith's claim that hegemony 'depends on the normalization of the idea that there is no alternative'.[224] More accurately for my purposes, scientific writings did not explicitly deny the existence of alternatives, but rather presented them as untenable. The representation of fringe states and behaviours allowed scientists to reinforce the norm. This norm was not only numerical, as the hermaphrodites were 'abnormal' because they were allegedly rare, but it was also and primarily ideological. It represented the ideal of sexuality, family and lifestyle in contemporaneous French society. Relations between men and women should contribute to the grandeur of France by allowing the male population to grow in terms of number and virility. It should be mentioned that, in scientific discourse, the physician became the new 'moral preceptor of society'.[225] Barbin's 'Souvenirs' [Memoirs] exemplify the replacement of the priest by the doctor. The hermaphrodite reports the words of the physician who examined them at the bishop's request: 'vous ne devez pas seulement voir en moi un médecin, mais un confesseur' [You should not only regard me as a doctor, but also as a confessor].[226] Scientists secularized Judeo-Christian morals rather than abolishing them. Like divine laws, natural laws could not be transgressed.

In the attitude of scholars, the heterosexual model, by implicit comparison with the 'abject' signs of pederasty so eagerly described by Tardieu, was the only one that was admissible because it was believed to reproduce natural laws. Similarly, Virey imposed a vision of the relations between men and women that was moral and family-centred, although he condemned 'vicious' practices by referring mostly to science and medicine. Masturbation, celibacy and polygamy were criticized because they were judged unhealthy and counter-procreative. Where polygamy was concerned, Virey concluded:

224 Anna Marie Smith, 'Hegemony Trouble: The Political Theories of Judith Butler, Ernesto Laclau and Chantal Mouffe', in Jeffrey Weeks, ed., *The Lesser Evil and the Greater Good: The Theory and Politics of Social Diversity* (London: Rivers Oram Press, 1994), 222–34 (230).

225 Laqueur, 'Orgasm', 29; Laqueur, *Making Sex*, 215. See also Gazalé, *Le mythe de la virilité*, 168, 275 and 281–82.

226 Barbin, 'Mes souvenirs', 89.

> Il en résulte cette vérité morale aussi bien que médicale, [à] savoir que l'état le plus heureux pour l'espèce humaine, le plus favorable à sa santé, le plus conforme à la raison, est de *suivre la nature* sans en abuser, soit par excès, soit par défaut.[227]
>
> [The result is this moral as well as medical truth, that is, the happiest state for the human species, the most favourable for its health, the most conform to reason, is *to follow nature* without over-indulging out of either excess or defect.]

Medicine and morality agreed in this case because they were purportedly supported by nature, which would condemn celibacy, encourage monogamous marriage and promote moderate and reproductive intercourse. Doctors' ideal model of sexuality and sexual identity coincided with the promotion of the patriarchal family, with the paterfamilias exercising soft domination over his wife, who was pleased with her lot. If readers had doubts regarding the 'natural' patriarchal system, Moreau de la Sarthe reassured them through a quote from Desmahis's 'Femme' article in the *Encyclopédie*: 'La nature semble avoir conféré aux hommes le droit de gouverner' [Nature seems to have given men the right to rule].[228]

Nature determined strict male and female categories, but homosexuals, eunuchs and masculine women, like individuals with intersex traits, could not belong to the female category and even less to the prestigious male category. According to Nye's interpretation of Virey's text, the latter was defined by chivalric honour.[229] This meant that the male biological sex would 'naturally' convey the values of bravery and liberality associated with masculine behaviour. Consequently, homosexuals, masculine women and hermaphrodites contravened the norm. When confronted with such non-normative figures, the texts displayed signs of social anxiety.[230] For Lauvergne, the convict Frédéric was an 'homme douteux' [dubious man] and he added: 'nous avouons l'erreur de la nature dans les êtres homologues de Frédéric' [We admit that nature has made a mistake in men like Frédéric].

227 Virey, *De la femme*, 130; emphasis in original.

228 Desmahis, 'Femme', in *Encyclopédie ou Dictionnaire raisonné des sciences, des arts et des métiers*, quoted by Moreau de la Sarthe, *Histoire naturelle*, I, 728. *Hommes* clearly refers to male individuals rather than human beings in general.

229 Nye, *Masculinity and Male Codes of Honor*, 62.

230 For instance, Cabanis's already quoted rhetorical question about the social position of masculine women (*Rapports*, I, 363).

By designating them as 'ces avortons de notre race, ces homoncules' [these runts of our race, these homuncules], Lauvergne believed homosexuals to be subhuman beings, as did Tardieu when assimilating them to animals and objects.[231] The use of the diminutive suffix –*cule*, the perception of homosexuals as criminals and the inclusion of intersexuality in essays dedicated to monsters constituted attempts to minimize their importance within the dominant discourse.

The rejection of men who did not conform to the male norm is best illustrated by Geoffroy Saint-Hilaire:

> La destinée de chaque enfant nouveau-né, du moment où son sexe est connu ou déclaré connu, se trouve donc réglée à l'avance pour les circonstances principales de sa vie: il est rangé dans l'une ou l'autre de ces deux grandes classes à laquelle appartiennent des fonctions non seulement différentes, mais presque inverses dans la famille aussi bien que dans la société. À cet égard, point d'intermédiaires; nos lois n'en admettent pas l'existence; n'en prévoient pas la possibilité.[232]
>
> [The fate of each newborn, from the moment their sex is known or declared known, is thus predetermined for the main circumstances of its life: he or she is placed in one or another of these great classes, to which not only different but almost contrary functions belong in the family as well as in society. In this respect, there are no intermediates; our laws do not admit their existence; neither do they allow for their possibility.]

For Geoffroy Saint-Hilaire and his fellow scientists, every baby should be sexually identified because society and the law attributed specific functions to each sex. From this perspective of strict sexual difference, the position of the masculine woman, the homosexual and the hermaphrodite seemed to be an aberration. They were devoid of a firm identity because they were split across two entities: male and female. As Geoffroy Saint-Hilaire's text points out, this contradiction was impossible to accept and had to be resolved by scientists through empirical observation. The teratologist and his colleagues attempted to normalize the hermaphrodite by arguing that there usually was a 'real' sex despite the ambiguous genitalia. Nonetheless, they could not deny that these individuals existed, and, by

231 Lauvergne, *Les Forçats*, 293 and 299.
232 Geoffroy Saint-Hilaire, *Traité de tératologie*, III, 573.

their simple existence, they questioned scientists' implacable insistence on sexual dichotomy.

This analysis of early nineteenth-century scientific writings focusing on sexual difference, masculine women, homosexuals and intersex individuals shows that scientists made their diagnosis primarily based on physical and biological characteristics, whose perception could be deformed through human bias. This approach reveals the obsession among the nineteenth-century French scientific community with separating the sexes. The examination of mannish, intersex and homosexual bodies reveals scientists' eagerness to turn them into 'deviant' creatures. Consequently, the scrutiny and the classification of these allegedly unconventional bodies excluded them from the 'normal' categories of maleness and femaleness.

The focus on bodies and genitalia, nonetheless, had its limits, as Murat points out: 'le troisième sexe oblige les deux autres à se penser. Il remet en question les normes, et en crée de nouvelles' [The third sex forces the other two to define themselves. It challenges the norms and creates new ones].[233] By admitting the existence of 'abnormal' bodies, scientists seemed to call into question the very sexual dichotomy on which their arguments were based. Accordingly, experts in intersexuality had to determine a predominant sex to adhere to the idealized binary sexual system. This process had to be supported by a psychological and behavioural approach. Tastes, habits and other psychological characteristics were considered to be as relevant as genitalia because they were believed to stem from the internal organization of the body itself. Rather than challenging the results of the focalization on genitalia, as is often the case nowadays (where a person's own sense of gender identity is taken into account), the psychological aspects only served to justify empirical observations. However, in the case of homosexuals, no reconciliation between physical and psychological characteristics seemed possible. Despite their alleged feminine tastes, their deformed anuses and their twisted penises, they remained men.

Like the scientific texts analysed in this chapter, literary narratives of the July Monarchy endorsed traditional distinctions between sexual and gender roles. This norm provided writers (who were often regular readers of

233 Murat, *La loi du genre*, 398.

scientific texts)[234] with a stable framework that was shared by their readers. It allowed them to refer to masculine and feminine characteristics without having to explain the meaning of these traits. Yet, also like scientific texts, the framework of these literary texts reveals that this binary system was restrictive and could hardly accurately represent male and female relations in French society. Many French writers endeavoured to depict, in their narratives, unconventional roles and relations that transgressed sexual and gender models. Accordingly, the next chapter analyses how hermaphroditic characters in literature became powerful vehicles through which the constraints of sexual and gender norms were discussed and even challenged.

234 Gutermann, 'Le désir et l'entrave', 70.

CHAPTER 2

The Hermaphrodite[1]

Hermaphroditus was the son of the god Hermes and the goddess Aphrodite. According to Ovid's version of the myth in the *Metamorphoses*, the god was born a boy and became the object of the nymph Salmacis's desire. When Hermaphroditus rejected her, she surprised him while he was having a bath in a fountain in Caria and wrapped herself around his body. She begged the gods to unite her forever to him. Her wish was granted; the two became one individual with both sexes: 'Thus, when in fast embrace their limbs were knit, They two were two no more, nor man, nor woman – One body then that neither seemed and both.'[2] Hermaphroditus further wished that any being who would bathe in that fountain would likewise turn into an intersex creature. This curse highlights the fact that the intersexuality of the god is realized against his own volition and experienced as a burden, or a 'fallen state', in Kari Weil's words, rather than a gift.[3]

Marie Delcourt, Frédéric Monneyron and Michael Groneberg consider that Ovid's version of the myth is unconventional, since the intersexuality of Hermaphroditus traditionally comes from birth and not from

1 Some material in this chapter, reframed around the subject of homosexuality and the beauty of the hermaphrodite, has been published as an article: Géraldine Crahay, '"Oh! non, je ne pourrai jamais te le dire … j'aime un homme!": Hermaphrodite et l'esthétisation de l'homosexualité dans *Mademoiselle de Maupin*', *Nineteenth-Century French Studies*, 45/1–2 (Autumn–Winter 2016–17), 17–32

2 Ovid, *Metamorphoses*, 4th book, A.D. Melville, trans., Kenney, E. J., ed. (Oxford: Oxford University Press, 2008), 85.

3 Weil, *Androgyny*, 18.

metamorphosis.[4] The cult of Hermaphroditus, which appeared in the Hellenic world in the fourth century BC, had obscure origins and was of little interest in Greece.[5] Delcourt highlights the paradox between Hermaphroditus's lack of fame during Antiquity and the later popularity of hermaphroditic or androgynous entities:

> L'étrange est que le mythe nouveau, qui semblait si bien fait pour servir de support à ces imaginations, n'en a pas attiré une seule. Les poètes n'ont prêté à Hermaphrodite aucune aventure, contents de broder un peu sur sa naissance. Et les cosmogonies tardives, hantées par le rêve de l'androgynie, n'en ont pas fait le propre du dieu le mieux fait pour la symboliser […]. C'est ainsi que, dans la littérature, Hermaphrodite est une idée plutôt qu'une personne.[6]
>
> [What is strange is that the new myth, which seemed so well made to support these imaginations, attracted none. The poets gave Hermaphroditus no adventure, content merely to elaborate on his birth a little. And late cosmogonies, haunted by the dream of androgyny, did not make it the main feature of the god best suited to symbolize it […]. This is how, in literature, Hermaphroditus is an idea rather than a person.]

Delcourt's statement resonates with the July Monarchy context. While Ovid's *Metamorphoses* had had, since Antiquity and especially during the Renaissance, considerable influence over Western arts and literature as a gateway to Greco-Roman mythology, their fame significantly declined in the nineteenth century.[7] Still, the figure of the hermaphrodite enjoyed great popularity in literary narratives of the 1830s.[8] A writer

4 Delcourt, *Hermaphrodite*, 81; Monneyron, *L'androgyne romantique*, 19; Michael Groneberg, 'Hermaphrodite's Voice: Dealing with the Either-Or Attitude in Science, Law, and the Arts', in *Transgender and Intersex*, 225–52 (234).

5 Delcourt, *Hermaphrodite*, 68–71; Monneyron, *L'androgyne romantique*, 18; Groneberg, 'Hermaphrodite's Voice', 235.

6 Delcourt, *Hermaphrodite*, 2.

7 On *Metamorphoses*'s success: John F. Miller and Carole E. Newlands, 'Introduction', in John F. Miller and Carole E. Newlands, eds, *Handbook to the Reception of Ovid* (Hoboken: John Wiley and Sons, 2014), 1–7 (3). On their decline: Melville, in *Metamorphoses*, xxxvii; Mechthild Fend, *Les limites de la masculinité: L'androgyne dans l'art et la théorie de l'art en France, 1750–1830*, Jean Torrent, trans. (Paris: La Découverte, 2011), 162.

8 Weil, *Androgyny*, 68–72; Monneyron, *L'androgyne romantique*, 8; Linton, *Prescribed Fictions*, 127

like Théophile Gautier showed the extent of his knowledge of the Latin poet's works in *Mademoiselle de Maupin*, for the fictional hermaphrodite borrowed more from classical mythology than from scientific readings. Hermaphroditus, nevertheless, was not depicted for himself, but personified the idea of sexual and gendered completeness with ambivalent perspectives. I argue in this chapter that French writers of the 1830s used the figure of the hermaphrodite as a means of depicting gender ambiguities symbolically, as well as challenging the issues that were believed to arise from these ambiguities.

Due to his intersexuality, Hermaphroditus belongs to the category of composite monsters defined by Pierre Jourde as 'assemblages hétérogènes des espèces, des règnes, des sexes, réunion de plusieurs êtres' [heterogeneous blends of species, kingdoms, sexes, combination of several beings].[9] This category also includes the Sphinx, the Minotaur and the Chimaera, who all combine different animal and/or human species. Composite monsters, Jourde explains, are usually considered to be ugly either because each part of them is unpleasant or because the assemblage itself is judged dissonant and unnatural.[10] The heterogeneity that results from the assemblage of the two sexes, rather than the unattractiveness of the parts, can explain the alleged monstrosity of the hermaphrodite, as each body incarnates the perfection of their sex. As will be seen, the eighteenth-century German art historian Johann Joachim Winckelmann (1717–68) was enthusiastic about the beauty of hermaphrodites and established a correlation between Hermaphroditus and ideal beauty in neoclassical aesthetics.[11]

Throughout this chapter, I rehabilitate this figure as complex rather than unequivocally monstrous. I use the term *hermaphrodite*, which emphasizes the physicality of the androgynous figure while sustaining its

9 Pierre Jourde, *Littérature monstre: Études sur la modernité littéraire* (Paris: L'Esprit des péninsules, 2008), 238.

10 Ibid., 239.

11 In contrast, Kari Weil highlights the notion of change that is contained in the concept of the hermaphrodite, suggesting that *hermaphrodite* recalls the Romantic changing nature of beauty (while *androgyne* is associated with classical art and beauty): 'Romantic Androgyny and its Discontents: The Case of *Mlle de Maupin*', *Romanic Review*, 78/3 (May 1987), 348–58 (349).

ambivalent connotations. Firstly, it refers to two entities: a god who is characterized by a corporeal metamorphosis and enjoyed great popularity in classical sculpture and a human being whose body has male and female characteristics. Secondly, the combination of classical beauty and erotic intersexuality renders the hermaphrodite appropriate for representing the ambivalence of the main protagonists of the literary narratives studied in this chapter. However, I will often designate these protagonists as hermaphroditic characters rather than hermaphrodites to highlight the fact that hermaphroditism does not refer to physical intersexuality exclusively, but also to other forms of sexual and gender indeterminacies, such as cross-dressing and the surgical alteration of the sex undergone by castrati. Whilst the hermaphroditic protagonists share several characteristics with masculine women and effeminate men, they differ from these figures because it is not only their gender that is called into question in these novels, it is also and especially their sexes that cast doubt. Moreover, although other terms such as *androgynous*, *trans* or *non-binary* may at times appear to be more appropriate, I prefer to use *hermaphroditic* as an umbrella term that encompasses these meanings because it emanates from the paradigmatic framework of the authors studied, for whom the reference to the god Hermaphroditus constitutes one of the few ways to conceive of gender non-conformity.[12]

This chapter focuses on two novels and one novella representing hermaphroditic characters: Latouche's *Fragoletta: Naples et Paris en 1799* (1829), a love story between a French officer d'Hauteville and Camille, an Italian hermaphrodite; Gautier's *Mademoiselle de Maupin* (1835) in which the poet d'Albert falls in love with a boy named Théodore, unaware of the fact that 'he' is really the cross-dressed girl Madeleine de Maupin; and Balzac's novella *Sarrasine* (1831), in which the sculptor Sarrasine falls in love with the castrato Zambinella because he mistakes him for a woman. Despite being written at the beginning of – or just before – the July Monarchy, none of these plots occur during that time: *Fragoletta* and the main plot

12 On debates over nomentaclures, see Sebastian Jansen, 'Transgender and Intersex: Unavoidable Essentialism and the Normative struggle for Recognition', in *Transgender and Intersex*, 115–40 (135). On the difference between *hermaphrodite* and *androgyne*, see the beginning of the section 'Hermaphroditic Monstrosity and Exclusion' in this chapter.

of *Sarrasine* take place during the eighteenth century, whilst *Mademoiselle de Maupin*, roughly inspired by the life of the adventuress and actress Julie d'Aubigny (1673–1707), occurs at the end of the seventeenth century.[13] The correlation between the timeframes of the stories and the context of their writing is also loose: *Mademoiselle de Maupin* is detached from historical references, whereas the narrator of *Sarrasine* stresses the difference between 'olden times' and the present day where the lot of castrati is concerned.[14] Only *Fragoletta* seems to follow the tenet of the historical novel, according to which the past may be used to depict and instruct the present,[15] as the Restoration of the Italian Bourbons recalls the Restoration of the French Bourbons during Latouche's time. However, as will be shown, connections between history and fiction in the novel are vague. *Fragoletta*, *Sarrasine* and *Mademoiselle de Maupin* distance themselves from historical accuracy and even show indifference towards historical events. Hermaphroditic protagonists are situated in the background of political and historical action. Despite this, or perhaps owing to this, the marginal hermaphroditic characters, I will suggest, allegorize the plight of 1830s individuals unable to conform to the sexual norm. Indeed, I will determine how some authors avoided challenging the sexual dichotomy, notably by depicting hermaphrodites as monsters. I will also show how other writers of this corpus redeem the figure of the hermaphrodite, notably by mediating hermaphroditic bodies through arts. Ultimately, by studying a hermaphroditic character endowed with a harmonious combination of moral qualities traditionally belonging to both genders, I will argue that hermaphroditism can be used as a means of both expressing masculine insecurity and surmounting this insecurity.

13 For more on the historical Mademoiselle de Maupin, see Monneyron, *L'androgyne romantique*, 119–20; Jennifer Waelti-Walters, *Damned Women: Lesbians in French Novels, 1796–1996* (Montreal: McGill-Queen's University Press, 2000), 27.

14 The narrator claims that castrati no longer exist in Italy. They nevertheless sang in the Sistine Chapel choir until 1898. Balzac was probably misled by the fact that Napoleon had intended to stop this practice, but it started again when the French withdrew from Italy: Yvonne Noble, 'Castrati, Balzac, and "BartheS/Z"', *Comparative Drama*, 31/1 (Spring 1997), 28–41 (34–35).

15 Jean Molino, 'Qu'est-ce que le roman historique?', *Revue d'Histoire Littéraire de la France*, 75/2–3 (March–June 1975), 195–234 (216–18); Laforgue, *L'éros romantique*, 59.

Hermaphroditism and Hybridity of Structure

How are hermaphroditic characters portrayed in July Monarchy narratives? The complexity of both the sex and the gender of these protagonists is elaborated through the construction of the narratives. *Fragoletta* is divided into two parts: the first part takes place in Naples, during the Italian campaign; and the second in Paris, before Napoleon's coup d'état. Following this division, the hermaphrodite Camille appears in Italy as an innocent and boyish girl, known as Fragoletta, and is introduced in France as a vicious and perfidious young man, 'her' brother Philippe Adriani.[16] In *Sarrasine*, the principle of *mise en abyme* that is adopted in the novella allows the narrator to depict Zambinella as an old man in the outer story and a young woman in most of the embedded story. While the feminine version of Zambinella incarnates for Sarrasine the ideal woman, the old Zambinella is perceived by other characters first as a supernatural and frightful creature, then as a decaying, ludicrous and anachronistic man because the castrato has lost his good looks and beautiful voice.

In other terms, as the hermaphroditic protagonists are intrinsically double, their portrayals are also two-fold. Each hermaphrodite is successively depicted as a man, and then as a woman (or conversely), rather than as simultaneously male and female. This is due to their hermaphroditism not being immediately, and sometimes never, identified. Camille is portrayed as both Fragoletta and her so-called twin brother Philippe. Madeleine de Maupin is introduced as both Madeleine and Théodore. Zambinella is both the *prima donna* and the mysterious old man.[17] This

16 In Latouche's novel, *Camille* refers to the female self of the hermaphrodite, whereas the name of their elder brother Philippe is adopted to introduce their male self. In this study, however, I use *Camille* to refer to the whole person of the hermaphrodite, while Camille's nickname *Fragoletta* designates the female side and *Philippe*, the male side.

17 According to Italian operatic terminology, a sopranist castrato was called *primo uomo*, whereas *prima donna* designated a female soprano: Patrick Barbier, *The World of the Castrati: The History of an Extraordinary Operatic Phenomenon*, Margaret Crosland, trans. (London: Souvenir Press, 1996), 2. The narrator's

does not mean that each gendered portrayal of hermaphrodites cannot contain stereotypical characteristics of the other gender, but rather that their sexes are not questioned in these descriptions despite androgynous characteristics. The double portrait of these hermaphroditic characters reveals the authors' difficulty in distancing themselves from the norm of the separation of the sexes, even when describing characters with dual gender and/or sex. As with the scientific texts seen previously, the descriptions of these hermaphrodites are not only grounded in references to physical characteristics. They also draw inspiration from psychological characteristics, such as tastes and activities, which look beyond the manifestation of a visible sex. However, whilst these interests supposedly disclose the 'real' sex of an individual in the opinion of Isidore Geoffroy Saint-Hilaire and his colleagues, these narratives demonstrate that relations between sex and gender are far more complex.

Moreover, the narrator is seldom omniscient in these narratives. It is either a heterodiegetic narrator, to use Gérard Genette's terminology, who adopts the perspective of a character (d'Hauteville in *Fragoletta*), or a homodiegetic narrator, as in *Sarrasine*, who espouses the viewpoint of Sarrasine in the embedded story.[18] The hermaphroditic protagonist, as a result, is mediated through the gaze of another character, who has only a partial – if any – understanding of their identity. This mediation manifestly serves the narration, as the revelation of their hermaphroditism constitutes an important element of the plot.[19] It thus directs and, at times, intentionally

mistake is therefore deliberately misleading. Roland Barthes calls *prima donna* a 'leurre' [lure; trick]: *S/Z* (Paris: Seuil, 1970), 117.

18 Pierre Citron offers a biographical interpretation of *Sarrasine* and suggests that the narrator and Sarrasine are closely related as they both embody two versions of Balzac: the sculptor is the young Balzac, whilst the narrator is an older Balzac: 'Interprétation de *Sarrasine*', *L'Année balzacienne*, 1st ser. (1972), 81–95 (87–90). About narratological techniques, see Gérard Genette, *Figures III* (Paris: Seuil, 1972) and *Nouveau discours du récit* (Paris: Seuil, 1983).

19 See what Barthes calls the hermeneutic code in *S/Z*: 'l'ensemble des unités qui ont pour fonction d'articuler, de diverses manières, une question, sa réponse et les accidents variés qui peuvent ou préparer la question ou retarder la réponse; ou encore: de formuler une énigme et d'amener son déchiffrement' (24) [all of the units whose function is to articulate, in different ways, a question, its answer and

misleads the reader's interpretation. However, it also objectifies the hermaphrodite through this third party, which, in contrast, bestows particular importance on the speeches of hermaphrodites.

The structure of *Mademoiselle de Maupin* being even more complex, I will shortly dwell on it. But first, a few plot details: d'Albert is a young poet in search of the perfect woman. Although he already has a mistress, Rosette, he falls in love with a beautiful young man named Théodore de Sérannes and regards his feelings as monstrous. He does not know that the young man is in fact Madeleine de Maupin, a girl wearing male attire to observe how men behave in the absence of women. Rosette, who is also ignorant of Théodore's 'true' identity, is equally infatuated with him but is rejected by him despite his obvious attraction to her. When the so-called young man plays the part of Rosalind in Shakespeare's comedy *As You Like It* (also a story of cross-dressing), d'Albert becomes aware of Théodore's 'real' sex. Later, Madeleine spends the night with him, before joining Rosette in her bedroom. She then leaves them forever but asks them in a farewell letter to continue loving each other and sometimes mention her name.

Gautier's novel is a hybrid creation, just like its female protagonist, since it combines letters, theatrical dialogues, aesthetic reflections on theatre and a performance of *As You Like It*.[20] It also incorporates three different narrators: one heterodiegetic narrator, who narrates three out of the seventeen chapters of the novel, and two homodiegetic narrators, d'Albert (the main character and the privileged viewpoint of the novel) and Madeleine de Maupin. The homodiegetic chapters are usually letters addressed to a

the diverse accidents that can either prepare the question or delay the answer; or: to formulate a riddle and bring out its deciphering].

20 The hybridism of *Mademoiselle de Maupin* has been analysed by Jean-Marie Roulin, 'Confusion des sexes, mélange des genres et quête du sens dans *Mademoiselle de Maupin*', *Romantisme*, 103 (1999), 31–40. See also Pierre Albouy, 'Le mythe de l'androgyne dans *Mademoiselle de Maupin*', *Revue d'Histoire Littéraire de la France*, 72/4 (July–August 1972), 600–08 (604); Anne Bouchard, 'Le masque et le miroir dans *Mademoiselle de Maupin*', *Revue d'Histoire Littéraire de la France*, 72/4 (July–August 1972), 583–99 (593–94); Laforgue, *L'éros romantique*, 210–11; Wing, *Between Genders*, 30; Maxime Foerster, *The Politics of Love: Queer Heterosexuality in Nineteenth-Century French Literature* (Durham, NH: University of New Hampshire, 2018), 115–16.

penfriend. Madeleine appears to be the beautiful young man Théodore in d'Albert's letters to Silvio, whereas Madeleine's own letters to her friend Graciosa reveal her sex. This illustrates what Genette calls *focalisation interne multiple* [multiple internal focalization], typical of epistolary novels, where the same event can be told by several characters,[21] or, as is the case here, the same protagonist's sex can be interpreted differently by several characters. This literary process, coupled with the intervention of a heterodiegetic narrator, allows the novel to delay the revelation of Madeleine's 'real' sex. When she starts writing to Graciosa, her mastery of masculine roles has already been asserted. D'Albert and Madeleine also send a letter to each other; the former to let the so-called Théodore know that he is no longer fooled by 'his' disguise (XIII); the latter to bid him farewell (XVII). Proper communication between the protagonists can only occur once gendered identity has been thought through (by d'Albert) and acknowledged (by Madeleine). However, whereas d'Albert's letter attempts to bring Madeleine back to the fold of gender conformity, Madeleine's embraces an almost macho tone. She sees women – and men – as sexual objects to be conquered; she expresses disdain for her latest conquest and longing for new ones.[22]

Finally, the timeline of the novel is disconcerting. Madeleine's letters only start in Chapter X (out of seventeen). They relate events that mostly occurred before her encounter with d'Albert and meet the poet's chronology only at the end of Chapter XV. Whilst the flash-back process is not unusual, what puzzles the reader is the lack of interest that Madeleine, the eponymous character of the novel, takes in its main narrator and protagonist. Madeleine, or rather Théodore, occupies the main (but by no means the only) subject of d'Albert's letters and thus plays a considerable role in his defining his own gendered identity, but the reverse is not true. Madeleine's encounters with Rosette are far more important. She grants d'Albert a purely utilitarian function in her sexual quest, something I will come back to later.

21 Genette, *Figures III*, 206–07.

22 See Donald L. Mosher and Mark Sirkin, 'Measuring a Macho Personality Constellation', *Journal of Research in Personality*, 18/2 (1984), 150–63 (151).

In brief, it appears that the variety of genres that characterizes *Fragoletta*, *Sarrasine* and above all *Mademoiselle de Maupin* reproduces the variety of genders within the stories. Narrative techniques, such as internal focalization on non-omniscient characters and the use of multiple narrators, facilitate the construction of gendered identity as a mystery that the narratives progressively unveil. Significantly, they also contribute to blurring the boundaries between 'real' and adopted gender and sexual identities.

Portraits of Hermaphrodites: Transgressing the Sexual Dichotomy?

The case of *Fragoletta* overwhelmingly reveals the difficulty of conceiving of a fully intersex individual. The novel's struggle with the complexity and the variety of sexes is reflected by its rather disjointed plot.[23] Marius d'Hauteville is a French officer during the Italian campaign in Naples who falls in love with the tomboy Camille (nicknamed Fragoletta). Although Fragoletta seems to like him, she rebuffs his advances. Separated from the young girl, the officer tries to find her, while the French army and their Italian allies fight the Royalists. D'Hauteville is later sent to Paris, where he meets Philippe, Fragoletta's alleged twin brother. Philippe courts d'Hauteville's sister Eugénie and rapes her in her brother's absence. D'Hauteville wants to avenge his sister and pursues Philippe to Naples. There, he finally meets Fragoletta who has found refuge in a convent: she rejects him once again but promises to hand her brother over to him. D'Hauteville kills Philippe in a duel. The novel ends with the physician examining Philippe's dead body and recommending that it should be moved to a women's convent. This claim implies that Camille and Philippe are in fact the same person: a hermaphrodite.

23 On *Fragoletta*'s flaws, see Busst, *The Image of the Androgyne*, 77; Michel Crouzet, 'Monstres et merveilles: Poétique de l'Androgyne, à propos de *Fragoletta*', *Romantisme*, 45 (1984), 25–41 (28–31); Laforgue, *L'éros romantique*, 89–92.

This summary of *Fragoletta* shows that the hermaphrodite is never presented as being simultaneously a boy and a girl despite their intersex body. Whilst Michel Crouzet considers that Camille is 'ni masculin ni féminin' [neither male nor female] and claims that the hermaphrodite thus prefigures Zambinella, I interpret the construction of the hermaphrodite differently.[24] Indeed, Camille is not *neither* male *nor* female; Camille is not *both* male *and* female; Camille, I claim, is *either* male *or* female. The heterodiegetic narrator first draws a detailed portrait of the hermaphrodite as a young girl:

> [Camille] n'avait rien changé à sa toilette, bien qu'elle en eût annoncé l'intention. Tout le désordre de son accoutrement capricieux accusait même de sa part une grande distraction ou une parfaite indifférence en matière de coquetterie. [...] Il lui [à d'Hauteville] semblait qu'elle avait quatorze ans à peu près; aucune grâce de la femme n'était développée en elle. Ses membres étaient forts, sa démarche alourdie; elle entrait dans ce rapide passage de la vie où l'on n'a point de sexe, qui n'est pas encore la jeunesse et qui pourtant n'est plus l'enfance.[25]

> [Camille had not changed her clothes, although she had intended to do so. All the disorder of her whimsical outfit even attested to her great distraction or her complete indifference to vanity. [...]. It seemed to d'Hauteville that she was about fourteen years old; she had developed no womanly grace. Her limbs were strong, her gait seemed heavy; she was entering this brief time of life when one has no sex, which is not yet youth and which is, however, no longer childhood.]

Whilst asserting her female sex, the narrator stresses her asexual appearance. He reiterates this claim shortly afterwards: 'Camille n'avait rien [...] des attraits qui soumettent nos sens' [Camille had none of the charms that subdue our senses].[26] Her asexuality is attributed to her age, since Camille is an adolescent. This explanation tallies with scientific studies on intersexuality. As shown in Chapter 1, scientists reported many cases of intersex children who were assigned a 'wrong' sex at birth because their sexual ambiguity was barely noticeable during childhood. The 'real' sex

24 Crouzet, 'Monstres et merveilles', 35.

25 Henri de Latouche, *Fragoletta: Naples et Paris en 1799*, 2 vols, Paris, Delloye, 1840, I, 13–14.

26 Ibid., I, 31.

became obvious with the development of secondary sexual characteristics at puberty. In contrast, the Ancients considered such cases to be changes of sex.[27] Scientifically speaking, Camille could be a prepubescent boy in Italy who reached puberty in Paris after a disappearance of several months, whereupon he developed male characteristics.

However, this empirical explanation is not consistent because, as François Kerlouégan points out, hermaphroditic narratives – and especially *Fragoletta*, I might add – express 'le refus du modèle naturaliste, c'est-à-dire du critère anatomique' [the rejection of the naturalist model, that is, the anatomic criterion].[28] Indeed, Latouche does not portray his character according to scientific coherence and, for the sake of the plot, constantly spreads confusion. For instance, the end of the novel annuls the clear interpretation of Camille as a male individual, since the doctor appears to have identified female genitalia in them. Latouche's strategy of depicting a hermaphrodite alternatively as either male or female does not challenge, but rather reinforces, the traditional dual system of sexes and highlights the artistic difficulty of creating a new gendered system.

The effeminacy of Camille's male side is as manifest as the tomboyishness of their female side. Certainly, the manliness of Philippe's traits and personality – a light moustache on his upper lip, a low voice and 'une assurance à peu près mâle et quelque chose de fier et de hardi assez inconnu à la femme' [an almost manly confidence and something proud and daring that is quite unknown to woman] – is highlighted.[29] However, the novel also points out his supposed feminine physical and behavioural characteristics: he looks identical to his so-called twin sister, he has white and effeminate hands, and Eugénie notices his skills in embroidery and hair-plaiting and his ability to talk to her as if he were one of her female friends.[30]

Unlike the case of Camille, that of Madeleine de Maupin suggests the possibility of a non-binary gender identity, although she is physically the least hermaphroditic protagonist of the corpus. Gautier's novel calls her gendered identity as a woman into question. Madeleine does not fully

27 Delcourt, *Hermaphrodite*, 66.
28 Kerlouégan, 'Une relecture de l'androgynie romantique', 42.
29 Latouche, *Fragoletta*, II, 30.
30 Ibid., II, 29, 45–46, 50, 98 and 230.

identify with feminine gender roles and feels more comfortable in typically masculine ones, including fighting, womanizing and bravado. *Fragoletta* shows that the depiction of a 'real' hermaphrodite has its limitations: it can confine them to a fixed, stereotypical representation of both sexes. *Mademoiselle de Maupin*, on the other hand, highlights how a symbolic hermaphrodite can illustrate gender fluidity and successfully question the sexual dichotomy. Both hermaphroditic characters, however, are similarly described at some point in the novels as physically female, but masculine in their behaviour and tastes. In this way, the writers make the hermaphrodites fathomable by portraying them as boyish girls. Barbara Creed observes: 'During the early stage, the tomboy [...] behaves like a "little man" enjoying boy's games, pursuing active sports, refusing to wear dresses or engage in feminine pursuits.'[31] Likewise, Fragoletta playfully attacks d'Hauteville, takes physical exercise, goes fishing, rides a horse and wears male clothes when she travels in the mountains and by the sea.[32] The young girl also seems indifferent to traditional female concerns. Her carelessness regarding her clothes and the threat that the sun presents to her pale complexion mirrors her lack of vanity. Similarly, Madeleine prefers masculine activities. She writes: 'j'aime les chevaux, l'escrime, tous les exercices violents, je me plais à grimper et à courir çà et là comme un jeune garçon' [I like horses, fencing, all violent pursuits, I enjoy climbing and running here and there like a young boy].[33]

While the adult Camille apparently changes sex in the narrative process and is depicted as a man, the adult Madeleine does not become a man but retains her masculine behaviour. She continues practising manly activities and is considered by men to be one of them. Indeed, she engages easily in homosocial bonding with men, which Eve Kosofsky Sedgwick has defined as 'social bonds between persons of the same sex'.[34] Like Philippe with

31 Barbara Creed, 'Lesbian Bodies: Tribades, Tomboys and Tarts', in Janet Price and Margrit Shildrick, eds, *Feminist Theory and the Body: A Reader* (Edinburgh: Edinburgh University Press, 1999), 111–24 (117).

32 Latouche, *Fragoletta*, I, 16–17.

33 Théophile Gautier, *Mademoiselle de Maupin*, Michel Crouzet, ed. (Paris: Gallimard, 'Folio Classique', 1973), 327.

34 Eve Kosofsky Sedgwick, *Between Men: English Literature and Male Homosocial Desire* (New York: Columbia University Press, 1985), 1.

Eugénie, Madeleine seduces Rosette. Overcoming her timidity, she lavishes intimate caresses on a willing Rosette. In other terms, she is incapable of embracing what Creed describes as the fate of a young tomboy: to renounce her masculinity and 'to capture a man whose job it is to "tame" her as if she were a wild animal'.[35] Madeleine does not relinquish her male clothes and does not marry a domineering man, just as Fragoletta refuses to marry d'Hauteville. She adopts a male identity and has sexual intercourse with another woman.[36] Creed claims that the tomboy who refuses to conform to the patriarchal model and to act as a feminine woman is stigmatized as a lesbian.[37] However, in Madeleine's case, it appears that sapphism is not an object of stigmatization. It is rather an expression of independence and agency, as it is the means through which Madeleine spurns d'Albert's romantic script in which she is his conquest and claims her own desires.

Cross-dressing is the means employed in *Mademoiselle de Maupin* to introduce sexual ambiguity and to critique the separation of the sexes. Whilst Camille is not a cross-dresser, because both sexes are constitutive of their sexual identity, Madeleine and the castrato Zambinella are cross-dressers because they wear clothes that are not typical of their physical sex. Cross-dressed characters were common in baroque literature (for instance, Honoré Urfé's *L'Astrée* [1607–27] and Isaac de Benserade's *Iphis et Iante* [1634]). Moreover, many cross-dressers populate Shakespeare's plays, including *As You Like It*, the comedy performed by d'Albert and Madeleine, in which Rosalind pretends to be a man. Marjorie Garber explains the function of female cross-dressing in Shakespeare's comedies as follows: 'Cross-dressing […] was playful and liminal – and also ameliorative and educational, whether for the "women" in the plays (Rosalind et al.) or for the audience.'[38] The principle of cross-dressing within the plays was reinforced by cross-casting, when male actors play female parts (the opposite is not as common), notably in the Elizabethan theatre and

35 Creed, 'Lesbian Bodies', 117.

36 Gautier, *Mademoiselle de Maupin*, 410–11. See also 366–67.

37 Creed, 'Lesbian Bodies', 118.

38 Marjorie Garber, *Vested Interests: Cross-Dressing and Cultural Anxiety* (London: Penguin Books, 1993 [Routledge, 1992]), 72.

the Roman opera.[39] In other words, male actors played female characters who were posing as men.

Garber has analysed the explanations of cross-dressing and judged 'unconvincing' and 'problematic' a common scenario that appears in many fictions (especially in films such as Sydney Pollack's *Tootsie* [1982], but also in Shakespeare's comedies). She calls this scenario the *progress narrative* and explains it as follows:

> Each [transvestite] is 'compelled' by social and economic forces to disguise himself or herself in order to get a job, escape repression, or gain artistic or political 'freedom'. Each, that is, is said to embrace transvestism unwillingly, as an instrumental strategy rather than an erotic pleasure and play space; [...] heterosexual desire is for a time apparently thwarted by the cross-dresser's assumed identity, so that it becomes necessary for him or her to unmask. The ideological implications of this pattern are clear: cross-dressing can be 'fun' or 'functional' so long as it occupies a liminal space and a temporary time period.[40]

From this perspective, cross-dressing is not the result of a willing choice, but is rather a temporary solution to cultural, social and financial issues. It constitutes 'progress' because it improves cross-dressers' living conditions and chances of success.[41] Garber criticizes this interpretation because it somewhat reduces cross-dressing to being solely the consequence of socio-economic need, as well as denying its eroticism.[42]

The *progress narrative* explanation cannot be used to interpret Madeleine's case. Her cross-dressing is neither motivated by socio-economic reasons, nor restricted to a specific time. An unspecified but long period separates Madeleine's first visit to Rosette's castle from the second one when she meets d'Albert. More significantly, her final departure (which occurs at least two months after first meeting d'Albert)[43] suggests that

39 On cross-casting in France, see Julia Prest, *Theatre under Louis XIV: Cross-Casting and the Performance of Gender in Drama, Ballet, and Opera* (Houndmills, Basingstoke and New York: Palgrave Macmillan, 2006).

40 Garber, *Vested Interests*, 70.

41 Ibid., 68–69.

42 Ibid., 70–71.

43 Gautier, *Mademoiselle de Maupin*, 305.

she does not plan to renounce her male attire. Furthermore, the necessity for her to 'unmask' (to borrow Garber's terminology) and appear in her female attire may be, at first glance, in line with the *progress narrative* interpretation, where characters must give up their costume in order to have heterosexual intercourse and *de facto* avoid homosexual relations. Her own unmasking might be appropriate in order to seduce d'Albert but is not required regarding her relationship with Rosette. Madeleine joins her in her bedroom in Rosalind's, not Théodore's, outfit, with the purpose of achieving the homoerotic relations that the *progress narrative* aims to deny.

Madeleine's strongest motivations for cross-dressing are curiosity and a thirst for knowledge, notably about *scientia sexualis*, as Christopher Rivers argues, using Foucauldian terminology.[44] Foucault defines a *scientia sexualis* civilization as one which '[a] développé au cours des siècles, pour dire la vérité du sexe, des procédures qui s'ordonnent pour l'essentiel à une forme de pouvoir-savoir rigoureusement opposée à l'art des initiations et au secret magistral' [has developed over the centuries, to tell the truth of sex, procedures which are essentially organized as a form of knowledge-power rigorously opposed to the art of initiations and the masterful secret].[45] Whereas an *ars erotica* civilization values the notion of pleasure, the *scientia sexualis* considers sex to be a secret that only experts can access. As a result, *savoir*, *connaître* [to know], *apprendre* [to learn] and their derivatives are frequently used in Madeleine's letters (especially at the beginning of her first letter to Graciosa). Her quest for knowledge of men, or, more accurately, for knowledge of the difference between the sexes, is expressed more than once, but nowhere more frankly than in the following sentence: 'je voulais étudier l'homme à fond, l'anatomiser fibre par fibre avec un scalpel inexorable et le tenir tout vif et tout palpitant sur ma table de dissection' [I wanted to study the man in detail, dissect him fibre by fibre with an inexorable scalpel and hold him all alive and fluttering on my dissecting table].[46] The desire for exhaustive and scientific knowledge is highlighted by the medical, even surgical, vocabulary. Madeleine seems to share with

44 Rivers, '*Inintelligibles pour une femme honnête*', 12–15 .

45 Foucault, *Histoire de la sexualité*, I, 78. For more on *scientia sexualis* and *ars erotica*, see 71–98.

46 Gautier, *Mademoiselle de Maupin*, 305.

nineteenth-century doctors their obsession for scrutinizing the sexuality of her fellow human beings. Ironically, as a female cross-dresser and a lesbian, she is the sexually unconventional individual, whereas her patients, all (heterosexual) men, comply with sexual norms. More significantly, Madeleine's sentence reveals the domineering, erotic dimension that is intrinsic to her thirst for knowledge about sexual difference. Willing to penetrate the mysteries of men already means defining herself as a sexual being who wants to exercise sensual domination over men. This erotic authority anticipates the end of the novel, when Madeleine uses d'Albert to discover the mysteries of sexuality.

The *progress narrative* seems *a priori* more convincing to explain the castrato Zambinella's cross-dressing, as the castration of a son represented a large sum of money for humble families or, at the very least, the possibility of improving his living conditions by becoming an accomplished singer.[47] In Balzac's novella, Zambinella himself has become immensely rich and famous thanks to his operatic career. Of course, the singer's cross-dressing is not immediately manifest in *Sarrasine*, owing to Balzac's narrative techniques (the story within a story and the adoption of Sarrasine's limited perspective). In the novella, an anonymous narrator tells the Marquise de Rochefide a story that aims to unveil the secret of a mysterious old man. The protagonist of the story within the story is the sculptor Sarrasine. Whilst attending an opera in Rome, he becomes obsessed with the *prima donna* Zambinella. Sarrasine accepts the singer's invitation to a party with her fellow actors, but he is abashed by her simultaneous flirtation and rejection of his love. The next evening, at a private concert, Sarrasine realizes that Zambinella is not a woman, but a castrato. He abducts Zambinella but fails to eliminate both the singer and the statue that he made in 'her' image. He is killed by the henchmen of the castrato's protector.

If the *progress narrative* persuasively clarifies Zambinella's cross-dressing (and cross-casting) on stage, it fails to explain what is at the core of *Sarrasine*: his cross-dressing offstage. The castrato justifies it as follows: 'Je n'ai consenti à vous tromper que pour plaire à mes camarades, qui voulaient

47 Garber, *Vested Interests*, 254. See also Barbier, *The World of the Castrati*, 21; Noble, 'Castrati', 31.

rire' [I agreed to deceive you only to please my friends who wanted to have fun].[48] His justification is certainly convenient because it clears him of the responsibility of the deceit. However, it does not explain why he is so proficient at incarnating a woman and, above all, why he accepts this mission so willingly. His name already casts doubt over his virility. *Zambinella* is not a female first name, contrary to what Sarrasine thinks, calling him *la Zambinella*. It is either his last name or, as was then common, a pseudonym. Yet, most castrati's stage names had a masculine ending in *–o* or *–i*. *Cortona*, the only pseudonym with a feminine ending in Patrick Barbier's listing of castrati's names, is explained by the fact that it designates the castrato's hometown.[49] The choice of a feminine pseudonym suggests that the castrato had claimed his feminine identity long before taking part in the hoax.

Zambinella, although male, appears to be the perfect representation of traditionally feminine behaviour, as the narrator claims: 'C'était la femme avec ses peurs soudaines, ses caprices sans raison, ses troubles instinctifs, ses audaces sans cause, ses bravades et sa délicieuse finesse de sentiment' [It was woman with all her sudden fears, her irrational whims, her instinctive worries, her groundless boldness, her bravado and her delicious delicacy of feeling].[50] His case goes beyond mere effeminacy and supports the concept of gender performativity that is developed by Judith Butler in *Gender Trouble*. The American philosopher argues that '[t]here is not gender identity behind the expressions of gender; that identity is performatively constituted by the very "expressions" that are said to be its results'.[51] In other words, gender does not exist *per se*; it is always an imitation of so-called naturalistic – in reality, cultural – behaviour coded as 'masculine' or 'feminine'. The characteristics that are considered to reveal the gender of an individual thus produce this very gender. They are unconsciously adopted to correspond to a cultural norm. Women perform feminine behaviour and men perform virile behaviour because it is what is expected of them.

48 Honoré de Balzac, *Sarrasine. Gambara. Massimilla Doni*, Pierre Brunel, ed. (Paris: Gallimard, 'Folio Classique', 1995), 75.

49 Barbier, *The World of the Castrati*, 83–85.

50 Balzac, *Sarrasine*, 71.

51 Butler, *Gender Trouble*, 25.

If gender is a performative act, as Butler claims, this means that a gender can be performed by a person whose sex is not traditionally associated with this gender.[52] Whilst being a man, Zambinella outstandingly performs femininity. This performance must not be mistaken for either the operatic show that he produces on stage or the appearance of the ridiculous epicene old man at the Lantys, nostalgic for his former years of beauty and fame. Zambinella's performance internalizes feminine behaviour and all those so-called feminine personality traits that make Sarrasine certain that he has fallen in love with a woman. Not only is Zambinella physically feminine due to his soft voice and his womanly beauty, but his behaviour covers a wide spectrum of femininity. Sandra Gilbert and Susan Gubar's book *The Madwoman in the Attic* has highlighted how nineteenth-century women were perceived and depicted as either angelic or demonic.[53] The castrato embodies these two equally reductive roles of either spiritual and passive goodness, or carnal and assertive malignity. On the one hand, Zambinella exhibits signs of sensitivity, delicacy, modesty and timorousness. On the other hand, he acts as a capricious, voluptuous and nonchalant seductress who crosses his legs and plays 'footsie' with Sarrasine. Most of all, he is a *coquette* – a trait that assimilates him to Béatrix de Rochefide, who is also a convincing performer of femininity, but in a heteronormative fashion.[54]

In the castrato's performance of femininity, it is, however, difficult to distinguish between Zambinella's intrinsic femininity, the externalized

52 Butler, however, has distanced her arguments from the numerous interpretations of gender performativity as relating exclusively to 'queer' performativity. On the reception and re-readings of *Gender Trouble*, see Jay Prosser, *Second Skins: The Body Narratives of Transsexuality* (New York: Columbia University Press, 1998), 24–27; Elisa Glick, 'Sex Positive: Feminism, Queer Theory, and the Politics of Transgression', *Feminist Review*, 64 (Spring 2000), 19–45 (35–37).

53 Sandra M. Gilbert and Susan Gubar, *The Madwoman in the Attic: The Woman Writer and the Nineteenth-Century Literary Imagination*, 2nd edn (New Haven: Yale University Press, 2000 [1979]), 27–44. For women's representation in French literature, see Ruth P. Thomas, 'The Death of an Ideal: Female Suicides in the Eighteenth-Century French Novel', in Samia I. Spencer, ed., *French Women and the Age of Enlightenment* (Bloomington: Indiana University Press, 1984), 321–31.

54 On Zambinella's feminine features, see Balzac, *Sarrasine*, 47, 57, 62, 64–66, 69, 71 and 74. On Béatrix de Rochefide's coquetry: 52.

femininity that is imposed by the script of the prank and Sarrasine's own bias. For instance, the fear that characterizes the singer's attitude towards the sculptor is interpreted by the latter as a vague sign of femininity, whereas it is an understandable feeling motivated by the apprehension that Sarrasine may find out the truth, later exacerbated by his violence.[55] The narrator hints at the diverse possible interpretations when a bottle of champagne is opened: 'Sarrasine lut dans les yeux de sa voisine une crainte assez vive de la petite détonation produite par le dégagement du gaz. Le tressaillement involontaire de cette organisation féminine fut interprété par l'amoureux artiste comme l'indice d'une excessive sensibilité' [Sarrasine read in his neighbour's eyes a fairly sharp fear of the small explosion made by the gas emission. The amorous artist interpreted the involuntary shudder of this feminine constitution as the sign of an excessive sensitivity].[56] Is the quivering real or only fantasized by Sarrasine, an unreliable witness? If it is real, is it artless or is it part of Zambinella's conscious performance? Later, the sculptor persists in imposing his vision of gender identity on the singer, when telling Zambinella, who is afraid of a grass snake: 'oserez-vous bien prétendre que vous n'êtes pas femme? [Do you dare pretend that you are not a woman?].[57] These examples show that Sarrasine is keen to interpret any supposedly feminine behaviour as proof of femaleness.

For Sarrasine, if an individual behaves according to specific gender patterns, he or she must belong to the sex traditionally associated with that gender. To put it simply, gender performativity always equates to sex performativity for him. His belief can be linked to Butler's claim in *Bodies That Matter* that sex is also performative; that it is also culturally constructed. If sex can only be accessed through its construction, it is therefore a kind of fantasy.[58] In other words, sex is often interpreted through gender displays. Individuals are compelled to perform the gender that is associated with their sex because they are determined by regulatory norms, which are those of compulsory heterosexuality.[59]

55 Ibid., 67–68.
56 Ibid., 66.
57 Ibid., 71.
58 Butler, *Bodies That Matter*, 5.
59 Ibid., 12.

The issue with Zambinella's sex is that it is twice socially constructed, first surgically, then as a fantasy. It thus escapes clear classification. Although his sex does not seem to correspond to the gender that is performed, it has been altered by the surgery that he endured in his boyhood and cannot be regarded as entirely male. As seen in Chapter 1, biological sex is nowadays defined according to different parameters, such as external and internal genitals, chromosomes and hormones (most of them unknown in Balzac's time). Zambinella's anatomical and chromosomal sex is male, but the operation of severing the ducts to his testes has modified his hormonal sex, since the absence of testosterone causes the overproduction of female hormones.[60] As a castrato, Zambinella might have developed secondary female sexual characteristics, such as the sought-after high-pitched voice, prominent breasts, deposits of fat on the hips and the thighs, and, conversely, he might not have grown male characteristics such as facial hair and an Adam's apple.

Roland Barthes certainly exaggerates when he claims that the text lies when asserting that Zambinella is a man, but it is not unreasonable to interpret the following excerpt as a suggestion as to Zambinella's female sex: 'Pendant une huitaine de jours, il [Sarrasine] vécut toute une vie, occupé le matin à pétrir la glaise à l'aide de laquelle il réussissait à copier la Zambinella, malgré les voiles, les jupes, les corsets et les nœuds de rubans qui la lui dérobaient' [For eight days, Sarrasine lived a whole life, spending the morning kneading the clay with which he was able to copy la Zambinella, despite the veils, the skirts, the corsets and the ribbons that concealed her from him].[61] As Thomas Pavel notes, the verb *copier* is ambiguous.[62] Manifestly, the sculptor does not copy Zambinella, since the female garments that allegedly prevent him from seeing the *prima donna*'s female sex actually conceal a key part of Zambinella's body: his penis. Female attire

60 On the effects of castration, see Barbier, *The World of the Castrati*, 13–14; Noble, 'Castrati', 31; Prest, *Theatre under Louis XIV*, 131.

61 Balzac, *Sarrasine*, 60; Barthes, *S/Z*, 48. Sandy Petrey's contestation of Barthes's claim also appears exaggerated: 'Speech Acts, and the Realist Differences: *S/Z* versus *Sarrasine*', *PMLA*, 102/2 (March 1987), 153–65 (156–58).

62 Thomas Pavel, 'Énergie et illusion: Sur les protagonistes de *Sarrasine*', *French Forum*, 21/3 (September 1996), 301–18 (308).

here acts as a fetish, that is, 'an object or a body part [endowed] with an unusual degree of [...] erotic pleasure'.[63] Indeed, in a reversal of Freudian interpretation, ribbons and laces do not substitute for the male sex that the singer actually has; they replace the female sex that the singer lacks. And yet, the opposition between the 'real' male sex and the fantasized female sex of Zambinella is not as clear as it seems. The secondary female sexual characteristics of his body make it impossible to be sure that Zambinella's corset does not hide a bosom. Barthes, however, is certain of the fact. Commenting on the phrase '[s]a poitrine, dont une dentelle dissimulait les trésors par un luxe de coquetterie' [her bosom, whose lace concealed the treasures by an abundance of coquetry], he claims: 'ce qu'il faut dissimuler, c'est qu'il n'y a rien' [what needs to be concealed is that there is nothing].[64] All the ambiguity of Zambinella's body lies here: his attire hides physical features that, as a man, he should not possess, but that he may nevertheless possess. The modification of Zambinella's male sex suggests that his performance of a feminine gender is also due to his physique and that this performance is therefore expected of him on stage as well as in his daily life. Indeed, as we have seen, Cabanis, Moreau de la Sarthe and Virey repeatedly asserted that a man deprived of his physical virility bordered on being a woman, and thus could not fully be considered a man.[65]

The case of Zambinella demonstrates that, while intuitively perceiving the principle of gender performativity, writers cannot conceive of an individual that can harmoniously combine characteristics of both sexes and both genders. The hermaphrodite is thus never constructed as a fully intersex individual, but rather through gendered alternatives that make the hermaphrodite more comprehensible for the reader, such as tomboyishness, effeminacy, cross-dressing and gender performativity. The inability to fathom an intersex individual outside of the categories of maleness or femaleness

63 'Fetishism', in Joseph Childers and Gary Hentzi, eds, *The Columbia Dictionary of Modern Literary and Cultural Criticism* (New York: Columbia University Press, 1995), 109.

64 Balzac, *Sarrasine*, 65; Barthes, *S/Z*, 148.

65 Cabanis, *Rapports*, I, 384–90; Moreau de la Sarthe, *Histoire naturelle*, I, 195–97; Virey, *De la femme*, 180, 210 and 231.

contributes to the ambiguous position of the writers when judging these characters as monstrous.

Hermaphroditic Monstrosity and Exclusion

The inherent ambivalence of hermaphrodites raises the question: Are they superior beings who have overcome the difference of sex, or are they monsters? We already know that the hermaphrodite is a monster in the eyes of scientists. After all, Geoffroy Saint-Hilaire's chapter on hermaphroditism is in a treatise about physical monstrosities. This physical monstrosity is linked to a supposedly moral one; both of which constitute the main difference between the hermaphrodite and the analogous concept of the androgyne, although these two words are sometimes regarded as being synonymous.[66] The myth of the androgyne is told by Aristophanes in Plato's *Symposium*. Human beings used to be double, round-shaped creatures: twice male, twice female, or both male and female, namely androgynous (*androgyne* means 'man-woman' in Greek). Zeus separated them when they conspired against the gods. Each half henceforth longed for its fellow and attempted to reunite through sexual encounter.[67] Unlike Hermaphroditus for whom the fusion of the male and female bodies is made against his will, it is the separation of sexed halves that is against the will of the androgynes.

The androgyne is often considered to embody a positive, Romantic representation of metaphysical intersexuality. Because of their spiritual connotation, androgynes are often connected to the ideal figures of angels.[68] Conversely, hermaphrodites are seen as negative, decadent and even corrupt

66 On the difference between hermaphrodite and androgyne, see Ramacciotti, 'L'esilio dell'androgino', 190; Weil, *Androgyny*, 17–21; Kerlouégan, 'Une relecture de l'androgynie romantique', 38–41. On the other hand, Busst (*The Image of the Androgyne*, 1) considers that the distinction between them is arbitrary.

67 Plato, *Lysis. Symposium. Gorgias*, W. R. M. Lamb, ed. and trans. (London: W. Heinemann-Macmillan, 1914), 132–51.

68 Joséphin Péladan, *De l'Androgyne: Théorie plastique* (Puiseaux: Pardès, 'Rebis', 1988), 37–67. An illustration of the philosophical and religious aspects of the myth

versions of intersexuality because of their sensual implications.[69] Their intersexuality is considered excessive eroticism. As Mircea Eliade explains:

> Chez les écrivains décadents, l'androgyne est compris uniquement comme un hermaphrodite dans lequel les deux sexes coexistent anatomiquement et physiologiquement. Il s'agit, non pas d'une plénitude due à la fusion des sexes, mais d'une surabondance des possibilités érotiques. Il n'est pas question de l'apparition d'un nouveau type d'humanité, dans lequel la fusion des sexes aurait produit une nouvelle conscience, apolaire, mais une soi-disant perfection sensuelle, résultant de la présence active des deux sexes.[70]
>
> [For decadent writers, the androgyne is understood solely as a hermaphrodite in which the two sexes coexist anatomically and physiologically. It is not a wholeness caused by the fusion of the sexes but an overabundance of erotic possibilities. It is not a question of the arrival of a new type of humankind in which the fusion of the sexes would have created a new a-polar consciousness, but of so-called sensual perfection, resulting from the active presence of the two sexes.]

In brief, the perfect completeness of androgynes is opposed to hermaphrodites' incompleteness, which is expressed through their impossible search for this absolute.[71] This interest in incomplete hermaphrodites occurs in earlier texts than is assumed by Eliade. Narratives by Latouche, in particular, but also Balzac and – to a lesser extent – Gautier, hint at the possibility of hermaphrodites' incompleteness and moral imperfection owing to their physical monstrosity. What deserves meticulous examination is how writers grant hermaphrodites the ability to give an ontological interpretation of their own situation. These moments of hermaphroditic speeches are even more significant since they occupy a relatively secondary position in the narratives and, as has been seen, are

---

of the androgyne during the July Monarchy times is Balzac's Swedenborgian novel *Séraphîta*, published in 1834 in *La Revue des deux mondes*.

69 Mircea Eliade, *Méphistophélès et l'androgyne* (Paris: Gallimard, 'Idées', 1962), 143–44; Busst, *The Image of the Androgyne*, 10–11.

70 Eliade, *Méphistophélès et l'androgyne*, 144.

71 On the opposition between androgyne/completeness and hermaphrodite/incompleteness, see Albouy, 'Le mythe de l'androgyne', 601, 605 and 607; Ramacciotti, 'L'esilio dell'androgino', 190; Martial Guédron, ed., *Monstres, merveilles et créatures fantastiques* (Paris: Hazan, 2011), 111.

usually mediated through an external narrator or another character. These discourses are fashioned by the tragic dimension of those who cannot identify with the sexual and gender dichotomy.

In *Fragoletta*, intersexuality is experienced as a curse.[72] Referring frequently to the Bible, the idea of a curse is expressed mostly by the male side of the hermaphrodite, who is perceived as more wicked than his female counterpart. It stems from the feeling of not belonging to Creation. Camille tells d'Hauteville before they fight: 'Qu'y a-t-il de commun entre moi et les créatures humaines? Je ne suis pas de leur espèce' [What do human creatures and I have in common? I do not belong to their species].[73] Camille defines themselves as an 'erreur, crime ou rebut de [la] nature' [nature's error, crime or waste], because their existence defies understanding.[74] In this respect, the discourse of Latouche's fictional hermaphrodite on curse, difference and isolation echoes the writings of historical hermaphrodites, such as Herculine Barbin. In their 'Souvenirs', the hermaphrodite asserts their cursed and lonely nature from the outset: 'J'ai beaucoup souffert, et j'ai souffert seul! seul! abandonné de tous! Ma place n'était pas marquée dans ce monde qui me fuyait, qui m'avait maudit' [I have suffered a lot and I have suffered alone! Alone! Forsaken by all! My place was not marked out in this world, which shunned me, which had cursed me].[75] Whether Barbin, an avid reader, actually read *Fragoletta* and identified with the character is not known, although they adopted the same gender-neutral first name,

72 Crouzet, 'Monstres et merveilles', 36; Ramacciotti, 'L'esilio dell'androgino', 195; Laforgue, *L'éros romantique*, 97.

73 Latouche, *Fragoletta*, II, 226–27. See also II, 225.

74 Ibid., II, 126. See also II, 194.

75 Barbin, 'Mes souvenirs', 9. See also: 'Et maintenant seul! … seul! … pour toujours! Abandonné, proscrit au milieu de mes frères! Eh! que dis-je! Ai-je le droit de donner ce nom à ceux qui m'environnent? […] Va, maudit, poursuis ta tâche! Le monde que tu invoques n'était pas fait pour toi. Tu n'étais pas fait pour lui' (111) [And now alone!... Alone!... Forever! Forsaken, banned among my brothers! Eh! What am I saying! Do I have the right to give this name to those who surround me? […] Go, cursed one, carry out your task! The world that you are invoking was not made for you. You were not made for it].

Camille, in their memoirs.[76] Similar experiences of isolation are also to be found in more recent accounts of intersex individuals that are probably not influenced by *Fragoletta*. The twentieth-century intersex activist Cheryl Chase similarly uses the vocabulary of monstrosity and solitude when recalling the discovery of her medical records: 'I was no longer a woman in my own eyes but rather a monstrous and mythical creature. [...] To myself, I was a freak, incapable of loving or being loved, filled with shame about my status as a hermaphrodite and about my sexual dysfunction. [...] I was in emotional agony, feeling utterly alone, seeing no possible way out.'[77] This suggests that isolation and the feeling of being cursed are intrinsic to the narrative framing of the hermaphrodite.

As their sexual difference leads to feelings of loneliness and rejection, the hermaphrodite holds a grudge against God for giving them an existence that differs from the existence of other creatures: 'Dieu lui-même est injuste aussi, car l'existence qu'il m'a donnée porterait malheur à qui placerait en moi une espérance d'attachement' [God Himself is also unfair, because the existence that He granted me would bring misfortune to anyone who would hope to be attached to me].[78] The curse is thus extended to anyone who dares to love them by condemning them to an impossible love, as the hermaphrodite's difference means that they are unable to find their opposite in the logic of the separation of the sexes. While hermaphrodites may symbolize the union of two bodies and thus exemplify androgynous completeness, their bodily difference, as Eliade suggests, is perceived as a form of incompleteness. Consequently, it prevents them from concretely realizing this fusion with another being. The burden of contradictory fates causes Camille to view God as an unfair and whimsical being that enjoys torturing His creatures in transgressing His own laws.[79]

76 Linton (*Prescribed Fictions*, 32) is more certain than me that Barbin had read *Fragoletta*. Gabrielle Houbre suggests other sources of inspiration, such as Saint Camille de Lellis, Virgil's *Aeneid* and Roman history: *Les deux vies d'Abel Barbin, né Adélaïde Herculine (1838–1868)* (Paris: Presses Universitaires de France, 2020), 138–39, endnote 38.

77 Cheryl Chase, 'Hermaphrodites with Attitude: Mapping the Emergence of Intersex Political Activism', in *The Transgender Studies Reader*, 300–14 (304).

78 Latouche, *Fragoletta*, I, 112.

79 Ibid., 125.

Camille therefore considers that their place is in Hell owing to the curse of which they are a victim: being excluded from God's love. They judge their life to be a torment, and, consequently, want to die.[80] Since the Catholic Church perceives suicide as a mortal sin, that is, a sin that cannot be absolved, a Catholic who commits suicide must consider her or his life worse than Hell.[81] According to this logic, by committing suicide, Camille could never be redeemed. The hermaphrodite is destined to damnation.

Their Inferno, however, is predominantly on earth and is linked to the original sin of knowledge: 'connaître est déjà un enfer' [knowledge is already hell].[82] Interestingly, Camille is both the victim of this sin and the tempter, as the hermaphrodite is compared to a snake.[83] This knowledge is that of the union of the sexes. As a hermaphrodite, Camille possesses this knowledge 'naturally', whereas other individuals usually gain it through sexual intercourse, as is the case with Madeleine. The obsession with knowledge assimilates them to Cain, Abel's brother and murderer in the Bible. Like Lord Byron's Cain (1821), the hermaphrodite is preoccupied by knowledge, although its object is different and expressed in reverse terms. Cain is obsessed with death, or rather with his current ignorance of death and with his ineluctable knowledge of death as a mortal being.[84] Conversely, Camille is haunted by their 'natural' knowledge of the union of the sexes as a hermaphrodite and by their ignorance of the union of the sexes with other individuals.

Moreover, both the biblical character and the hermaphrodite bear a mark on their bodies. Cain bears a mark of God's protection, while Camille has a fruit-shaped spot on their right arm, which explains their nickname *Fragoletta*, or 'little strawberry' in Italian. Above all, both are inexplicably

80 Ibid., II, 124 and 228–30.

81 Rodney Stark, Daniel P. Doyle and Jesse Lynn Rushing, 'Beyond Durkheim: Religion and Suicide', *Journal for the Scientific Study of Religion*, 22/2 (June 1983), 120–31 (121).

82 Latouche, *Fragoletta*, II, 194.

83 Ibid., II, 98.

84 George Gordon Byron, 'Cain', in *Lord Byron's Cain: Twelve Essays and a Text with Variants and Annotations*, Truman Guy Steffan, ed. (Austin and London: University of Texas Press, 1968), 153–258 (for instance, Act I, vv. 268–299, 174–75 and vv. 451–460, 184–85).

excluded by God. In *The Curse of Cain*, Regina Schwartz argues that God is responsible for the first biblical murder by favouring Abel over Cain and, consequently, is as criminal as Cain: 'What kind of God is this who chooses one sacrifice [Abel's] over the other [Cain's]? [...] who excludes some and prefers others, who casts some out[?]'[85] Schwartz's reading of Cain undermines the representation of the biblical character as a criminal and considers him to be a victim. It thus supports the Romantic rehabilitation of this character in the works of authors such as Byron and Hugo. Likewise, Camille's case – cast out and unable to understand the reasons for their exclusion – denounces God's unfairness because He has mischievously modified His own Creation to include a being who is incapable of bonding with other humans. Latouche's novel highlights the fact that, although Camille is a potential criminal due to their counter-nature, the hermaphrodite is also a victim of God's whim and abandonment.

Fragoletta presents the hermaphrodite as both an innocent girl and an insensitive, cruel and violent man. Although their portrait is little nuanced, Latouche's novel outlines the ontological crisis that individuals who do not fit the binary gender and sexual system may face. This crisis, whether it appears as malaise, fear or revolt, is also present in other hermaphroditic narratives.

Like Camille, Zambinella is a hermaphroditic character who struggles with his creator. Zambinella is not simply a creature of God. Although virtually no information is given about his 'making', it can be assumed that he has been fashioned by three other creators: an anonymous surgeon barber who operated on the castrato in his boyhood, his mentor Prince Chigi and Sarrasine. The singer's beautiful voice and looks have been brought about by human intervention, as Chigi reveals to Sarrasine at the end of the story. The old man was pointedly called an 'homme artificiel' [artificial man] and a 'création artificielle' [artificial creation] in the frame story.[86] Later, through the gaze of the sculptor, the face of the *prima donna* is perceived as worthy of a painting and 'her' body is seen as that of a Greek statue of

85 Regina M. Schwartz, *The Curse of Cain: The Violent Legacy of Monotheism* (Chicago and London: University of Chicago Press, 1997), 3.

86 Balzac, *Sarrasine*, 40 and 47.

Venus.[87] The word *création*, which had just been used to refer to classical art, is employed again to refer to Zambinella; in Sarrasine's eyes, the singer is not only a work of art, but 'she' embodies artistic perfection: 'C'était plus qu'une femme, c'était un chef-d'œuvre!' [She was more than a woman; she was a masterpiece!].[88] Sarrasine's observation is intended to be hyperbolical; he ignores that this is literally true: Zambinella is more than a woman because he is an artificially created castrato.

Zambinella is also perceived as a 'creature'. The *Dictionnaire de l'Académie française* indicates that the word *créature* is used to designate 'une personne qui tient sa fortune ou son élévation d'une autre' [a person who gets their wealth or ascension from another], which is precisely Zambinella's position towards Chigi, whilst also explaining that *créature* 'se dit aussi par mépris' [is also used out of scorn].[89] Consequently, Zambinella-the-creature infringes on Zambinella-the-masterpiece. Zambinella himself attempts to assuage Sarrasine's fervour by calling himself 'une créature maudite' [a cursed creature].[90] He expresses the same feeling of being cursed as Camille did. Like the hermaphrodite, he feels excluded from the world because no other creature can bond with him and, importantly, he cannot bond with anyone. His words 'Je n'ai pas de cœur!' [I have no heart!] highlight his isolation.[91] Accordingly, he turns his feelings of seclusion into hatred: 'J'abhorre les hommes encore plus peut-être que je ne hais les femmes' [I loathe men perhaps even more than I hate women].[92] Zambinella's misanthropy mirrors the hermaphrodite's anger against God since men are responsible for his exclusion. God, nevertheless, is not spared because He provided him with his feminine beauty: 'Fatale beauté!' [Fatal beauty!], the castrato whispers while looking at the sky, and thus at Heaven, with horror[93]. If the castrato's beauty is due to human intervention, its perfect outcome may be due to Fate.

87 Ibid., 58.
88 Ibid. See also 57.
89 'Créature', in *Dictionnaire de l'Académie française*, I, 447.
90 Balzac, *Sarrasine*, 69.
91 Ibid., 70.
92 Ibid., 69.
93 Ibid., 70.

Not only does the word *créatures* signify inferiority, it is also a common euphemism to designate castrati. Chigi unveils the enigma by saying: 'Et ne savez-vous pas par quelles créatures les rôles de femme sont remplis dans les États du pape?' [And don't you know which creatures fulfil female roles in the Papal States?], whilst the narrator finishes his story by saying: 'On n'y [en Italie] fait plus de ces malheureuses créatures' [These unfortunate creatures are no longer made in Italy].[94] Incidentally, the nature of the old man in the frame story may have been hinted at by the narrator calling him 'cette créature bizarre' [this strange creature].[95] Unlike a *chef-d'œuvre* [masterpiece], the term *créature* does not connote the ideal beauty and the exceptional voice of the castrato but refers to the physical human being and his surgical operation. Euphemism is also used in *Fragoletta* to allude to the process of castration. One reads on a sign: 'Qui si castrano i puti miravigliosamente' [literally: Here boys are castrated wonderfully], which Latouche freely – and ironically – translates as: 'Ici on *perfectionne* les petits garçons' [Here little boys are *perfected*].[96] The perfection embodied by castrati is denied by the narrator, who instead mocks a tradition that has been little accepted in France.[97] This so-called perfection is also questioned by Béatrix de Rochefide, when admiring the portrait of Adonis inspired by Zambinella: 'Un être si parfait existe-t-il?' [Does such a perfect being exist?].[98] The novella suggests that beauty is possible only through artificiality and performance.

The castrato's physical perfection is mainly challenged by the revelation of his 'real' sex, which leads to Sarrasine's violent acts and an offensive speech in which he calls the castrato a monster.[99] Sarrasine threatens the singer by claiming: 'Tu n'es rien. Homme ou femme, je te tuerais! mais …' [You are nothing. If you were a man or a woman, I would kill you! But …].[100] The use of the conditional was already a feature of Zambinella's

94 Ibid. 73 and 78.

95 Ibid., 42.

96 Latouche, *Fragoletta*, II, 178 (emphasis in original).

97 Barbier, *The World of the Castrati*, 186–201; Prest, *Theatre under Louis XIV*, 137–38 and 146–48.

98 Balzac, *Sarrasine*, 49.

99 Ibid., 76.

100 Ibid., 75.

speech, when hinting at his 'true' identity.[101] The castrato assumed that the sculptor would not take the deceit lightly, were he to know the truth. This conditional both confirms and infirms the singer's assumption, as Sarrasine knows the truth and thinks about killing him, but ultimately renounces it. So, while the conditional and the conjunction *mais* diminish the impact of the threat, Sarrasine's assertion nevertheless reflects the anger arising from his inability to fathom Zambinella's sexual identity. The sculptor denies the singer's identity, first openly, then by stating an alternative construction to which Zambinella does not belong because it follows the common belief in the separation of the sexes. Sarrasine struggles to conceive of an individual with no determinate sex (and a gender suitable to that sex) and his threat is explained by his fear of his own potential sexual and gender indeterminacy, as well as of sterility, as is evident from the use of the indirect object pronoun *me*: 'toi qui ne peux donner la vie à rien, tu m'as dépeuplé la terre de toutes ses femmes' [You who can give life to nothing, you have depopulated the Earth of all its women for me].[102] His fear echoes the anxiety of Virey and other scientists relating to the depopulation and degeneration of French society. Just as they dreaded the feminization of Frenchmen, so the sculptor perceives himself as lacking virility and holds Zambinella's sterility (caused by his emasculation) responsible for his own sexual ambiguity. Barthes summarizes: 'Sarrasine contemple en Zambinella sa propre castration' [Sarrasine contemplates in Zambinella his own castration].[103]

However, whilst the castrato's sterility mirrors the most extreme cases of intersexuality, it does not deprive him of descendants. Certain aspects of his patrimony have been transmitted to his family. His great-nephew Filippo has inherited his effeminate looks: the narrator compares him to Antinous, Hadrian's ephebic lover.[104] More significantly, Zambinella has given his love for music and his beautiful voice to his great-niece Marianina. The young woman's portrait emphasizes her lyrical gift that can stem only

101 Ibid., 67 and 69.

102 Ibid., 76.

103 Barthes, *S/Z*, 113. See also 42–43, 171, 181, 204–08 and 216–18. On castration in *Sarrasine*, see also Jean Reboul, '*Sarrasine* ou la castration personnifiée', *Cahiers pour l'analyse*, 7 (1967), 91–96.

104 Balzac, *Sarrasine*, 38.

from her great-uncle's own musical perfection.[105] The infertility of hermaphrodites is thus overcome by the ability to transmit an aesthetic heritage. Art, as will be discussed later, becomes the sanctuary of hermaphroditic characters by counterbalancing their monstrosity. The old Zambinella may have lost his perfect voice and hearing, but what prevents him from being ridiculous is his never-ending passion for music.[106]

Valeria Ramacciotti has established a connection between Camille, Zambinella and Madeleine as being victims of exclusion.[107] Although the hermaphrodite and the castrato's rejection is unquestionable, it is exaggerated in the case of the young woman. *Mademoiselle de Maupin* differs from *Fragoletta* because the suffering protagonist is not Madeleine, the hermaphroditic character, but d'Albert, the man who falls in love with such a character. The insecure young man shares similar sentiments of difference, loneliness and incompleteness. His incompletion as a man and an artist explains his quest for an ideal partner.[108] However, this quest for beauty differentiates him from other men and prevents him from bonding with them: 'les autres hommes ne sont guère pour moi que des fantômes, et je ne sens pas leur existence' [other men are only ghosts to me and I cannot feel their existence].[109] His thoughts are typical of Romantic heroes suffering from the *mal du siècle*.[110] At the beginning of Chapter XI, d'Albert expresses his ontological *ennui*.[111] His long letter to Silvio highlights his feeling of inadequacy in modern life:

> Ah! malgré l'étreinte furieuse dont j'ai voulu enlacer le monde matériel au défaut de l'autre, je sens que je suis mal né, que la vie n'est pas faite pour moi, et qu'elle me repousse; je ne puis me mêler à rien: quelque chemin que je suive, je me fourvoie; l'allée unie, le sentier rocailleux me conduisent également à l'abîme.[112]

105 Ibid., 37–38.
106 Ibid., 45 and 48.
107 Ramacciotti, 'L'esilio dell'androgino', 189.
108 Weil, 'Romantic Androgyny', 353–54.
109 Gautier, *Mademoiselle de Maupin*, 124. See also 280.
110 Monneyron, *L'androgyne romantique*, 100.
111 Gautier, *Mademoiselle de Maupin*, 269.
112 Ibid., 277–78.

> [Ah! Despite the furious embrace with which I wanted to hold the material world for lack of anything else, I feel that I was ill-born, that life is not made for me and rejects me; I can blend into nothing: whatever path I follow, I err; both the even and the stony paths lead me to the abyss.]

This maladjustment, a constant topic in his letters, is coupled with his inability to take responsibility for his actions. D'Albert describes his impression of sleepwalking through his life and his lack of self-awareness. Accordingly, the young man claims his inability to fulfil an artistic career and to achieve anything concrete in life.[113]

On the contrary, Madeleine does not suffer from loneliness, exclusion or passivity. Her quest for knowledge is not experienced as a burden and, unlike Zambinella and Camille, her destiny is not marked by decay and brutal death. She is capable of assuming her identity and of embracing the world instead of running away from it, whereas d'Albert broods over questions of beauty and art and shuns the rest of the world. I must therefore dispute Albert Smith's and Ramacciotti's interpretations of Madeleine's departure.[114] The former claims that Madeleine seeks a (male) lover who is entirely devoted to love; her disappearance is a sign of her failure to find durable happiness with such a partner. For the latter, Madeleine's departure represents her annihilation as a hermaphrodite since she acknowledges the impossibility of living beyond d'Albert's erotic fantasy. However, these interpretations misread the purpose of her quest. Madeleine does not define herself through d'Albert's gaze. She seeks knowledge rather than happiness, especially if it is defined as heteronormative conjugal bliss. Accordingly, she leaves the castle because she has learned all that she could from d'Albert and Rosette and needs to go elsewhere to put her knowledge into practice.

As 'products' of ontological anguish, hermaphrodites feel as though they do not belong to humanity due to their hermaphroditism, which has been fashioned by their creators, either God or men. Their physical

113 Ibid., 280–81 and 286–90. See the section 'The Weak Man and the Aspiring Artist' in the following chapter.

114 Albert B. Smith, 'Gautier's *Mademoiselle de Maupin*: The Quest for Happiness', *Modern Language Quarterly*, 32 (1971), 168–74 (171–72); Ramacciotti, 'L'esilio dell'androgino', 198.

monstrosity causes isolation, which leads to sentiments of incompleteness and even hatred against those who have rejected them. Their monstrosity is also judged to be immoral as they are suspected of homosexuality due to their intersexuality.

Hermaphroditism and Homosexuality

The previous chapter showed that homosexuality was judged a physical and psychological monstrosity in the nineteenth-century medical field. It also demonstrated that the connection between same-sex relations and intersexuality was concurrently highlighted and denied in scientific writings. Hermaphrodites were in fact compelled to choose partners belonging to the opposite sex of their own predominant sex in order to meet the heterosexual norm. However, the correlation between homosexuality, monstrosity and gender ambiguity is not limited to medical texts. Homosexuality and intersexuality have also been connected in society, generally in order to render these 'deviant' behaviours or physical characteristics more intelligible. As Garber observes with regard to cross-dressing, 'if there is a difference (between gay and straight), we want to be able to *see* it, and if we see a difference (a man in women's clothes), we want to be able to *interpret* it'.[115] Similarly, when nineteenth-century novels suggest, or even depict, homoerotic relations, they often connect homosexuality to hermaphroditism.

The importance of homosexuality in these narratives is inversely proportional to the degree of intersexuality of the protagonists, as though the link between the two notions were self-evident and did not need to be stressed. Homosexuality constitutes a cornerstone of *Mademoiselle de Maupin*, whereas it is seldom mentioned in *Fragoletta*. Camille's sexuality appears to be the most intelligible. As an intersex individual, they display feelings for members of both sexes, depending on the sexual identity

115 Garber, *Vested Interests*, 130 (emphasis in original). See also Prest, *Theatre under Louis XIV*, 5.

that they assume. As a girl, Fragoletta unintentionally provokes the love of d'Hauteville; as a man, Philippe rapes Eugénie. The passivity of the female side of the hermaphrodite can be interpreted in one of two ways. Either it corresponds to the feminine, 'natural' and highly recommended inclination to modesty concerning relations with men, coupled with the hermaphrodite's youth and ignorance in the domain of love; or it must be regarded as contempt for all men. If one follows Geoffroy Saint-Hilaire's claim about a *hermaphrodite bisexuel*'s impossible inclination to both sexes and the necessity of choosing an appropriate partner for their predominant sex, Camille's choice of a female partner would mean that their predominant sex must be male.[116] This conclusion, however, challenges the observation of the doctor who perceives Camille's dead body as female, once again creating sexual confusion and preventing the reader from interpreting Camille's sex as unique. Owing to the construction of the novel and the bad press of sapphism in *Fragoletta*, it is equally dubious that Camille should be interpreted as a lesbian character. The narrative construction of *Fragoletta* separates the female side from the male side of the hermaphrodite. As previously shown, the female side portrayed in the first part of *Fragoletta* is positively connoted, since it is determined by innocence and purity. In contrast, the only explicit sapphic relationship depicted in the novel appears immoral and stereotypical. This is the relationship between the Messalina-like Queen Caroline and Emma Hamilton (who also happens to be the mistress of Napoleon's enemy, Admiral Nelson). These women are portrayed as vicious, manipulative and careless, since they inconsiderately talk about politics and the predicaments of the Neapolitan people whilst lasciviously having a bath together.[117] Accordingly, the 'good' Fragoletta cannot be associated with such concupiscent individuals. Like their body and morals, the hermaphrodite's sexuality is depicted in an *either/or* perspective. It follows the scientific view that a hermaphrodite should be attracted to the sex that is opposite their predominant sex and avoids making claims that contradict compulsory heterosexuality.

116 Geoffroy Saint-Hilaire, *Traité de tératologie*, II, 171.

117 Latouche, *Fragoletta*, I, 177–86.

In *Sarrasine*, the connection between homosexuality and gender indeterminacy is more clearly affirmed. Sarrasine's strong reaction to the revelation of Zambinella's 'real' sex (he abducts and threatens to kill the castrato) is due to the singer's feminine performance calling the sculptor's own masculinity into question. Throughout the embedded story, the sculptor Sarrasine is presented by the narrator as hypermasculine. Dana Oswald defines hypermasculinity as 'a category marked by inflated physical traits, as well as performances of aggression and domination'.[118] Sarrasine indeed exhibits a violent virility in which the traditional characteristics of male behaviour are turned into flaws. His restlessness, lack of coquetry, taste for fighting, impiety and even his ugliness are exaggerated and appear to be evidence of his manliness.[119]

His manhood, however, is questioned by his name, which has feminine and oriental connotations. Let us not forget that the Orient was seen as feminine in the nineteenth-century Western imagination. The suffix –*ine* is a common one for French female names and *sarrasin* designates a 'musulman d'Orient, d'Afrique ou d'Espagne, au Moyen Âge' [Eastern, African or Spanish Muslim, in the Middle Ages].[120] Importantly, Sarrasine originally lacks one of the components that define hypermasculine behaviour according to Donald Mosher and Mark Sirkin: 'calloused sex attitudes

118 Dana Oswald, 'Monstrous Gender: Geographies of Ambiguity', in Asa Simon Mittman and Peter J. Dendle, eds, *The Ashgate Research Companion to Monsters and the Monstrous* (Farnham, Surrey and Burlington, VT: Ashgate, 2012), 343–63 (347).

119 Balzac, *Sarrasine*, 53–56 and 68.

120 'Sarrasin, ine' in *Le Nouveau Petit Robert: Dictionnaire de la langue française* (Le Robert-SEJER, 2004), 2364. On the femininity of 'Sarrasine', see also Reboul, '*Sarrasine* ou la castration personnifiée', 93; Barthes, *S/Z*, 24; Citron, 'Interprétation de *Sarrasine*', 90. Citron also underlines the orientalism of *Sarrasine* in 'Le rêve asiatique de Balzac', *L'Année balzacienne*, 1st ser. (1968), 303–36 (310–11). For his part, Jean Seznec suggests that Balzac found this name in Diderot's *Salon de 1767* which mentioned a sculptor who lived in the seventeenth century and was named Jacques Sarrazin: 'Diderot et *Sarrasine*', *Diderot Studies*, 4 (1963), 237–45 (239–40).

toward women'.[121] Indeed, he knows nothing about love and his virility has not been proven by female conquests.[122] He had only one mistress, the dancer Clotilde, and only for a brief period. Sarrasine feels threatened by any woman who confidently embodies the masculinity that he attempts to imitate and therefore rejects 'une femme forte, une Sapho, courageuse, pleine d'énergie, de passion' [a strong woman, a brave Sappho, full of energy and passion].[123] Accordingly, as Barbara Johnson rightly shows, the sculptor affirms his manliness in choosing a partner among women who are 'his symmetrical opposite[s]', namely very feminine women.[124] He does not conceal the fact that he loves Zambinella *because* 'she' is weak, thus apparently harmless.[125]

Like those men whose 'style of life [...] protest[s] its masculinity just a little too much not to convey an underlying insecurity', as Lynne Segal observes about Ernest Hemingway, Sarrasine's virility is faltering.[126] As a result, he is compelled to adopt over-compensatory behaviour to assert his manliness and conceal a fear of his own effeminacy. This perhaps constitutes the ultimate form of homophobia: the 'secret fear of one's own homosexual wishes'.[127] Sarrasine's involuntary attraction to another man suggests 'que c'est sa propre identité qui n'est pas sûre et que son erreur n'en est peut-être pas une' [that his own identity is not certain and that his mistake may not be a mistake].[128] He cannot accept being regarded as effeminate and homosexual by Zambinella's friends in spite of his active

121 Mosher and Sirkin, 'Measuring a Macho Personality Constellation', 151. Other components are 'a conception of violence as manly' and 'a view of danger as exciting'.

122 Balzac, *Sarrasine*, 56.

123 Ibid., 72.

124 Barbara Johnson, *The Critical Difference: Essays in the Contemporary Rhetoric of Reading* (Baltimore and London: The Johns Hopkins University Press, 1980), 10.

125 Balzac, *Sarrasine*, 71–72.

126 Lynne Segal, *Slow Motion: Changing Masculinities, Changing Men* (London: Virago Press, 1990), 112.

127 Ibid., 158. See also Katherine Kolb, 'The Tenor of *Sarrasine*', *PMLA*, 120/5 (October 2005), 1560–75 (1570).

128 Éric Bordas, '*Sarrasine* de Balzac: Une Poétique du Contresens', *Nineteenth-Century French Studies*, 31/1–2 (Fall–Winter 2002–03), 41–52 (41).

performance of masculine identity. More importantly, he cannot admit that they might be right, that the deception almost occurred of his own volition, because he repeatedly rejected the clues to Zambinella's maleness. After all, did the singer not offer to be '*un* ami dévoué' [a devoted (male) friend], and did he not even suggest 'Si je n'étais pas une femme?' [What if I was not a woman?]?[129] Sarrasine goes as far as to explain to the singer that, as an artist, he cannot be fooled by her feminine body. Sarrasine prefers the security of his blindness over clarification.

As Sarrasine's greatest insecurity is his wavering masculinity, the resemblance with the castrato that he cannot help but perceive is judged to be appalling: 'Tu m'as ravalé jusqu'à toi' [You have brought me down to your level].[130] This reluctant identification results in attempts to eliminate Zambinella. The elopement that Sarrasine had planned turns into the vengeful abduction that the castrato had feared. Yet, whilst Sarrasine fears being mistaken for a homosexual due to his attraction to, and resemblance with, Zambinella, the castrato's sexuality has not been clearly defined by the narrator. Certainly, like many young castrati in Rome, he is the protégé of an ecclesiastic man, the jealous Cardinal Cicognara, with whom he may have had sexual relationships.[131] Still, no conclusion can be drawn about the exact nature of their relationship. Sarrasine's interpretation of the sartorial difference that he perceives (Zambinella's cross-dressing) as an unacceptable sexual difference is his own and echoes his insecurity.

Homosexuality is as monstrous and unbearable to d'Albert in *Mademoiselle de Maupin* as it is to Sarrasine. D'Albert notices Théodore's beauty at their first encounter and is puzzled because Théodore looks like his ideal woman: 'Voilà donc enfin un des types de beauté que je rêvais réalisé et marchant devant moi! Quel dommage que ce soit un homme, ou quel dommage que je ne sois pas une femme!' [Here is finally one of those types of beauty of which I dreamt fully made and walking before me! What a pity

129 Balzac, *Sarrasine*, 69 (my emphasis) and 70.

130 Ibid., 76. On the similarities between the sculptor and the singer (sensitivity, ignorance, fear of the realities of life and even attire), see Kolb, 'The Tenor of *Sarrasine*', 1570–71.

131 Barbier, *The World of the Castrati*, 153.

that it is a man, or rather, what a pity that I am not a woman!].[132] Or, more accurately, he reinterprets his ideal after this meeting.[133] Confronted with his growing attraction to the young man, d'Albert first adopts a strategy of denial in his letters to Silvio. Certainly, he confesses his attraction to another man, but he attempts to reject his feelings by considering them to be 'impossible' and by highlighting his inner confusion.[134] Contrary to Pierre Albouy's claim that d'Albert never really had doubts about the sex of Théodore,[135] it is unlikely that he considered Théodore to be a woman before the production of *As You Like It*. Admittedly, d'Albert asserts: 'Il faut que Théodore soit une femme déguisée; la chose est impossible autrement' [Théodore must be a woman in disguise; this thing is impossible otherwise], but this claim is made when he has no reason to suspect that Théodore is actually a woman.[136] His desperation to stress the necessity of the so-called young man's femaleness shows that he is trying to convince himself, not Silvio.

The confession itself is made reluctantly, as the ellipses indicate: 'Enfin, à travers tous les voiles dont elle s'enveloppait, j'ai découvert l'affreuse vérité … Silvio, j'aime … Oh! non, je ne pourrai jamais te le dire … j'aime un homme!' [Finally, through all the veils in which it wrapped itself, I have discovered the awful truth … Silvio, I love … Oh! No, I shall never be able to say it to you … I love a man!].[137] The lexical fields of shame, guilt and monstrosity are used to designate d'Albert's passion. This passion is 'insensée, coupable et odieuse' [insane, guilty and odious] and 'une honte

132 Gautier, *Mademoiselle de Maupin*, 178.

133 As Carol Rifelj shows, the woman of his dreams is blonde (*Mademoiselle de Maupin*, 84), whereas Théodore has brown hair (178 and 294): 'Nobody's Perfect: Hair and Sexual Identity in *Mademoiselle de Maupin*', *Dalhousie French Studies*, 69 (Winter 2004), 33–41 (36).

134 Gautier, *Mademoiselle de Maupin*, 213, 217, 220 and 223.

135 Albouy, 'Le mythe de l'androgyne', 602.

136 Gautier, *Mademoiselle de Maupin*, 223. Similarly, Sarrasine desperately attempts to convince himself that Zambinella is a woman *after* he finds out that the singer is a castrato: 'C'est une femme […]. Il y a là-dessous quelque intrigue secrète. Le cardinal Cicognara trompe le pape et toute la ville de Rome!' [She is a woman […] There is some secret intrigue behind all this. Cardinal Cicognara is deceiving the Pope and the whole city of Rome!] (Balzac, *Sarrasine*, 74).

137 Gautier, *Mademoiselle de Maupin*, 220.

dont la rougeur ne s'éteindra jamais sur mon front' [shame whose blush will never disappear from my forehead].[138] The object of this guilty love is regarded as demonic. Théodore's beautiful hand, for example, is described as the claw of Satan.[139] Once d'Albert realizes that Théodore is a woman, he can explicitly call his supposed same-sex love monstrous because he knows that his passion no longer threatens morals: 'Je sentis s'évanouir l'horreur que j'avais de moi-même, et je fus délivré de l'ennui de me regarder comme un monstre' [I felt the horror at myself vanish and I was delivered from the trouble of regarding myself as a monster].[140]

D'Albert's avowal shows that he is obsessed with the concepts of guilt and monstrosity, and these have a clear sexual connotation. Sexual monstrosity is even more explicit when he becomes certain that Théodore is a woman, as though these words could be voiced only when the crisis is resolved:

> Aimer comme j'aimais d'un amour monstrueux, inavouable, et que pourtant l'on ne peut déraciner de son cœur; être condamné à garder le silence le plus profond, et n'oser se permettre ce que l'amant le plus discret et le plus respectueux dirait sans crainte à la femme la plus prude et la plus sévère; se sentir dévoré d'ardeurs insensées et sans excuses, même aux yeux des plus damnés libertins; que sont les passions ordinaires à côté de celle-là, une passion honteuse d'elle-même, sans espérance, et dont le succès improbable serait un crime et vous ferait mourir de honte? Être réduit à souhaiter de ne pas réussir, à craindre les chances et les occasions favorables et à les éviter comme un autre les chercherait, voilà quel était mon sort.[141]
>
> [To love as I did with a monstrous, guilty love that one cannot remove from one's heart; to be condemned to remain fully silent and not to dare to allow oneself what the most discreet and respectful lover would fearlessly say to the most modest and severe woman; to feel devoured by insane and inexcusable ardour, even in the eyes of the most damned libertines; what are ordinary passions compared with this, a passion ashamed of itself, hopeless and whose unlikely success would be a crime and make you die of shame? To be reduced to not wanting to succeed, to fearing chances and favourable opportunities and to avoiding them as another would seek them, this was my lot.]

138 Ibid.
139 Ibid., 217.
140 Ibid., 294.
141 Ibid., 296.

Here, the attraction to Théodore is depicted as a sexual inclination. D'Albert contemplates the possibility of intercourse with Théodore in terms that immediately condemn any same-sex relationship through the association between the negatively connoted adjective 'monstrueux' and the concepts of shame and crime, as well as the subsequent necessity to remain silent, given that sodomy was a crime punishable by death at the time at which d'Albert is supposed to be writing. The text also stresses the fact that d'Albert considers that his homoerotic feelings are unlikely to be shared ('sans espérance', 'le succès improbable'), which may be for the best in his opinion ('souhaiter de ne pas réussir'), since, if the relationship were consummated, this would lead to emotional, social – and physical – death.

Later, the poet recalls his youth. This period, which is traditionally perceived as a time of innocence, is experienced differently by d'Albert:

> Cette monstruosité remplaçait pour moi les fraîches et pudiques illusions du bel âge; mes rêves de tendresse si doucement caressés, le soir, à la lisière des bois, par les petits sentiers rougissants, ou le long des blanches terrasses de marbre, près de la pièce d'eau du parc, devaient donc se métamorphoser en ce sphinx perfide, au sourire douteux, à la voix ambiguë et devant lequel je me tenais debout sans oser entreprendre d'expliquer l'énigme![142]
>
> [This monstrosity replaced for me the fresh and modest illusions of the beautiful age; my dreams of tenderness, which I had nursed so softly, in the evening, at the edge of the woods, through the blushing paths, or along the white marble terraces, near the pool in the park, had to turn into this perfidious sphinx, with an enigmatic smile, an ambiguous voice and before which I stood without daring to try to explain the riddle!]

The alleged innocence of his youth is highlighted by the reference to white and bucolic landscapes ('blanches terrasses de marbre') and by words connoting virtue ('pudiques'). The rhetorical figure of the hypallage ('les petits sentiers rougissants') endows nature with his own feelings, whilst bestowing the latter's own purity upon him. However, this fantasized adolescence and his dreams of tenderness are illusions. The innocence of its heterosexual desires is highlighted in order to emphasize by contrast

142 Ibid., 296–97.

the so-called hideousness of his current homoerotic desire, designated through the vocabulary of deception ('perfide', 'douteux', 'ambiguë').

In the phrases 'cette monstruosité' and 'ce sphinx perfide', neither demonstrative adjective has a direct grammatical referent in the text. This grammatical imprecision is also present in the sentence preceding the quoted paragraph: 'c'était chez moi la première effervescence de jeunesse, la première pâquerette de mon printemps d'amour' [it was for me the first tumult of youth, the first daisy in my springtime of love].[143] What is the meaning of this *ce*? We must seek the antecedent outside the text. Jean-Marie Roulin considers that 'ce sphinx […] à la voix ambiguë' refers to Madeleine.[144] It is nonetheless unlikely because d'Albert does not slander his ideal woman once he is sure about Théodore's 'real' sex. On the contrary, Madeleine in Rosalind's costume is described in the preceding pages with unadulterated admiration.[145] It would be more appropriate to correlate the demonstrative adjectives, not with a person, but with homoerotic desire. D'Albert hates the emotions that Théodore has awakened in him, and not their instigator. This grammatical rupture through the absence of a textual referent underlines the difficulty of expressing such love in a heteronormative context.

As his love seems monstrous, even to him, d'Albert needs to confess his secret, which makes him a genuine product of Foucault's *scientia sexualis* civilization. This practice of sexual discourse stems from the Catholic confession, especially the confession of sins of the flesh.[146] Foucault has shown the progressive expansion of this practice, which became exercised in the frameworks of family, school, justice and medicine.[147] Confession does not only relate to sexual acts, but also to thoughts, images and desires, that is, to sexual fantasies. Likewise, d'Albert's confession does not cover his actions since his love remains platonic until he discovers Théodore's 'real' sex. The young man rather recounts his thoughts and anxiety regarding his homoerotic desires.

143 Ibid., 296.
144 Roulin, 'Confusion des sexes', 31–32.
145 Gautier, *Mademoiselle de Maupin*, 294–95.
146 Foucault, *Histoire de la sexualité*, I, 27–30.
147 Ibid., 78–80.

Furthermore, Foucault has suggested that a confession is a power relationship in which the one who holds the power is the one who listens because he or she can judge, punish and forgive.[148] The interlocutor in this case is Silvio, who may be a double for the reader and eager to condemn d'Albert's morals. My interpretation of this narratee, or recipient, as a substitute for the reader is based on Genette's *Nouveau discours du récit* [*New Narrative Discourse*]. Although the narration in *Mademoiselle de Maupin* is different from a traditional intradiegetic tale, in which a story is told within another story (like in *Sarrasine*), owing to its epistolary structure, it shares with the latter the transmission of a story to an identified fictional recipient. For Genette, the assumed reader of an intradiegetic story cannot be mistaken for the narratee of the story, as the narrator makes specific observations that can only be addressed to the narratee. However, some commentaries are still relevant to the reader.[149] In the case of d'Albert's letters, the narratee is Silvio, an indubitably silent recipient. Not only are his responses not shown, but he is also hardly acknowledged by d'Albert's letters, as direct addresses to him are rare.[150] In brief, he is deprived of his status as an interlocutor. The boundary between him and the assumed reader becomes blurred. Silvio is thus particularly fit to embody the supposed homophobic prejudices of the reader. He may represent the internalization of the moral law, inherited from the bourgeois society that rules d'Albert. Consequently, the young man must be able to convince Silvio and himself that his feelings are innocent in making them culturally acceptable. The next section thus explores how d'Albert endeavours to redeem himself in the eyes of his interlocutor by substituting the topic of monstrosity for that of beauty. His 'dreadful' same-sex love quest is turned into an aesthetic quest.

148 Ibid., 82–83. For more on the fear of not being forgiven, see Martine Reid, 'La confession selon Musset', *Littérature*, 67 (1987), 53–72 (56).

149 Genette, *Nouveau discours du récit*, 91 and 103. See also *Figures III*, 265–66.

150 The first sentence of the novel, 'Tu te plains, mon cher ami, de la rareté de mes lettres' [You complain, my dear friend, of the scarcity of my letters] (Gautier, *Mademoiselle de Maupin*, 71), which refers to a reciprocal correspondence between the two friends, constitutes an exception.

A Quest for Hermaphroditic Beauty

Mademoiselle de Maupin first appears to follow the Romantic aesthetic principle that dominates the nineteenth century, even though the plot is supposed to occur in the seventeenth century. According to Romantic aesthetics, beauty is essentially a female attribute, whereas male beauty is considered to be non-existent or may raise suspicion regarding a man's sexuality.[151] A beautiful man risks being suspected of effeminacy, marginality and foppery in the Romantic period.[152] This belief in beauty as a female attribute is seemingly shared by Madeleine and d'Albert. Madeleine, for instance, claims: 'l'homme ne me tente pas beaucoup; car il n'a pas la beauté comme la femme' [Men do not really tempt me because they have no beauty like women do].[153] D'Albert, even before avowing his homoerotic feelings, conceives of Théodore as the embodiment of feminine beauty, thoroughly admired and frequently described with superlative phrases, despite his apparent male sex:

> [C]e qui me charme le plus est un jeune cavalier qui est arrivé depuis deux ou trois jours; – il m'a plu tout d'abord, et je l'ai pris en affection, rien qu'à le voir descendre de son cheval. Il est impossible d'avoir meilleure grâce […]. […] Il est muni d'une paire d'yeux les plus beaux et les plus noirs du monde, qui ont une expression indéfinissable et dont il est difficile de soutenir le regard; mais, comme il est fort jeune et n'a pas d'apparence de barbe, la mollesse et la perfection du bas de sa figure tempèrent un peu la vivacité de ses prunelles d'aigle […]. Cet Adonis, qui, à sa belle figure, joint un esprit très vif et très étendu, jouit encore de ce privilège d'avoir à mettre au service de

151 The concept of female beauty, conceived in opposition to male sublimity, is already present in Edmund Burke's *Philosophical Enquiry into the Origin of our Ideas of the Sublime and the Beautiful* (1757): Anne K. Mellor, *Romanticism and Gender* (New York and London: Routledge, 1993), 107; Alex Potts, *Flesh and the Ideal: Winckelmann and the Origins of Art History* (New Haven and London: Yale University Press, 1994), 113–14.

152 Lawrence R. Schehr, *Parts of an Andrology: On Representations of Men's Bodies* (Stanford: Stanford University Press, 1997), 78.

153 Gautier, *Mademoiselle de Maupin*, 383.

> ses bons mots et de ses plaisanteries une voix d'un timbre argentin et mordant qu'il est difficile d'entendre sans être ému. – Il est vraiment parfait.[154]
>
> [The one who pleases me the most is a young rider who arrived two or three days ago; – I liked him right away and I became fond of him just by seeing him climbing from his horse. It is impossible to be more graceful. […]. […] He owns the two finest and blackest eyes in the world, which have an undefinable expression and whose gaze is impossible to meet; but, as he is very young and beardless, the smoothness and the perfection of the lower part of his face temper a little the intensity of his eagle gaze […]. Besides his beautiful face, this Adonis also has a very bright and very extensive mind. To serve his puns and jokes, he is also blessed with the privilege of having a voice with such a silvery and acerbic tone that one can hardly hear it without being moved. – He is truly perfect.]

Whilst d'Albert struggles at times to find the words to describe Théodore's beauty, he does not conceal the confusion that the latter's physical perfection raises in him. He also immediately notes the so-called man's sexual ambiguity in terms that are reminiscent of Madame de Rochefide's observation about the portrait of Adonis in *Sarrasine*. Whilst d'Albert states that '[l]e seul défaut qu'il [Théodore] ait, c'est d'être trop beau et d'avoir des traits trop délicats pour un homme' [Théodore's only flaw is that he is too beautiful and his features are too delicate for a man], Béatrix criticizes Adonis by saying: 'Il est trop beau pour un homme' [He is too beautiful for a man].[155] Théodore and Adonis's male (but feminine) beauty appears disconcerting and unexplainable in the Romantic context.

Accordingly, d'Albert must adopt an alternative perspective to explain Théodore's attractiveness. He explicitly argues that male beauty belongs to another time; only ancient men, heroes and gods were able to reach such a degree of physical perfection.[156] Indeed, a significant aesthetic shift occurred between the neoclassical and Romantic periods and Gautier makes his character refer to the neoclassical aesthetics that were defined at the end of the eighteenth century under the influence of Johann Joachim Winckelmann. The German art historian established a neoclassical model through the

154 Ibid., 178.

155 Ibid.; Balzac, *Sarrasine*, 49.

156 Gautier, *Mademoiselle de Maupin*, 237 and 354.

study of ancient Greek statues and their Roman copies according to which beauty was seen as the prerogative of men, rather than women.

The nature of this male beauty, however, varies depending on historians and art historians' interpretations. According to George Mosse, Winckelmann's rediscovery and interpretation of Greek sculpture played a decisive role in the creation of the stereotype of ideal masculinity that remained stable throughout the next two centuries.[157] Winckelmann highlighted the beauty of male bodies, the harmony of their proportions and the regularity of their features; he interpreted Greek statues, notably that of Laocoön, a model of male heroism and restraint, as a mixture of 'noble simplicity and tranquil grandeur'.[158] Put another way, the physical beauty of Greek men was thought to reflect the moral values that Winckelmann's society wanted to promote: power, virility and self-control.

The art historian Abigail Solomon-Godeau nevertheless declares that not one, but two aesthetic male archetypes concomitantly existed in French neoclassical art: 'a heroic, virile, and purposeful manhood understood as active and dominating' (such as the Horatii in David's painting and Laocoön's statue) and 'a typically younger model – adolescent or ephebic – whose sensual and erotic appeal derives at least in part from its relative passivity'.[159] Mechthild Fend also highlights the increasing popularity of beautiful, sexually ambivalent adolescents as incarnations of the ideal antique beauty at the end of the eighteenth century.[160] In other words, the conception of male beauty that predominated at that time was intrinsically hermaphroditic. It possessed attributes such as softness and passivity, which were generally judged to belong to women. These traits were nonetheless

157 Mosse, *The Image of Man*, 29–39.

158 For more on Johann Joachim Winckelmann's designation of Greek statues, see 'Thoughts on the Imitation of the Painting and Sculpture of the Greeks' (1755), in *German Aesthetic and Literary Criticism: Winckelmann, Lessing, Hamann, Herder, Schiller, Goethe*, H. S. Nisbet, trans. and ed. (Cambridge: Cambridge University Press, 1985), 32–54 (42 and 44). On Laocoön, see Mosse, *The Image of Man*, 29 and 32–33.

159 Abigail Solomon-Godeau, *Male Trouble: A Crisis in Representation* (London: Thames and Hudson, 1997), 26. See also 88.

160 Fend, *Les limites de la masculinité*, 43.

embodied by, and appreciated in, male bodies.[161] Winckelmann, for instance, described Apollo and Bacchus as combining 'the forms of prolonged youth in the female sex with the masculine forms of a beautiful young man'; these male gods were therefore 'plumper, rounder, and softer' than other male figures.[162] The male beauty of a sensual and passive ephebe in neoclassical art is illustrated in *Sarrasine* by the fictional painting of Adonis and the real painting *Le Sommeil d'Endymion* [*The Sleep of Endymion*] by Girodet (1791), which is mentioned as being inspired by Zambinella.[163] In the neoclassical period, this beautiful, erotic and effete ephebe did not promote homosexuality and, accordingly, must not be interpreted through a homoerotic or homosexual reading, but through a homosocial one, claims Solomon-Godeau.[164] Reviews of *Salons* at the end of the eighteenth century indicate that spectators were unaware of, or unwilling to notice, a homoerotic dimension to the paintings.[165]

With his slimness, smoothness, silvery voice and long curly hair, as well as his feminine hands and feet, Théodore incarnates the second model of male beauty, that of beautiful and sexually ambivalent young men. He therefore belongs to neoclassical aesthetics. The text is particularly interested in Théodore's walk: 'il a quelque chose de moelleux et d'onduleux dans la démarche […] qui est on ne peut plus agréable' [there is something smooth and sinuous in his gait […] that is excessively agreeable].[166] This focus once again reveals the importance of one's walk as a gendered

161 Mosse, *The Image of Man*, 29–39; Solomon-Godeau, *Male Trouble*, 26 and 88; Fend, *Les limites de la masculinité*, 43.

162 Johann Joachim Winckelmann, *History of the Ancient Art: Among the Greeks*, Giles Henry Lodge, trans. (London: Chapman, 1850), 56.

163 Balzac, *Sarrasine*, 77. For more on the function of the painting of *Endymion/Adonis* in *Sarrasine*, see Fend, *Les limites de la masculinité*, 211–26.

164 Solomon-Godeau, *Male Trouble*, 48–50.

165 On Girodet's painting, for instance, see Solomon-Godeau, *Male Trouble*, 66–68. See also the comments on Guérin's painting Aurore et Céphale (1810) (161) and Granger's *Apollon et Cyparissus* (1817) (167). See also Fend, *Les limites de la masculinité*, 121–22; Francis Moulinat, 'Les Amours grecques: Homosexualité et représentations, du *Léonidas* de Jacques-Louis David (1799–1814) au *Swimming Hole* de Thomas Eakins (1885)', *Romantisme*, 159 (2013), 73–83 (77).

166 Gautier, *Mademoiselle de Maupin*, 178.

characteristic in the nineteenth century. Moreover, the constant description of Théodore's body as undulating corresponds to what Fend calls, referring to the Winckelmann's aesthetics, *ligne ondoyante* [undulating line].[167] It designates a sensual and almost aquatic fluidity of the contours that is supposed to define feminine and androgynous bodies in classical art, as Winckelmann observed of the sexually ambiguous god Bacchus: 'The forms of his limbs are soft and flowing, as though inflated by a gentle breath.'[168] Incidentally, Madeleine frequently describes Rosette's body as undulating: 'ses bras pendaient au long de son corps aussi ondoyants et aussi moelleux que des écharpes dénouées' [her arms hung by her body, as undulating and smooth as untied scarves].[169] This portrayal highlights the feminine roots of the undulating reference.[170]

By claiming an objective and deserved admiration for Théodore's beauty, d'Albert positions himself not as a lover but as an aesthete. His strategy is to justify his 'monstrous' feelings for a man by constantly referring to the classical canon of beautiful and sexually ambivalent men. The approach that Gautier attributes to d'Albert is not innovative. Many homosexuals of the nineteenth century attempted to legitimize their lifestyle by establishing a continuity between the Greeks' morals and their own. They notably referred to famous historical or mythological, alleged or recognized, homosexuals, such as Achilles, Patroclus, Plato and Michelangelo.[171] Likewise, d'Albert justifies this passion for another man by citing the Roman poets Catullus, Tibullus, Martial and Virgil, as well as their 'sérail monstrueux' [monstrous harem].[172] This list manifestly allows him to claim his passion for beauty and refinement. However, d'Albert

167 Ibid., 295. Fend, *Les limites de la masculinité*, 50–62. See also Potts, *Flesh and the Ideal*, 170. Gautier was familiar with Winckelmann's writings: Raymond Giraud, 'Winckelmann's Part in Gautier's Perception of Classical Beauty', *Yale French Studies*, 38 (1967), 172–82.

168 Winckelmann, *History of the Ancient Art*, 93. On flowing beauty, see also 44 and 47–49.

169 Gautier, *Mademoiselle de Maupin*, 337. See also 338 and 345.

170 Roulin, 'Confusion des sexes', 38.

171 Mosse, *The Image of Man*, 88; Robb, *Strangers*, 144.

172 Gautier, *Mademoiselle de Maupin*, 225–26.

mainly reminds his reader of the pederastic customs of this admired civilization. This list also attests to the unwillingness of nineteenth-century French society to acknowledge this type of love. The male names of the poets' lovers were banned from schools because they were feminized in translation. D'Albert himself is not freed from self-censorship, since the use of 'monstrueux', as well as of the previously analysed expressions of monstrosity, shame and crime, suggests that he shares this repugnance for homosexuality. Like Sarrasine, he fears being homosexual, which constitutes the epitome of homophobia.

The most efficient strategy to make his passion for a man more acceptable is by recalling the figure of Hermaphroditus.[173] Hermaphroditus is not known as a homosexual god, but, because of his 'natural' androgyny, he embodies the ideal of the sexually ambivalent ephebe celebrated by Greco-Roman and neoclassical arts. Although the reference to Hermaphroditus is not always directly connected to Madeleine, d'Albert establishes a strong textual correlation between the Greek god and Théodore on several occasions.[174] In the part of his confession to Silvio in which he intends to justify his love for an effeminate man, he highlights the popularity of the mythological monster: 'l'hermaphrodite est [...] une des chimères les plus ardemment caressées de l'antiquité idolâtre' [the hermaphrodite is [...] one of the most ardently cherished fantasies in idolatrous antiquity].[175] Like the Latin poets and their 'monstrous' harem, he considers himself not as an outsider, but as a mere follower of sophisticated classical culture.

173 In contrast, Madeleine does not use aesthetic reasoning to justify her attraction to Rosette, although she refers to women's beauty. Unlike d'Albert, she feels no guilt in loving another woman. Rivers (*'Inintelligibles pour une femme honnête'*, 15) highlights the difference between her speech and that of Suzanne in Diderot's *La Religieuse* [*The Nun*], referring to theology and sin.

174 The word *Hermaphrodite* appears for the first time even before the existence of Théodore is revealed. D'Albert explains his attempts to merge his body and that of Rosette in a sexual embrace and compares himself to 'l'antique Salmacis, l'amoureuse du jeune Hermaphrodite' [the ancient Salmacis, young Hermaphroditus's lover] (Gautier, *Mademoiselle de Maupin*, 128). Interestingly, the young man compares himself not to the young male god, but to his female counterpart, the naiad Salmacis.

175 Ibid., 237.

D'Albert's description of Hermaphroditus is extensive and rich in laudatory vocabulary praising the monster's sensuality and harmony, the latter being an essential quality in the conception of classical beauty.[176] The original monstrosity of Hermaphroditus, due to his intersexuality, becomes a perfection that exceeds the beauty of the individual:

> C'est en effet une des plus suaves créations du génie païen que ce fils d'Hermès et d'Aphrodite. Il ne se peut rien imaginer de plus ravissant au monde que ces deux corps tous deux parfaits, harmonieusement fondus ensemble, que ces deux beautés si égales et si différentes qui n'en forment plus qu'une supérieure à toutes deux, parce qu'elles se tempèrent et se font valoir réciproquement.[177]
>
> [This son of Hermes and Aphrodite is indeed one of the smoothest creations of the pagan genius. Nothing more ravishing can be imagined than these two equally perfect bodies, harmoniously merged together, than these two beauties so alike and so different which become a being superior to both of them, because they temper and highlight each other reciprocally.]

This description highlights the harmonious, composite and therefore perfect beauty of Hermaphroditus. Significantly, Gautier's definition of Hermaphroditus's beauty reverses the argument of the scientists of his time. Whereas they deemed 'real' hermaphrodites' infertility to be monstrous, the god's sterility illustrates Gautier's defence of the uselessness of art and beauty in the Préface. Gautier criticizes the journalists who expect literature and the arts to be useful and contribute to the moralization, wellbeing and progress of humankind.[178] Conversely, he claims that beauty cannot support utilitarian functions: 'Il n'y a de vraiment beau que ce qui ne peut servir à rien' [Only what is useless can be truly beautiful].[179] Hermaphroditus is therefore the paragon of beauty because he is not 'médicalement bien conformé, en état de faire des enfants' [medically fit and able to bear children], to use Gautier's words.[180] He embodies the eroticization of beauty, the object of Gautier's quest according to Michel

176 Mosse, *The Image of Man*, 29.
177 Gautier, *Mademoiselle de Maupin*, 237.
178 Ibid., 50.
179 Ibid., 54.
180 Ibid., 53.

Crouzet. The latter rightly notes that Gautier links beauty and fruitless sensual pleasure in his Préface, which suits the portrayal of the sensual but sterile Greek god.[181]

Hermaphroditus's beauty is grounded in the contradictory yet harmonious combination of male and female beautiful traits belonging to the young god and the nymph:

> Le torse est un composé des monstruosités les plus charmantes: sur la poitrine potelée et pleine de l'éphèbe s'arrondit avec une grâce étrange la gorge d'une jeune vierge. Sous les flancs bien enveloppés et d'une mollesse toute féminine, on devine les dentelés et les côtes, comme aux flancs d'un jeune garçon; le ventre est un peu plat pour une femme, un peu rond pour un homme, et toute l'habitude du corps a quelque chose de nuageux et d'indécis qu'il est impossible de rendre, et dont l'attrait est tout particulier.[182]

> [The torso is composed of the most charming monstrosities. Above the plump and full chest of the ephebe rise with a curious grace the round breasts of a young virgin. Beneath the fleshy sides, whose smoothness is entirely feminine, one guesses at the serratus and the ribs, like those of a young boy. The belly is a little too flat for a woman, a little too round for a man, and the whole bent of the body has something fluffy and indeterminate that is very difficult to convey and whose attraction is very peculiar.]

This portrait of Hermaphroditus recalls, once again, the principle of the undulating line of the contours, the softness and the roundness of female and effeminate bodies, and the firmness of male bodies. Although harmonious, such an assemblage is conceived of as strange and unique. Oxymoron ('monstruosités les plus charmantes') seems to be the only figure of speech able to express the complexity and rarity of such a creature. Such singularity is precisely the reason why d'Albert is entitled to undertake an aesthetic quest. He originally feared that his quest would make him 'tomber à la fin dans l'impossible ou dans le monstrueux' [fall in the end into the impossible or the monstrous], that is, develop homoerotic desires.[183] His later rhetoric, however, reverses the meaning of this

181 Michel Crouzet, '*Mademoiselle de Maupin* ou l'Éros romantique', *Romantisme*, 8 (1974), 2–21 (3–5).

182 Gautier, *Mademoiselle de Maupin*, 238.

183 Ibid., 213.

quest. The alleged monstrosity of homosexuality is now positively connoted as the rarity of ideal neoclassical beauty.

Despite seeing Théodore as the embodiment of Hermaphroditus's physical perfection, d'Albert expresses reservations about Théodore's masculinity, as his female features are overwhelming: 'Théodore serait à coup sûr un excellent modèle de ce genre de beauté; cependant je trouve que la portion féminine l'emporte chez lui, et qu'il lui est plus resté de Salmacis qu'à l'Hermaphrodite des Métamorphoses' [Théodore would undoubtedly be an excellent model for this kind of beauty. However, I think that the feminine part in him takes over and that more of Salmacis remained in him than in the Hermaphroditus of the *Metamorphoses*].[184] This observation must be linked to his comment about Théodore's excess of beauty. Interestingly, the dominance of the female side in the Hermaphroditus incarnated by Théodore seems to call into question the whole association with the young god. However, the feminine characteristics that d'Albert highlights are those observed by Winckelmann about hermaphrodites and Greek gods. In other words, what was conceived of as evidence of both feminine and male beauty in the classical canon is considered by d'Albert to be a clue as to Théodore's femaleness. Consequently, he interprets Salmacis's side in Théodore as a challenge to his sex when he tells 'Rosalind': 'Mais vous êtes une femme, nous ne sommes plus au temps des métamorphoses; – Adonis et Hermaphrodite sont morts, – et ce n'est plus par un homme qu'un pareil degré de beauté pourrait être atteint' [But you are a woman, we no longer live in the age of metamorphoses; – Adonis and Hermaphroditus are dead, – and no man could reach such a degree of beauty].[185] A shift is therefore perceptible in d'Albert's reasoning which mirrors the aesthetic shift that took place at the turn of the century, as Solomon-Godeau notes:

> In the course of the four decades between the French Revolution of 1789 and the establishment of the July Monarchy in 1830, the heroic male nude, alpha and omega of history painting, gradually lost its privileged position *in practice*, to be increasingly eclipsed, and ultimately supplanted, by the female nude.[186]

184 Ibid., 238.

185 Ibid., 354.

186 Solomon-Godeau, *Male Trouble*, 43 (emphasis in original). See also 22.

Accordingly, d'Albert does not refer to the neoclassical model of beauty, but to the Romantic model, according to which beauty is exclusively feminine. He therefore presents Théodore as an aberration. The novel has come full circle: the reference to Hermaphroditus becomes useless and Romantic aesthetics are once again the norm from the moment that d'Albert does not have to justify a desire that is no longer 'répréhensible'.[187]

In brief, the case of d'Albert has shown that a discourse about male homosexuality solely based on emotions and morals is fruitless because the speaker can only repeat his horror of homosexuality, as well as his shame and guilt in a society in which homosexuality is a crime or, at least, an offence to morals. This moralistic discourse must therefore be replaced by an aesthetic discourse based on a neoclassical model that promotes the beauty of effeminate young men. The discourse about Hermaphroditus in *Mademoiselle de Maupin* has proved to be the only possible way to discuss homoerotic feelings from a morally non-condemnable perspective. Although the novel does not depict a 'real' homosexual man, its reasoning has shown its effectiveness in turning a monster into an object of praise.

Hermaphrodites, Heterotopias and Art

Given the Romantic obsession with artistic self-sufficiency (as exemplified by the famous 'l'art pour l'art' [art for the sake of art] motto), it comes as no surprise that art is a key medium in the narratives explored here in order to express the beauty of hermaphrodites' bodies and diminish their alleged physical and moral monstrosities. Art is omnipresent throughout the novels: sculpture in *Fragoletta*, *Sarrasine* and *Mademoiselle de Maupin*, music and opera in *Sarrasine*, and Shakespearean comedies in *Mademoiselle de Maupin*. Art, arguably, constitutes one of these spaces described by Foucault as *hétérotopies* [heterotopias]. These are:

187 Gautier, *Mademoiselle de Maupin*, 358.

> des lieux réels, des lieux effectifs, des lieux qui ont [été] dessinés dans l'institution même de la société, et qui sont des sortes de contre-emplacements, sortes d'utopies effectivement réalisées dans lesquelles les emplacements réels [...] sont à la fois représentés, contestés et inversés, des sortes de lieux qui sont hors de tous les lieux, bien que pourtant ils soient effectivement localisables.[188]
>
> [real places, actual places, places that have been designed within the very institution of society and that are sorts of counter-places, of effectively realized utopias in which real places [...] are at once represented, challenged and reversed, sorts of places that are outside all places, although they are actually locatable.]

In the hermaphroditic narratives, the utopia of art is embodied in material spaces: the works of art and the places of art. The Museum of Naples (which houses the artistic treasures from the newly discovered cities of Pompeii and Herculaneum), the Roman opera house, the salon of the ambassador where Zambinella sings, Sarrasine's studio, the Museum Albani containing the marble copy of la Zambinella's statue,[189] the stage on which d'Albert, Madeleine and Rosette play *As You Like It*, the room in which d'Albert writes his letters to Silvio, even the statues of Hermaphroditus sculpted by Polykles at the Museum of Naples and of la Zambinella fashioned by Sarrasine, can be seen as heterotopias. In these spaces, beauty comes to life; homosexuality is confessed; fusion between the sexes is fully realized.

The liberation of alternative expressions of sex, gender and sexuality is possible through various means, notably the theatre and the museum. These examples of heterotopias are given by Foucault. A theatre illustrates the capacity of heterotopias to juxtapose several spaces in one, whilst a museum is a space in which time is perpetually accumulated and suspended.[190] These functions appear in the hermaphroditic narratives. As d'Albert himself notes

188 Michel Foucault, 'Des espaces autres', in *Dits et écrits: 1954–1988*, 4 vols, Daniel Defert and François Ewald, eds (Paris: Gallimard, NRF, Bibliothèque des sciences humaines, 1994), IV, 755–56.

189 Incidentally, Winckelmann (*History of the Ancient Art*, 58–59) mentions the statue of a hermaphrodite in this museum.

190 Foucault, 'Des espaces autres', 758–59. On the importance of the theatrical space in the construction of gender, see Kerlouégan, 'Une relecture de l'androgynie romantique', 43–47.

with regard to Baroque theatre: 'Ce pêle-mêle et ce désordre apparents se trouvent, au bout du compte, rendre plus exactement la vie réelle sous ses allures fantasques que le drame de mœurs le plus minutieusement étudié' [This apparent jumble and disorder, in the end, depicts real life in all its original more exactly than the most minutely studied moral drama].[191] Indeed, the superposition of a 'real' story and of a theatrical plot highlights their similarities. Shakespeare's comedy and the personal drama of d'Albert and Madeleine coexist in the same space. Rosalind and Orlando's apparently impossible love (owing to the former being mistaken for a man by the latter) echoes Madeleine and d'Albert's own situation. D'Albert himself points out: 'les paroles que j'avais à prononcer répondaient si bien à ma situation qu'elles me semblaient plutôt inventées par moi qu'apprises par cœur dans un livre' [the words that I had to utter echoed my own situation so well that I felt as if I had invented them rather than learned them by heart from a book].[192] As for museums as heterotopias, the Museum of Naples in *Fragoletta* exemplifies how the past is conserved and protected in such a space. Visitors can admire ancient masterpieces and discuss ancient traditions. Whether they identify with them (as does Camille) or criticize them (as does d'Hauteville), they are still relevant to them.

Heterotopias, nevertheless, exceed these functions in hermaphroditic narratives. As Stefan Horlacher points out, 'art creates spaces in which ludic, creative, and experimental thinking becomes possible, in which humankind can transcend itself and create new ways of living and alternative concepts of understanding, also of one's own gender or sexual identity'.[193] Indeed, in these narratives, artistic heterotopias aim to allow the hermaphrodites to fathom or reveal their 'true' nature. Camille understands what they are when discovering the statue of Hermaphroditus and can evaluate the difference between their own condition and ancient hermaphrodites, either real or fantasized. Similarly, d'Albert comprehends the 'real' sex of the impersonator of Rosalind while rehearsing *As You Like It* and Sarrasine

191 Gautier, *Mademoiselle de Maupin*, 275.

192 Ibid., 297. See also 297–98, 302 and 308.

193 Stefan Horlacher, 'Transgender and Intersex: Theoretical, Practical, and Artistic Perspectives', in *Transgender and Intersex*, 1–27 (14).

realizes that Zambinella is a man when hearing him sing in male attire at a private concert.

Heterotopias also have the function of emphasizing how 'real' spaces are illusory and/or disorganized.[194] The heterotopias in *Fragoletta*, *Sarrasine* and *Mademoiselle de Maupin* have a similar function to the Puritan and Jesuitical colonies in America as described by Foucault: they are ideal places due to their flawless organization of space and activities and, therefore, show by contrast the disorganization and deception of 'real' spaces.[195] Heterotopias in hermaphroditic narratives reveal that sexual dichotomy and compulsory heterosexuality are deceiving norms, as Murat points out: 'En introduisant le théâtre dans le roman, et le travestissement dans le travestissement, Gautier parvient à rendre caducs tous les référents habituels' [In introducing theatre within the novel, and cross-dressing within cross-dressing, Gautier manages to cancel out all the usual referents].[196] While Murat is referring to Madeleine's cross-dressing, this observation can also be applied to d'Albert. His performance as Orlando in Shakespeare's play enables him to express homoerotic feelings that he must conceal in daily life. Likewise, in the Museum of Naples and only there, intersexuality is celebrated; elsewhere hermaphrodites are despised and even condemned to death. On the stage of the opera house, Zambinella is a masterpiece, whereas he is a ridiculous foppish old man at the Lanty hotel.

The bodies of the hermaphrodites also function as crucial heterotopias in the narratives. Like other heterotopias, these spaces exist; they represent, contest and invert human sexes. In them, sexual ambiguity is emphasized and the connection between sex and gender is called into question. Being male and masculine (or female and feminine) is neither evident nor natural, as shown by the hermaphroditic individuals who combine diverse genders and sexes. The transformation of their bodies into heterotopias is achieved through the mediation of, or the conversion into, works of art. They become utopian and similar to Hermaphroditus, the ultimate embodiment of ideal beauty:

194 Foucault, 'Des espaces autres', 761.

195 Ibid.

196 Murat, *La loi du genre*, 86.

> [Art] united the beauties and attributes of both sexes in the figures of hermaphrodites. The great number of hermaphrodites [...] shows that artists sought to express in the mixed nature of the two sexes an image of higher beauty; this image was ideal. [...] [E]very artist cannot have an opportunity of seeing so rare a deviation of nature; and hermaphrodites, like those produced by sculpture, are probably never seen in real life. All figures of this kind have maiden breasts, together with the male organs of generation; the form in other respects, as well as the features of the face, is feminine.[197]

Winckelmann presents the sculptures of hermaphrodites as improvements of 'real' Asian eunuchs, claiming that the artistic approach is in continuity with the traditions of Asia Minor.[198] In both cases, the goal is to create a beautiful and youthful individual, combining the virility of a man with the soft traits of a girl, someone who is, as Winckelmann puts it, 'intermediate between the two'.[199] The art historian, nonetheless, notes no correspondence between the 'real' intersex individuals and the artistic hermaphrodites. Although he reluctantly admits the past existence of hermaphrodites, he doubts that artistic representations of hermaphrodites could be inspired by 'real' ones.[200] We thus find in Winckelmann the same inclination as we found in the scientific texts studied in Chapter 1: to separate the artistic being considered to be the canon of beauty from the human being seen as a monster.

The discrepancy between 'real' and artistic hermaphrodites is highlighted in *Fragoletta* when d'Hauteville and Camille visit the Museum of Naples. Their guide, the Italian poetess Éléonore Pimentale, reveals to the officer that the statue made by Polykles that fascinates him represents Hermaphroditus. The description of the sculpture corresponds to the so-called *Sleeping Hermaphroditus*, lying on a bed sculpted by the baroque artist Bernini, now at the Louvre Museum:

> Quelle est [...] la déesse ou la mortelle couchée sur cette espèce de lit dont la forme est toute moderne ? [...] Appuyée sur un bras plein de souplesse, on la dirait dans un demi-sommeil. La tête, tournée en sens inverse de la pose du corps, semble exprimer

197 Winckelmann, *History of the Ancient Art*, 56–57.
198 Ibid., 53–54.
199 Ibid., 54.
200 Ibid., 56.

à la fois le sourire et la tristesse. Une des jambes croisées relève un pied charmant, et l'autre pied est engagé avec grâce dans les plis d'un manteau qui laisse cependant presque toute cette beauté sans voile.[201]

[Who is [...] the goddess or mortal woman who is lying on this sort of bed whose shape is very modern? [...] Leaning on a very flexible arm, she looks half-asleep. The head, facing in the opposite direction to the pose of the body, seems to express both smile and sadness. One of the crossed legs lifts a charming foot, and the other foot is gracefully trapped inside the pleats of a coat that still leaves almost all this beauty unveiled.]

The sight of this statue is an excuse for Pimentale's panegyric on the beauty of hermaphrodites. She attempts to convince the sceptical d'Hauteville not only of the existence but, above all, of the perfection of hermaphrodites. Her considerable erudition and aesthetic enthusiasm recall d'Albert's in *Mademoiselle de Maupin* and make her a substitute for Winckelmann.

Éléonore first refers to Greek philosophy, especially to the myth of the androgyne, then to the Bible, suggesting that Adam and Eve constituted one being before the female sex was extracted from the first man. Afterwards, her speech borrows arguments from scientific discourses and tallies with the observations of the eighteenth-century scientists who examined cases of animal and vegetal hermaphroditism.[202] However, her catalogue of metaphysical and scientific facts does not convince d'Hauteville. He is unable to renounce his belief in clear sexual identity, which Éléonore attributes to his northern origins.[203] He continues to consider 'real' hermaphrodites to be monstrous and reluctantly mentions their destiny, as they were pushed into the sea in Athens and into the Tiber in Rome – Camille will later follow this ancient tradition and attempt to throw themselves into the sea.[204]

201 Latouche, *Fragoletta*, I, 60. The presence of the *Sleeping Hermaphroditus* in Naples is rather surprising, as the statue was moved directly from the Villa Borghese into the Louvre in 1807. Although Latouche, according to classical tradition, attributes the statue to the Greek sculptor Polykles, this paternity is uncertain. For more on the contestation of Polykles's paternity, see Lodge (Winckelmann's translator), *History of the Ancient Art*, 57, footnote k; Delcourt, *Hermaphrodite*, 97.

202 Latouche, *Fragoletta*, I, 61–63. See Monneyron, *L'androgyne romantique*, 54.

203 Latouche, *Fragoletta*, I, 61 and 63.

204 Ibid., 62.

Among this collection of proof of the existence of hermaphrodites, Éléonore makes a point that she expresses in an anaphorical repetition. She claims that the hermaphrodite incarnates both the beauty and the union of the sexes. She states the Winckelmannian principle that the beauty of both sexes embodies an ideal beauty characterized by completeness: 'Vous demandez ce qu'a voulu l'artiste en composant ce chef-d'œuvre? Combiner [...] la beauté de Camille et la vôtre, réunir dans une figure adorable tout ce que la nature avare n'aurait du reste séparé que pour nous' [You want to know what the artist meant to do when composing this masterpiece? Combine [...] Camille's beauty with yours, uniting in one adorable figure what miserly nature would have separated only for us].[205] In addition to referring implicitly to the German art historian, she also alludes to Plato's myth of the androgyne, as she claims that the hermaphrodite also symbolizes the physical love between a man and a woman: 'vous demandez ce qu'a voulu Polyclès? Personnifier l'union des corps, représenter cette alliance de deux êtres que l'amour précipite en un seul' [You want to know what Polykles meant to do? Personify the union of the bodies, represent this alliance of two beings that love throws into one].[206] This corresponds to the role of Hermaphroditus, ancient patron of marriage, as analysed by Groneberg. Indeed, the god personified the matrimonial union of man and woman, 'based on gender dualism and anchored in what we nowadays call heteronormative practice', rather than embodying intersexuality.[207]

Camille's body is thus mediated through Polykles's Hermaphroditus, especially as their father was a sculptor. He fashioned Camille as much as Polykles sculpted Hermaphroditus. The hermaphrodite becomes the living embodiment of an ancient and unquestionably admired sculpture since their beauty equals that of the statue. Here, however, the similarity ends because Camille's fate is different from that of Polykles's creation. The hermaphrodite's astonishment when discovering the statue and their embarrassment during Éléonore and d'Hauteville's discussion is caused by the discrepancy between their perception of their own body and the

205 Ibid.

206 Ibid., 63.

207 Groneberg, 'Hermaphrodite's Voice', 235.

idealization of this body in the statue.[208] The statue of Hermaphroditus acts as a mirror, as it supports an oxymoronic experience (to borrow Jenijoy La Belle's terminology).[209] This statue reflects both the self and the other. It reveals Camille's 'true' identity, while making them appear as a stranger, even to themselves. In other words, it is the means through which Camille becomes able to conceive of their 'alternative' sexual identity and to define themselves as an intersex individual.

Sculpture is also a salient element of the plot of *Sarrasine*, as the eponymous character, like Camille's father, is a sculptor. From Zambinella's first appearance, Sarrasine sees 'her' as the embodiment of ideal beauty, a feeling that is expressed through a laudatory repertoire that is similar to d'Albert's descriptions of Théodore and employs artistic references.[210] The sculptor adopts an artistic approach analogous to that of the legendary painter Zeuxis, who realized a painting of Helen of Troy by combining the features of several beautiful girls because he was unable to find a woman who would embody his ideal female beauty.[211] Likewise, Zambinella's body is composite for the sculptor, because it combines the most attractive parts of otherwise 'ignoble' individuals. According to the narrator, Sarrasine

> demand[ait] à un modèle, souvent ignoble, les rondeurs d'une jambe accomplie; à tel autre, les contours du sein; à celui-là, ses blanches épaules; prenant enfin le cou d'une jeune fille, et les mains de cette femme, et les genoux polis de cet enfant, sans rencontrer jamais sous le ciel froid de Paris les riches et suaves créations de la Grèce

208 Latouche, *Fragoletta*, I, 61, 62 and 65. See Monneyron, *L'androgyne romantique*, 57. For Fabio Vasarri, the discovery of Polykles's statue expresses Camille's transition from childhood to adulthood: 'I nomi dell'ermafrodito: L'esempio di *Fragoletta*', *Rivista di letterature moderne e comparate*, 44/2 (April–June 1991), 120–29 (123–24). For more on the hiatus between the mythical and historical hermaphrodite, see Laforgue, *L'éros romantique*, 105.

209 La Belle, *Herself Beheld*, 42.

210 Balzac, *Sarrasine*, 58.

211 For more on Zeuxis and *Sarrasine*, see Helen O. Borowitz, 'Balzac's *Sarrasine*: The Sculptor as Narcissus', *Nineteenth-Century French Studies*, 5/3–4 (Spring–Summer 1977), 171–85 (177); Patrick O'Donovan, 'Avatars of the Artist: Narrative Approaches to the Work of the Painter', in Peter Collier and Robert Lethbridge, eds, *Artistic Relations: Literature and the Visual Arts in Nineteenth-Century France* (New Haven and London: Yale University Press, 1994), 222–36 (223).

antique. La Zambinella lui montrait réunies, bien vivantes et délicates, ces exquises proportions de la nature féminine si ardemment désirées.[212]

[asked from one model, often awful, the roundness of an impeccable leg; from another one, the contours of her breast; from the next, her white shoulders; finally taking a young woman's neck and another woman's hands and a child's smooth knees, without ever encountering under the cold Parisian sky the rich and smooth creations of ancient Greece. La Zambinella showed him, united together, alive and delicate, those exquisite proportions of female nature, which he had desired so ardently.]

Both the extract from *Sarrasine* and the legend of Zeuxis indicate that physical perfection does not exist as a whole in nature. This beauty must be mediated by works of art, specifically the art of ancient Greece, in order to exist. It is therefore not surprising that Zambinella's perfect beauty emanates from an artificial being.

As 'real' beauty is inherently artificial or artistic in *Sarrasine*, the sculptor attempts to transform Zambinella into a tangible work of art. However, the drawings and the sculpture that are supposed to copy 'her' beauty do not imitate nature. Sarrasine fails to realize that the composite external beauty of Zambinella, which assembles the most perfect parts of previous models, reflects the ambiguity of the singer's sex. The sculptor fashions a new being by interpreting the face and especially the voice of his model, whose body is concealed by layers of fabric. The female statue of Zambinella becomes the epitome of the castrato's femininity. Like Polykles's Hermaphroditus, it embodies ideal beauty, here not only feminine, but also female, obeying Romantic rather than classical aesthetics. The sculptor creates a woman according to his desires, which makes him a modern Pygmalion. However, unlike another Pygmalion-like figure, Prince Chigi, who shaped his ideal Zambinella for profit (the castrato bears a grudge against him for this reason), Sarrasine loves his creation.

The mention of the mythological sculptor is not associated with the act of creation in the novella, but appears in a periphrasis that designates Galatea, Pygmalion's statue, and metaphorically refers to Zambinella: 'Sarrasine dévorait des yeux la statue de Pygmalion, pour lui descendue de son piédestal' [Sarrasine was devouring with his eyes

212 Balzac, *Sarrasine*, 57–58.

Pygmalion's statue descended from its pedestal just for him].[213] The text recalls the metaphor of the devouring – and *de facto* objectifying – male gaze. Harkness rightly shows that the myth of Pygmalion is connected to male dominance.[214] Drawing Zambinella is Sarrasine's way of possessing the singer when he cannot see 'her'. Whereas the medium of art promoted Zambinella as a living embodiment of ideal beauty, it also objectifies him, as he is reduced to being Sarrasine's plaything. The castrato becomes prisoner of his own status as a masterpiece. Incidentally, d'Albert similarly submits women to his male gaze, perceives them as works of art and therefore reifies them: 'J'ai pour les femmes le regard d'un sculpteur et non celui d'un amant' [I have for women the gaze of a sculptor, not that of a lover].[215]

Referring to the theme of Pygmalion in *Sarrasine*, Per Nykrog claims that Sarrasine is not a Pygmalion, but a naive artist, since he does not accept the fact that art is not a reflection of reality.[216] To be more accurate, Sarrasine is an involuntary Pygmalion because, despite his naivety, he has been able to create a perfect representation of feminine beauty and give life to his work. He created *la* Zambinella when sculpting 'her' statue. Whereas the singer was objectified by the sculptor, the statue has become an independent subject whose destiny does not depend on its creator. Unlike sterile hermaphrodites and castrati, Zambinella's sculptural representation is able to generate 'offspring'. Cardinal Cicognara acquired the statue and had a marble copy made. Then, the Lantys commissioned the artist Vien to paint his Adonis from the marble statue. Finally, Girodet allegedly found his inspiration for *Endymion* in Vien's work. The statue thus not only escapes

213 Balzac, *Sarrasine*, 58. See also 59.

214 Harkness, *Men of their Words*, 136 and 139. He also points out that Madame de Rochefide, the narratee, is almost a victim of this objectification, but escapes it by refusing the contract that is offered to her: 'Resisting Petrification in George Sand's *Lélia* and Balzac's *Sarrasine*', *French Studies*, 59/2 (2005), 159–72 (168 and 170–71). Madame de Rochefide protests: 'vous me faites à votre goût. Singulière tyrannie! Vous voulez que je ne sois pas *moi*' [You are fashioning me to your own taste. Peculiar tyranny! You do not want me to be *myself*] (Balzac, *Sarrasine*, 52 [emphasis in original]).

215 Gautier, *Mademoiselle de Maupin*, 227. See also 29, 277 and 281.

216 Per Nykrog, 'On Seeing and Nothingness: Balzac's *Sarrasine*', *Romanic Review*, 83/4 (November 1992), 437–44 (442).

the 'death' that its creator had planned for it when throwing a hammer at it but is also brought to life in different forms and, more importantly, different sexes. Regarded as a woman by Sarrasine, it recovers the 'real' sex of its model in Girodet's painting. These successive changes of sex illustrate that Zambinella cannot be constrained by a male or a female identity. They also reveal that the body of the castrato, like that of the hermaphrodite, struggles to achieve acceptance outside of artistic thinking.

Art in *Fragoletta*, *Sarrasine* and *Mademoiselle de Maupin* occupies an essential, if ambiguous, place in the hermaphroditic narratives in their heterotopian capacity. Works of art, and especially sculptures, mediate the bodies of hermaphrodites and therefore allow them to express their 'true' sex and gender. They also denounce the illusory norm of the separation of the sexes. Only art, because of its ability to create as well as to represent, can make these bodies fathomable and acceptable, whereas 'real' intersex characters are confined to monstrosity. Some of our chosen narratives, nevertheless, highlight the limits of artistic creation – and potentially their own limits – as artworks are misunderstood (the statue of Hermaphroditus) and even destroyed (the statue of Zambinella), as their own meta-artistic reflections can be misconstrued.

The Renewal of Masculinity

The previous sections have explored the variations in the representation of hermaphroditic characters from physical and moral aberrations to incarnations of ideal beauty. Here, I develop the issue of masculinity and oppose the insecure virility of male characters with the possibility of masculine regeneration through a hermaphroditic protagonist: Madeleine de Maupin.

Embodiments of manhood in the three literary narratives are compromised at best. D'Albert is vain and narcissistic, and he chooses isolation over action. With his overcompensating hypermasculine behaviour, Sarrasine is not spared from insecurity. He demonstrates that adopting a blatant form of virility is unsustainable because it cannot conceal men's weaknesses and leads to doubt and death; ultimately, he lets Cardinal Cicognara's henchmen kill him. Even Napoleon, who makes a brief appearance in *Fragoletta* at the onset of his coup d'état, is described as frail and unimpressive by the narrator.[217]

Certainly, d'Hauteville, at first glance, seems unaffected by masculine insecurity, as he combines numerous virile qualities. He can be regarded as a noteworthy incarnation of the Napoleonic officer. He is brave, chivalrous and educated. He despises cowardice and displays magnanimity towards his enemies – unlike the Italians' alleged obsession with revenge.[218] He shows honour when refusing to take advantage of Fragoletta and avenging Eugénie's dishonour, but, standing by his principles of generosity, he first offers his sister's hand in marriage to Philippe.

D'Hauteville is, nevertheless, oblivious to the 'true' nature of Camille. In the same way that Sarrasine refuses to see that the woman whom he loves is a castrato, the French officer does not realize that the girl with whom he fell in love and his enemy are the same person. This obliviousness also affects his involvement in historical events. Whilst playing a significant role in the war against the Italian royalists in the Neapolitan plot, he refuses to take part in the coup of 18 Brumaire. As Laforgue points out, 'Latouche met d'Hauteville littéralement en dehors de l'histoire pour lui faire assouvir sa destinée de personnage de fiction en le lançant à la poursuite de Fragoletta sur les routes de l'Italie' [Latouche literally positions d'Hauteville outside history to make him satisfy his destiny as a fictional character by setting him off in pursuit of Fragoletta on the Italian roads].[219] The young man renounces his political destiny to embrace a sentimental fate outside the public sphere.

This refusal to become involved in historical events is not specific to d'Hauteville. Despite their adoptive father's interest in politics and his support for the French side, Camille is interested neither in political events nor in history. The hermaphrodite disappears during the war between the French and the royalist army, and leaves Paris just before Napoleon's

217 Latouche, *Fragoletta*, II, 136–37.

218 Ibid., I, 116.

219 Laforgue, *L'éros romantique*, 92.

coup. It is rather surprising that Latouche, a moderate republican who denounces the Restoration of the Italian Bourbons and the rise of Napoleon in *Fragoletta*, does not explicitly link the situation of his protagonists to this historical period.[220] Attempting to connect the myth of the androgyne to the historical motif, Crouzet asks: 'Faut-il mettre en rapport la quête par d'Hauteville de l'être et de l'amour impossibles et la quête de la liberté ?' [Should we link together d'Hauteville's quest for impossible being and love and the quest for freedom?][221] Nevertheless, he rightly stresses the lack of coherence in Latouche's novel, which prevents him from validating this suggestion; the only similarity is disbelief in both historical events and the human (hermaphroditic) condition.[222] As for Laforgue, he argues that the hermaphrodite represents the disorder of historical events and meaninglessness.[223] The indeterminate, double gender identity of the hermaphrodite mirrors the confusion of political events. However, it seems as though Laforgue, like Crouzet, takes the shortcomings in Latouche's writing as an excuse not to address this issue. The co-presence of military operations and the dwelling on an impossible romance due to ontological singularity cannot be overlooked. I argue that the hermaphrodite embodies young men's lack of desire to fight and lack of political involvement in historical events. Interestingly, Latouche divides the figure of the apolitical young man into two characters: d'Hauteville and Camille. The former is originally involved in political events and even rescues the victims of an earthquake but is too obsessed with Camille by the end of the novel to care about making history. When invited to change the fate of France, he tells Macdonald's aide-de-camp: 'Je ne sais rien et n'ai guère de temps à perdre' [I know nothing and have no time to lose].[224] Camille has never been interested in the future of their country, as if to echo the lack of interest that history shows in unconventional beings. This indifference to history and politics also appears in *Sarrasine*, as the sculptor does not know that women performing on stage was banned in the Papal States.

220 Crouzet, 'Monstres et merveilles', 41; Laforgue, *L'éros romantique*, 103.

221 Crouzet, 'Monstres et merveilles', 31.

222 Ibid.

223 Laforgue, *L'éros romantique*, 96.

224 Latouche, *Fragoletta*, II, 137.

The protagonists' lack of participation evokes young men's increasing rejection of political actions against the regime that limited their freedom. Few students participated in the revolution of 1848. Maurice Aguilhon's table of the number of insurgents who were arrested in June 1848 (based on Roger Price's study) shows that students represented only 0.33 per cent of them; *ouvriers* [manual workers], on the other hand, constituted 62.7 per cent of all arrests.[225] The disillusionment caused by the wavering of the ideals of the July Revolution seemed too strong for this generation to act.

The conclusion that might be drawn from these hermaphroditic narratives is that young men appear too disconnected from reality to be regarded as suitable rulers. This is the hypothesis that will be further explored in the next chapter. What I claim for the moment is this: the result of this male withdrawal from public life is that the regeneration of masculinity cannot be realized by young male characters, but rather by hermaphroditic individuals who reconcile virile and feminine qualities. The ambiguities relative to Camille's and Zambinella's moral principles are overcome by Madeleine in order to embody an ideal form of masculinity. Théodore's transformation into Hermaphroditus is not the result of d'Albert's fantasies exclusively; Madeleine also perceives herself as a hermaphroditic being, although she does not explicitly refer to the Greek god – even if she is aware of his existence.[226] She embraces, to put it simply, gender ambiguity, as she is certain that she does not belong to the feminine gender. From the beginning of the tale of her adventures, she stresses the lack of 'romanesque' [romance] in her personality, an adjective with a sentimental, and thus feminine, connotation.[227] She also despises feminine activities:

> il m'ennuie de me tenir assise les deux pieds joints, les coudes collés au flanc, de baisser modestement les yeux, de parler d'une petite voix flûtée et mielleuse, et de faire

225 Maurice Aguilhon, ed., *Les quarante-huitards* (Paris: Gallimard-Julliard, 1975), 32. See also Magraw, *France 1815–1914*, 88; Broglie, *La Monarchie de Juillet*, 166.

226 Gautier, *Mademoiselle de Maupin*, 333.

227 Ibid. 244. This meaning stands in opposition to another use of 'romanesque' in the novel (249), which refers to the adventures told in a novel and thus has a masculine connotation that corresponds to Madeleine's personality. See 'Romanesque', in *Dictionnaire de l'Académie française*, II, 672.

passer dix millions de fois un bout de laine dans les trous d'un canevas; – je n'aime pas à obéir le moins du monde, et le mot que je dis le plus souvent est: – Je veux.[228]

[It bores me to sit with my feet together, my elbows glued to my side, to lower my eyes modestly, to speak with a little, fluted and unctuous voice and to push a bit of wool ten million times through the holes of a canvas; – I do not like in the least to obey, and the word that I say most often is: 'I want'.]

Her representation of feminine activities highlights the imperative of modesty that was imposed on women and denounces Rousseau's claim that women must be subservient to men and cannot express their will.[229] 'Je veux' reveals her assertive temperament, therefore coded as manly.

From this perspective, Madeleine regards her masculine disguise as her natural clothes.[230] Cross-dressing enables her to reveal her true self, unlike social and cultural norms, and to feel erotic pleasure in the opposition between her masculine outfit and behaviour, and her female body. Madeleine delights in engaging in erotic activities with Rosette while dressed in male clothes: 'je sentais sa gorge, demi-nue et révoltée, bondir contre ma poitrine, et ses doigts enlacés se crisper dans mes cheveux. – Un frisson me courut tout le long du corps, et les pointes de mes seins se dressèrent' [I was feeling her half-naked and revolted bosom leaping against my chest and her intertwined fingers tensing in my hair. A shiver ran through my whole body and my nipples hardened].[231] Her pleasure, she confesses, is augmented by the equivocation of her situation: being found in a risqué position with another woman who does not know her 'real' sex.[232] Such erotic pleasure

228 Gautier, *Mademoiselle de Maupin*, 327.

229 Many examples of the imperative of women's submission can be found in *Émile*, for instance: 'Il résulte de cette contrainte habituelle une docilité dont les femmes ont besoin toute leur vie, puisqu'elles ne cessent jamais d'être assujetties ou à un homme, ou aux jugements des hommes, et qu'il ne leur est jamais permis de se mettre au-dessus de ces jugements' [It results from this usual constraint a docility that women need all their lives because they never stop being subjected either to a man or to men's judgements, and they are never allowed to put themselves beyond these judgements] (Rousseau, *Émile*, 534).

230 Gautier, *Mademoiselle de Maupin*, 327.

231 Ibid., 339–40.

232 Ibid., 339.

is also stimulated by manipulating a sword, an obvious phallic object, when she fights in a duel against Alcibiade, Rosette's brother: 'ce bruit et ces éclairs tourbillonnants de l'acier m'enivraient et m'éblouissaient' [this noise and the whirling flashes of steel intoxicated and blinded me].[233] The young woman highlights the rapture that she feels in fighting Alcibiade, which is increased by both her adversary's strength and her own superiority.

Not only does Madeleine reject her behavioural femininity and prefer masculine conduct, but she is also able to verbalize her experience (showing that she can display 'viriles pensées' [virile thoughts]).[234] She opposes the sex of the body to the sex of the soul – that is, sex (femaleness) to gender (masculinity). In so doing, she shows acute awareness of the difference between biological sex and cultural gender at a time in which the modern meaning of *genre* did not exist. She claims: 'Il arrive souvent que le sexe de l'âme ne soit point pareil à celui du corps' [It is often the case that the sex of the soul does not match the sex of the body].[235] Assimilating herself to the masculine gender does not impede her from criticizing the male sex along with representatives of her own sex. She does not envy the lot of men because of their alleged ugliness, rude behaviour and stupidity and is not eager to renounce her femaleness, even though it does not suit her masculine mind.[236] If she wants to be a man, this is only because she does not understand how a woman can physically love another woman, although her experience with Rosette teaches her alternatives to heterosexual love.[237]

Bisexual, or even lesbian, Madeleine can also be interpreted as a non-binary or trans person. A person with a transidentity is defined as 'anyone who does not feel comfortable in the gender role they were attributed at birth, or who has a gender identity at odds with the labels "man" and

233 Ibid., 371.

234 Ibid., 327.

235 Ibid.

236 Ibid., 328, 378–80 and 383–84.

237 Ibid., 339 and 383–85. Rivers ('*Inintelligibles pour une femme honnête*', 3–4, 6 and 10–11) stresses her ignorance of lesbian sex and points to the role of d'Albert in this revelation: 'Ironically, the truth of homosexuality seems to come to Madeleine through her (hetero-)sexual initiation by d'Albert' (21).

"woman" credited to them by formal authorities'.[238] In this respect, hermaphroditism constitutes a powerful literary and artistic tool for depicting the stigma of intersex and trans individuals who are forced to comply with the binary sex system.[239] Madeleine is a transgender person who identifies with neither of the two gender labels, as shown in the following excerpt:

> À force d'entendre tout le monde m'appeler monsieur, et de me voir traiter comme si j'étais un homme, j'oubliais insensiblement que j'étais femme [...]. [...] Moi, par exemple, si je n'avais pas pris cette résolution, folle en apparence, mais très sage au fond, de renoncer aux habits d'un sexe qui n'est le mien que matériellement, et par hasard, j'eusse été fort malheureuse.[240]
>
> [Through hearing everyone calling me sir and treating me like a man, I was gradually forgetting that I was a woman [...]. [...] For instance, if I had not taken this apparently insane but actually very wise decision to renounce the clothes of a sex that is mine only materially and by accident, I would have been very unhappy.]

Madeleine points out here that she did not choose to put on male attire because she was uncomfortable in the gender of her birth, but that her cross-dressing revealed a malaise that she had long felt. Her so-called insane decision to change sex seems insane only to cisgender people. It makes sense to trans people who can thus claim their true gender through a name, a pronoun, looks and activities that may be associated with the opposite sex.

By refusing both gender labels, Madeleine deconstructs the binary system of sexes and genders. She discards both sexes and claims to belong to a third one:

> En vérité, ni l'un ni l'autre de ces deux sexes n'est le mien; je n'ai ni la soumission imbécile, ni la timidité, ni les petitesses de la femme; je n'ai pas les vices des hommes, leur dégoûtante crapule et leurs penchants brutaux: – je suis d'un troisième sexe à part qui n'a pas encore de nom: au-dessus ou au-dessous, plus défectueux ou supérieur: j'ai le corps et l'âme d'une femme, l'esprit et la force d'un homme, et j'ai trop ou pas assez de l'un et de l'autre pour me pouvoir accoupler avec l'un d'eux.[241]

238 Stephen Whittle, 'Foreword', in *The Transgender Studies Reader*, xi–xvi (xi). This definition highlights the external imposition of a gender identity on an individual.

239 See Costello, 'Intersex and Trans* Communities', 91–92.

240 Gautier, *Mademoiselle de Maupin*, 327.

241 Ibid., 393–94.

> [In truth, neither of these two sexes is mine; I have neither the stupid submissiveness, nor the timidity nor the pettiness of women; I do not have mens's vices, disgusting dishonesty and brutal inclinations: – I belong to a separate third sex, which does not yet have a name: higher or lower, more defective or superior. I have a woman's body and soul, and a man's mind and strength. And I have too much or not enough of either of them to be able to pair up with one of them.]

Once again, women are depicted as subservient, timid and petty. Although containing an 'embryon de révolte féminine' [embryo of female revolt], to borrow Chantal Bertrand-Jennings's words, *Mademoiselle de Maupin* is not a feminist novel advocating women's empowerment.[242] Gautier's female character never blames men for women's submission. Their oppression is justified in the novel, notably through Madeleine, by their numerous flaws. Yet, men's portrayal is not flattering either, as they are depicted as debauched and brutal.

Rejecting each pole of the sexual dichotomy equally, Madeleine adopts a dialectical reasoning by claiming to belong to a third sex. *Troisième sexe* refers here to a new sex (and its own gendering) that combines one physical and one moral quality of each sex: the beauty and the soul of women, and the strength and the mind of men. D'Albert himself ultimately corroborates this representation of a third gender, as he had stated that Madeleine has the body and the features of a woman, and the mind of a man, just before referring to her fencing talent.[243] Physically, Madeleine merges two aesthetic ideals: a virile and strong beauty, and a feminine and delicate beauty. Where her moral qualities are concerned, Madeleine's words seem to contradict a previous statement according to which she does not have a woman's soul.[244] It can be assumed, however, that the word *âme* highlighted women's sensitivity, whereas it refers to a more general moral composition in this excerpt. As innovative as this overcoming of sex and gender conventions is, it does not fully break away from a binary framework. The third sex conceived by Madeleine is only made up of elements taken from the other two.

242 Bertrand-Jennings, *Un autre mal du siècle*, 125.

243 Gautier, *Mademoiselle de Maupin*, 305.

244 Ibid., 327.

Madeleine pretends to doubt the superiority of this third sex to which she belongs but does not mean it. She judges her sex to be superior to traditional sexes, since she emphasizes men's and women's flaws, whereas she combines the physical, moral and intellectual qualities of both sexes, thus creating a new, perfect individual. Madeleine also underlines the inaccuracy of language to express hermaphroditism. She stresses the fact that few Western cultures recognize the existence of hermaphroditic beings and, accordingly, few have a proper name to designate them. The narrator of *Sarrasine* and the real intersex individual Herculine Barbin make a similar observation when the former designates the old Zambinella as 'cette créature sans nom dans le langage humain' [this creature with no existing name in human language] and the latter calls themselves 'un être sans nom' [a nameless being].[245] Their judgements, nevertheless, are mainly negative, as they refer to creatures regarded as either ridiculous or cursed.[246] In contrast, Madeleine deplores this lexical lacuna because she advocates the superiority of the hermaphroditic sex. Her words contrast with those of the physician Lauvergne who expressed his concerns about effeminate men, calling them 'sexes douteux' [dubious sexes].[247] For him, people of ambiguous sex are outcasts, like Camille. For Madeleine, their physical, moral and intellectual qualities render them superior. She can thus reverse the curse of hermaphroditic creatures by questioning the separation of the sexes and assuming the qualities of each.

Madeleine also refuses d'Albert's process of objectification. Like Sarrasine with Zambinella as well as d'Hauteville with Camille, the young man has put her on a pedestal. He appreciates her as the embodiment of ideal beauty but neglects her as a human being. He never considers the possibility that Madeleine might not love him. This is indeed the case, as she only mentions him towards the end of her final letter to Graciosa and casually writes: 'si je ne l'aimais pas avec passion, il me plaisait assez pour ne point le laisser sécher d'amour sur pied' [I did not love him passionately,

245 Balzac, *Sarrasine*, 45; Barbin, 'Mes souvenirs', 113.

246 The position of Barbin (Ibid.) is complex because they claim the superiority of their angelic nature in a fit of what Burgelin calls *megalomania* ('Le cas Herculine B.', 99).

247 Lauvergne, *Les Forçats*, 288. See also Cabanis, *Rapports*, I, 363.

but I liked him well enough not to let him pine away out of love].[248] In response to his reification, she uses him to discover the mystery of sexual intimacy, before abandoning him. She employs this new knowledge to have sexual intercourse with Rosette, then leaves the castle to seek new adventures. Her resourcefulness and eagerness to be part of the world not only contrast with d'Albert's passivity and seclusion, but also demonstrate Madeleine's ability to overcome gendered and sexual oppositions imposed by social rules. She can be in turn a seducer, an unparalleled fencer, a generous master and a clever individual.

In brief, Madeleine de Maupin, like young men and other hermaphroditic protagonists, shows a lack of interest in historical events and her story appears to be situated outside history. She nonetheless represents historical dynamism and embodies a new type of active masculinity that is grounded in her hermaphroditic nature. The latter enables her to refuse the confinement imposed on women by patriarchal societies and to explore new sexual territories.

It is clear from what precedes that the hermaphrodite is at the centre of the aesthetic and social preoccupations of July Monarchy literature. The representation of intersex individuals is a means for writers to examine the correlation between sex and gender. By their very existence, hermaphrodites mirror the sexual or gender insecurity of men. The most effective approach of the literary narratives presented here is to aestheticize the hermaphrodite, for instance by turning a shameful homoerotic desire into a quest for beauty. Art can transform the 'monstrous' nature of hermaphrodites into ideal beauty. Their bodies become real but idealized spaces that reject the Western norm of sexual difference and demonstrate the fusion of the sexes.

This emphasis on art highlights the origin of the hermaphrodite as a product of classical art and Greek mythology, which were widely regarded as universal at the time. Balzac justified the choice of a mythological figure to depict a contemporary malaise by mentioning the evocative power of myth: 'Il est déplorable, au xix [*sic*] siècle, d'aller chercher les images de la Mythologie grecque; mais je n'ai jamais été si frappé que je le suis de la puissante vérité de ces mythes' [It is deplorable in the nineteenth century

248 Gautier, *Mademoiselle de Maupin*, 396.

to use images of Greek mythology, but I have never been so struck by anything as I am by the powerful truth of these myths].[249] The supposed eternal dimension of the hermaphrodite suggests that the sexual and gender indeterminacy that this figure embodies is timeless. Just as hermaphrodites are outside of society, they are outside of history. Indeed, the medium of mythology and the absence of a definite connection between the writing context and the fictional context of the characters show that writers cannot distance themselves from the separation of the sexes and conceive of a fully intersex individual. Camille, the 'real' hermaphrodite, is depicted as either a man or a woman, and the castrato Zambinella justifies his perfect embodiment of femininity by the scam organized by his colleagues against Sarrasine. Only Madeleine de Maupin fully assumes her non-binary identity. The positive light in which her symbolic intersexuality is cast contrasts with the often-criticized sexual ambiguity of young men in the novels analysed in the next chapter.

249 Balzac, *Lettres à l'Étrangère*, quoted by Borowitz, 'Balzac's *Sarrasine*', 174.

CHAPTER 3

The Effete

In Balzac's novel *Splendeurs et misères des courtisanes* (1838–47), Vautrin, the famous criminal in *La Comédie humaine* [*The Human Comedy*], works tirelessly to secure the success of his protégé, the poet Lucien de Rubempré.[1] The young man is engaged to Clotilde de Grandlieu, but he must raise a substantial amount of money to buy his ancestors' land, an act which would seal their marriage. In order to do so, Vautrin prostitutes Esther, Lucien's mistress, to the banker Nucingen. After numerous twists in the plot, including the murder of a secret agent, the rape of the latter's daughter, Esther's suicide and the disappearance of Esther's money, Vautrin and Lucien are arrested and cross-examined by the *juge d'instruction* [examining magistrate] Camusot. While Vautrin prevails against the judge, he fears that Lucien would yield to pressure and reveal incriminating information about the identity and the plans of his protector, who hides behind the clothes of the Spanish priest Carlos Herrera. Vautrin's fears are indeed grounded, as Lucien falls to pieces and betrays him, before committing suicide. In order to convince Camusot not to examine Lucien, the criminal had told him: 'Je connais Lucien, c'est une âme de femme, de poète et de méridional, sans consistance ni volonté' [I know Lucien. He has the soul of a woman, a poet and a Southerner, with neither consistency nor willpower].[2] Lucien is indeed – or at least fancies himself – a poet. Native to Angoulême, in the southwest of France, he can be described as a Southerner. Both

1 Vautrin is mainly called Carlos Herrera (his fake identity) and Jacques Collin (his real name) in the novel, but Vautrin (his fake identity in *Le Père Goriot*) is the name by which he is mostly remembered.

2 Honoré de Balzac, *Splendeurs et misères des courtisanes*, Patrick Berthier, ed. (Paris: Librairie Générale Française, 'Le Livre de Poche Classiques', 2008), 498.

terms convey feminine connotations: the soul of a poet is made of aesthetic sensibility and reverie, and southern people were seen as more feminine.[3] Vautrin thus indicates in three different ways that Lucien is like a woman. This comparison is relevant in the criminal's speech because it entails the feminine features that have been highlighted in the first chapter, namely his weakness, pusillanimity and inconstancy, which enable Vautrin to prove the young man's insignificance in the eyes of the Law.

Balzac's novel illustrates a similar awareness of generational malaise as that described in the second chapter of Musset's *La Confession d'un enfant du siècle* (1836). In fact, Octave's disease, which is at the heart of the plot, does not correspond to the *mal du siècle* that is depicted in the second chapter of the novel. It consists of love sickness, infidelity, debauchery and, above all, jealousy. Nevertheless, as Gilles Castagnès rightly argues, the construction of the novel encourages the reader to interpret Octave's ailment as the collective and generational disease of the century in order to justify it.[4] The narrative is constructed in such a way that Octave's pathological jealousy seems to result from the 'maladie du siècle' that would affect his generation.[5] Octave's malady and Lucien's so-called feminine soul both reveal a common disbelief in young men's social abilities rooted in the social, political and economic changes that occurred during the July Monarchy. Among the many male characters displaying insecurity in narratives of this period (Sténio in *Lélia*, Calyste du Guénic in *Béatrix*, Fabrice del Dongo

3 Waller (*The Male Malady*, 41–42) has shown how René's femininity, in Chateaubriand's novel, was revalorised because it indicated aesthetic sensibility and poetic reverie.

4 Gilles Castagnès, '*La Confession d'un enfant du siècle*: Un romantisme de façade?', in Sylvain Ledda and Frank Lestringant, eds, *Musset: Un romantique né classique* (Toulouse: Presses Universitaires du Mirail, 2009), 87–102 (88–91). Maxime Foerster, however, supports the close connection between Octave's malady and that of the century, as he considers the latter to be (apparent in) 'dysfunctional heterosexual couples and a loss of faith in love' (*The Politics of Love*, 102).

5 For Castagnès ('*La Confession d'un enfant du siècle*', 101), the causality is reversed: the reference to the *mal du siècle* would be Musset's retrospective attempt to justify his own morbid jealousy.

in *La chartreuse de Parme*, to mention just a few), Octave and Lucien – the latter also appears in *Illusions perdues* (1837–43) – constitute two paragons of this masculine malaise, with their insecurity, effeminacy and immaturity.

This chapter studies the discourse that seemingly differs from the narratives that have just been examined. Whereas the previous chapter analysed the literary renegotiation of sexual difference through hermaphroditic characters and the somewhat positive connotations of their sexual ambiguity, this chapter explores how this often-negative opinion of young men's effeteness was embedded in doubts regarding social gender norms. Writers depict feminine and fragile young men in order to express their concern that sexual dichotomy may not be the stable framework that it was supposed to be and may be overthrown by gender ambiguities. Although male characters such as Lucien and Octave, unlike hermaphrodites, are not considered to be obvious challenges to maleness, they nevertheless call the bearings of traditional masculinity into question.

By taking as its starting point the problematic relationship between young men and the unachievable masculine model of the Napoleonic soldier, this chapter analyses how society ascribes demeaning feminine traits to these men in order to highlight their inability to exercise effective social roles. Young men can no longer distinguish themselves as soldiers and reject bourgeois professions, such as lawyers and doctors. Only a few fictional young men enter politics (Eugène de Rastignac is one of them). Young men's 'flaws' include puerility, moral weakness, effeminate beauty and homosexuality. I must also stress Balzac's complex attitude towards male homosexuality, which is paradoxically used as a vector to criticize men's alleged social incompetence and simultaneously to praise the combination of masculine and feminine qualities. The figure of a homosexual such as Vautrin, not Lucien, constitutes an alternative to both masculine insecurity and a more traditional model of virility.

The Son of the Napoleonic Soldier

During the first half of the nineteenth century, Napoleon and his soldiers embodied an exemplary – if inaccessible – virile model. They promoted qualities such as valour and honour that were expected from all men. Accordingly, soldiers of the Great Army stirred both the admiration and the anxiety of the generation of young men reaching adulthood in the 1830s, often including their own sons. However, young men of the Restoration and the July Monarchy could not express their manliness in the same way as their fathers owing to the military history of these periods – or rather the lack of fighting opportunities. These young men did not experience continuous wars. With the exception of the conquest of Algeria, which had started under the reign of Charles X, the military action of Louis-Philippe's regime was considerably limited.[6]

Stendhal's heroes, Julien Sorel and Fabrice del Dongo, are two fictional examples of this fascination for the figures of Napoleon and his soldiers of the Great Army. Brian Joseph Martin sees in these characters the incarnations of, respectively, *wannabes* and *latecomers*. These were young men who had not personally experienced the Napoleonic era, but who grew up lulled by the stories of imperial grandeur, who idealized the Empire and who yearned to equal the soldiers of the Great Army in bravery.[7] Many characters from novels written during the July Monarchy period do not long for military glory so explicitly. They are, nevertheless, fashioned by the idealization of virile values that echo those of the revolutionary and Napoleonic era.

Fathers held the same appeal for this young generation as the Napoleonic soldiers themselves. Writers such as Hugo, Musset and Sand were the offspring of soldiers and civil servants of the Empire. In the second chapter of *La Confession d'un enfant du siècle*, Musset depicts the discrepancy between

6 For more on the conquest of Algeria, see Bertaud, 'La virilité militaire', 183; Broglie, *La Monarchie de Juillet*, 330–34.

7 Martin, *Napoleonic Friendship*, 103–25 (esp. 105–08 and 115–16).

the admiration shown for the previous generation and the weakness of the generation following the heroes of the Napoleonic wars:

> Pendant les guerres de l'Empire, tandis que les maris et les frères étaient en Allemagne, les mères inquiètes avaient mis au monde une génération ardente, pâle, nerveuse. Conçus entre deux batailles, élevés dans les collèges aux roulements des tambours, des milliers d'enfants se regardaient entre eux d'un œil sombre, en essayant leurs muscles chétifs. De temps en temps leurs pères ensanglantés apparaissaient, les soulevaient sur leurs poitrines chamarrées d'or, puis les posaient à terre et remontaient à cheval.[8]

> [During the wars of the Empire, while their husbands and brothers were in Germany, anxious mothers had given birth to an ardent, pale and nervous generation. Conceived between two battles, raised in colleges with drum rolls in the background, thousands of children were looking at each other with a gloomy gaze, while trying out their frail muscles. From time to time, their blood-stained fathers appeared, held them to their chests adorned with gold and then put them down and got back on their horses.]

Two characteristics of this new generation emerge from this excerpt: their weakness and their ambition. The paleness and nervousness of these boys convey their physical and psychological fragility. Their emblematic lack of strength is precisely the hurdle preventing them from realizing their ambition, a masculine trait according to French scientists.[9] Raised in a bellicose context, these children long to prove their worth to their classmates, who impersonate potential enemies, and above all to their fathers – those distant figures who appear as glorious, bloody war gods – but Musset's novel and other contemporary narratives demonstrate that they cannot stand up to this model. Later, in the same chapter, these children, by then grown up, are portrayed as nostalgic for a fantasized past that they have never experienced. Imperial times are presented as an epic period in which the generation of their fathers was enthusiastically offered as a tribute to the demigod Emperor.[10] Invited by Napoleon, the first 'enfant du siècle' [child of the century], to avenge him, the younger generation is unable to comply with his request. Arguably, this may have been due to the anxiety caused by feelings of inadequacy when confronted with the

8 Musset, *La Confession d'un enfant du siècle*, 60.

9 Menville de Ponsan, *Histoire médicale*, I, vi.

10 Musset, *La Confession d'un enfant du siècle*, 60 and 61.

overwhelming myth of the Emperor, or at least with the attenuated but still threatening form of the myth as embodied by their fathers.[11]

The weakness of Octave's generation is interpreted as a sign of puerility in Musset's novel. As these young men are the children of Napoleonic soldiers, it is not surprising that they are depicted as children, notably in Musset's quoted extract. The symbolic link between paternity and authority is concretized in the Civil Code, which, in 1804, asserted the authority of the father over his wife and children.[12] Virile and endowed with moral and legal authority, the paterfamilias has his domination apparently ensured by the Code, since, as Allan points out, it transfers power from a national father, namely the King, to individual fathers.[13]

The Civil Code thus corresponds to what Jacques Lacan calls the *Nom-du-Père*, the association of the paternal function with the symbolic function as representative of the Law. As Lacan explains: 'C'est dans le *nom du père* qu'il nous faut reconnaître le support de la fonction symbolique qui, depuis l'orée des temps historiques, identifie sa personne [celle du père] à la figure de la loi.' [It is in the *name of the father* that we must acknowledge the support of the symbolic function, which, since the dawn of historical times, identifies the Father with the figure of the Law.][14] Needless to say, Napoleon, the father of the Civil Code, is a clear embodiment of the Law, and his soldiers perpetuate his symbolic and legal heritage. Accordingly, Octave, along with Lucien, appears as a child forced to comply with the *Nom-du-Père*, the Napoleonic Code and an unfair social system. Mary Orr notes that '[t]he Code can be seen to be the Law of the New Fathers determined to keep its children [...] in perennial childhood to prevent

11 Goruppi, 'Gli eredi', 101–04. Napoleon calls himself 'enfant du siècle' in Las Cases's *Mémorial de Sainte-Hélène* (quoted in Goruppi, 'Gli eredi', 101), as well as claiming: 'Quelle jeunesse [...] je laisse après moi! C'est pourtant mon ouvrage!' [What a youth [...] I leave behind me! It is, however, my work] (102).

12 On conjugal duties and paternal authority, see articles 212–26 and 371–87: *Code civil des Français, édition originale et seule officielle* (Paris: Imprimerie de la République, 1804), 53–55 and 92–95. On paternal power and the Civil Code, see Martin, 'Fonction paternelle et Code Napoléon'.

13 Allan, *Writing the Self*, 105–06.

14 Jacques Lacan, *Écrits* (Paris: Seuil, 1966), 278.

a repetition of Revolution'.[15] Indeed, the paterfamilias is encouraged to support the State in transmitting civic values and play a policing role on a domestic scale.[16]

In *La Confession d'un enfant du siècle*, the term *enfant* not only highlights Octave's affiliation to his time; it also emphasizes his condition as a lost and dominated child.[17] He displays insecurity because he does not understand the codes of the world in which he lives. This is noticeable in the way that Octave and, beyond the fictional character, Musset as a self-proclaimed spokesperson for the men of his generation portray themselves. Male writers, in contrast to female writers, were, indeed, often reluctant to write about childhood memories in autobiographies because a child's experiences could seem trivial.[18] In *Illusions perdues*, Lucien's childishness is perceptible to Vautrin when they first meet, as the escaped convict tells him: 'Vous êtes un enfant, vous ne connaissez ni les hommes, ni les choses' [You are a child, you know neither men nor things].[19] In both protagonists, puerility is characterized by ignorance of the things of life.

In Musset's novel, ignorance goes beyond innocence and purity, which are often associated with childhood, notably in the Romantic tradition.[20] Maija Lehtonen has noted that the word *enfant* is not only used in the novel to connote innocence, especially when Octave applies it to himself in conversations, but also to emphasize his lack of experience and his

15 Mary Orr, *Madame Bovary: Representations of the Masculine* (Bern, Berlin, Brussels, Frankfurt a. M., New York and Vienna: Peter Lang, 1999), 23.

16 Martin, 'Fonction paternelle et Code Napoléon', 468–70 and 473–74.

17 Rosemary Lloyd, somewhat surprisingly, considers that the notion of the child that is supported by the *enfant du siècle* is hardly compatible with more traditional definitions of the substantive: *The Land of Lost Content: Children and Childhood in Nineteenth-Century French Literature* (Oxford: Clarendon Press, 1992), viii.

18 Ibid., 33–37. Simone de Beauvoir, *Le deuxième sexe*, 2 vols (Paris: Gallimard, 'Idées', 1949), II, 358–60.

19 Honoré de Balzac, *Illusions perdues*, Maurice Ménard, ed. (Paris: Librairie Générale Française, 'Le Livre de Poche Classiques', 1983), 508. Vautrin also paternalistically calls him 'enfant' [child], 'jeune homme' [young man] and 'mon Petit' [my little one] (504, 509, 511–12, 514 and 519). Such expressions are also frequently employed by Lélia and Félicité to designate their lovers.

20 Lloyd, *The Land of Lost Content*, esp. 65.

irresponsibility.[21] The Octave of the first part of the novel appears as a carefree but unhappy man who endeavours to forget about his mistress's infidelity by giving in to a life of debauchery. This depraved existence ends when his father dies. He leaves Paris and settles in the countryside. There, he takes up residence in his father's house and, since his way of life has been so far unsatisfactory, he adopts his father's lifestyle by wearing the latter's clothes and imitating his habits. Although he pretends to worship independence,[22] the young man's experience in his father's house proves that he needs to follow a paternal model to be able to fashion his life. Only by abandoning his autonomy and, as Ronald Grimsley rightly suggests, by returning to the past, can Octave temporarily find peace and happiness.[23] Rather than promoting autonomy, his actions in fact demonstrate his need for protection and support.

In short, Octave is immature. Immaturity is characterized by insouciance, that is, the refusal to acknowledge one's own responsibility. This characteristic is judged to be customary and even positive in children, as they are objectively exempt from responsibilities, but it is noxious for Octave. After all, as an adult, he should have responsibilities, be it in a military career, in a liberal profession or as his father's heir. Far from the idealized purity of a child, Octave's insouciance degenerates into mischievousness and even cruelty. In his father's hometown, Octave encounters Brigitte Pierson, a young widow with whom he falls in love. Although at first rejecting his advances out of decency, she finally admits her feelings for him. Their romantic bliss is, however, short-lived. When Brigitte tells him a white lie – about a composition of her invention that she attributes to the famous composer Stradella in order to hear Octave's sincere opinion – the young man becomes insanely jealous. He observes: 'J'avais commencé par me montrer insouciant; j'en vins bientôt à me montrer méchant' [I had started with being carefree; I soon became unkind].[24] He gratuitously provokes Brigitte,

21 Maija Lehtonen, *Essai sur la 'Confession d'un enfant du siècle'* (Helsinki: Suomalainen Tiedeakatemia, 1982), 25–26.

22 Musset, *La Confession d'un enfant du siècle*, 93.

23 Ronald Grimsley, 'Romantic Emotion in Musset's *Confession d'un enfant du siècle*', *Studies in Romanticism*, 9/2 (Spring 1970), 125–42 (132).

24 Musset, *La Confession d'un enfant du siècle*, 269. See also 247.

just as a child would, as he knows that she is faithful to him. His offences against her consist of comparing her with other women and even forcing her to comply with his expectations by imitating his former cheating mistress. The paroxysm of his need to humiliate her is reached when he treats her even less respectfully than he would a prostitute.[25] Such cruelty aims at concealing Octave's own fragility, which shows through the narrative despite his struggle to admit it.

In Brigitte, as in his former mistress, Octave seeks substitutes for his absent mother.[26] Both of them are older than him and are widows, a marital status that allegedly gives them independence, as they are not subservient to a husband or a father.[27] His case is not isolated in July Monarchy narratives. Anxious sons of absent fathers and/or mothers seek a substitute paternal or maternal figure, a protector who will take charge and make all the vital decisions for these carefree protagonists, whilst providing them with love and emotional security. Roger Bellet points out that Octave's generation not only consists of 'enfants du Siècle' [children of the Century], but, specifically, 'de la Femme' [of the Woman].[28] The search for a maternal figure is thus as constitutive of a young man's identity as is his relation to his father. Lisbeth Fischer takes the main decisions regarding Wenceslas Steinbock's career and adulterous relationships in *La Cousine Bette*, whilst Félicité des Touches encourages Calyste du Guénic to discover culture in *Béatrix*. Brigitte has manifest motherly virtues towards Octave: 'je ne suis pas votre maîtresse tous les jours; il y en a beaucoup où je suis, où je veux être votre mère' [I am not your mistress every day; there are many days

25 Ibid., 270.

26 Grimsley, 'Romantic Emotion', 135; Anthony Rizzuto, 'Octave in Alfred de Musset's *La Confession d'un enfant du siècle*', *Kentucky Romance Quarterly*, 24/1 (1977), 83–94 (87); Lehtonen, *Essai sur la 'Confession d'un enfant du siècle'*, 26 and 52; Susan M. Levin, *The Romantic Art of Confession: De Quincey, Musset, Sand, Lamb, Hogg, Frémy, Soulié, Janin* (Columbia, SC: Camden House, 1998), 50–51. The absence of Octave's mother has been noted by Castagnès (*'La Confession d'un enfant du siècle'*, 96–98) and Reid ('La confession selon Musset', 65).

27 Beauvoir, *Le deuxième sexe*, II, 11.

28 Roger Bellet, 'La confession politique des enfants malades du siècle', *Europe: Revue littéraire mensuelle*, 583–84 (November–December 1977), 126–35 (131–34).

when I am, and I want to be, your mother].[29] She is nonetheless unable to provide Octave with a sense of security because of his jealousy.

Despite his quest for a nurturing maternal figure, the young man rejects attempts to be freed from his insecure state. The relationship between Octave and Brigitte is shaped by alternation between quarrels and reconciliation. Octave's jealousy highlights the fact that he needs to doubt Brigitte's fidelity in order to define his identity. The puerile man prefers ambiguity over certainty, because the latter would force him to engage with responsibility and commit to social functions. This uncertainty takes the form of deferring decisions. For instance, when Octave and Brigitte reconcile and decide to travel to Switzerland, Octave keeps postponing this trip: 'nous ne l'avions pas décidé encore [le lieu où nous allions nous ensevelir], et nous trouvions à cette incertitude un plaisir si vif et si nouveau, que nous feignions, pour ainsi dire, de ne pouvoir nous fixer sur rien' [We had not yet decided on the place in which we were going to bury ourselves, and we found such a great and new pleasure in this uncertainty that we pretended, so to speak, that we were unable to make up our minds on anything].[30] As Castagnès points out, Octave cannot be content with the real world and therefore prefers making travel plans over the concretization of the journey.[31] By changing his mind and delaying his decisions, the young man enjoys a kind of irresponsible freedom.

Octave's love for a woman who embodies in many ways the ideal mother constitutes only one side of his Oedipus complex. The novel bathes in a Freudian atmosphere, as according to Freud a young boy not only desires his mother but also expresses hostility towards his father, deemed to be a rival.[32] However, the novel complexifies this Oedipal relationship because the identity of the fatherly rival is symbolic, since none of the 'real' male figures (Octave's father, Brigitte's deceased husband, her friend Dalens, Abbot Mercanson and Henri Smith) constitute convincing rivals: it is the

29 Musset, *La Confession d'un enfant du siècle*, 258.

30 Ibid., 275.

31 Castagnès, '*La Confession d'un enfant du siècle*', 93.

32 Sigmund Freud, 'An Autobiographical Study', in *The Freud Reader*, Peter Gay, ed. (London: Vintage Books, 1995), 2–41 (22); 'The Ego and the Id', 628–58 (640).

Nom-du-Père.[33] The conflict shifts towards a social perspective in which what is at stake is not only the possession of the mother, as would be the case in Freud's 'family romance',[34] but also the possession of the authority to rule society. Yet, merely by his imitation of his father's lifestyle, Octave avoids assuming the kind of responsibility that is supposed to characterize a man.

In *Illusions perdues*, Balzac also depicts the relationship between a man and, in this case, a fatherly mentor. Vautrin, embodying a subversive *Nom-du-Père*, spontaneously gives his protection to the orphan Lucien and organizes the latter's life so that Lucien can access the aristocratic life of which he always dreamt, thus giving financial security and discipline to his protégé. Lucien remains passive, agreeing to marry a girl whom he does not love and to let his mistress prostitute herself to establish his capital. Previously, in *Illusions perdues*, his friend David Séchard, his sister Ève and his mistress Louise de Bargeton in Angoulême, then the poet of the Cénacle Daniel d'Arthez, the editor Finot, Étienne Lousteau and the clique of journalists in Paris, all acted as Lucien's mentors, which reveals the poet's desire for protection and his inability to take control of his life.[35]

Childlike, narcissistic characters do not always recognize the merits of their protectors. Lucien suffers from his mother and sister rejecting his help when he comes back to Angoulême after failing in Paris. He interprets their gentle rejection as them no longer loving him and being the victims of their bourgeois lifestyle: 'Elles sont bourgeoises, elles ne peuvent pas me comprendre' [They are so middle-class, they cannot understand me].[36]

33 On the resemblance between Octave and Henri Smith, see Reid, 'La confession selon Musset', 69; Castagnès, '*La Confession d'un enfant du siècle*', 99; Franziska Meier, 'La "Mort de l'auteur" dans l'écriture autobiographique romantique: À propos du "jeune" François-René de Chateaubriand (*René*) et d'Alfred de Musset (*La Confession d'un enfant du siècle*)', *French Studies*, 67/3 (July 2013), 323–39 (336).

34 Freud, 'Family Romances', in *The Freud Reader*, 297–300.

35 Julia Chamard-Bergeron calls the relationships between Lucien and David, d'Arthez, Lousteau and Vautrin *friendships*. However, she highlights the inequality within them, at least with David, as one gives and the other receives. Therefore, it is preferable to consider these relationships to be based on protection: ' "Vous croyez aux amis": L'amitié dans *Illusions perdues*', *L'Année balzacienne*, 3rd ser., 10 (2009), 283–313 (287).

36 Balzac, *Illusions perdues*, 466.

Their reluctance to accept his help derives in fact from their lucidity towards his unreliability. Lucien's ungratefulness also takes the shape of selfish indifference where the sacrifices of others are concerned, such as the case of Esther selling her body to Nucingen to provide money so that her lover can marry Clotilde. His ingratitude towards the women who love him is sustained by his ingratitude towards his spiritual father Vautrin, whom he betrays when questioned by the magistrate Camusot – who is blamed for his 'interrogations captieuses' [specious questions].[37] Whilst he admits being ungrateful in his farewell letter, he nonetheless accuses Vautrin of being responsible for his suicide.[38] Discussing the concept of gratitude in the eighteenth century, Patrick Coleman points out that this virtue was thought to reveal one's education and nobility of birth as well as nobility of character.[39] Lucien's ungratefulness reveals his failure to be an accomplished man, just as the revocation of his testimony shows his incapacity to assume his own words.

To writers such as Balzac, the absolute admiration for the father figure appears to be exaggerated given that it is based on symbolic rather than existing qualities. This fascination, in the form of the Napoleonic cult, is both depicted and challenged in a novel that will be analysed in detail in the next chapter but that I will discuss briefly now: *La Cousine Bette*. In the Hulot family, the father is worshipped by his wife Adeline, his daughter Hortense and especially his son Victorin: 'Quant à Hulot fils, élevé dans l'admiration du baron, en qui chacun voyait un des géants qui secondèrent Napoléon, il savait devoir sa position au nom, à la place et à la considération paternelle' [As for Hulot Junior, brought up in admiration of the Baron, in whom everyone saw one of the giants who assisted Napoleon, he knew that he owed his own position to his father's name, position and the esteem in which he was held].[40] The reputation of Hulot d'Ervy, a compulsive womanizer who squanders his fortune on his mistresses, stems from his position

37 Balzac, *Splendeurs et misères des courtisanes*, 530.

38 Ibid. and 531. On Lucien's letter, see 'The Reconciled Man' in this chapter.

39 Patrick Coleman, *Anger, Gratitude and the Enlightenment Writer* (Oxford: Oxford University Press, 2011), 20.

40 Honoré de Balzac, *La Cousine Bette*, Roger Pierrot, ed. (Paris: Librairie Générale Française, 'Le Livre de Poche Classiques', 2000), 46.

as one of the Emperor's lieutenants (and from the renown of his brother, Marshal Hulot de Forzheim, a hero of the revolutionary wars), rather than from his personal merits. The use of 'en qui chacun voyait', rather than 'qui était', implies that this reputation is exaggerated. Likewise, Victorin Hulot, a young lawyer, enjoys a respectable professional situation as a result of his father's good name. This extract echoes Saint-Paulien's observation concerning many officers of the Great Army in *La Comédie humaine*, including Baron Hulot; they became drunkards, crooks, paupers, ridiculous men and even criminals because they were unable to do anything but fight wars.[41] By dint of spending all his money on his mistress Valérie Marneffe, Hulot ruins his own family, then lives with a young working-class girl and ends up marrying the cook when his wife dies.

Balzac ridicules another Napoleonic soldier who appears in *Illusions perdues*: Giroudeau, the uncle of the editor Finot and the handyman of the editorial offices. Giroudeau is nostalgic for the Empire. His uniform-like clothes, his repeated 'broum! broum!' noise (reminiscent of a military march or the noise of a cannon), his digressions on Napoleonic times and the tales of his memories as an officer in the Great Army point to this nostalgia.[42] Whilst this almost senile man is mocked, he mostly represents the difficulties experienced by former soldiers in post-Napoleonic times. They were persecuted and humiliated during the Restoration, often deprived of half of their pension (which earned them their nickname of *demi-soldes* [half-pays]), put on inactive reserve duty and were unlikely to have a military career.[43] *Illusions perdues* depicts the former soldier as having no choice but to work as a cashier for his nephew, highlighting the paradox between the former heroes' high symbolic virile capital and their low economic and professional capital during Balzac's times.

However, Giroudeau is not devoid of common sense, unlike Lucien. He declares:

41 Saint-Paulien, *Napoléon, Balzac et l'Empire de 'la Comédie humaine'* (Paris: Albin Michel, 1979), 279. See also 37 and 352.

42 Balzac, *Illusions perdues*, 190–91.

43 Hazareesingh, *The Legend of Napoleon*, 236; Martin, *Napoleonic Friendship*, 152–57.

> Quant aux rédacteurs, c'est de singuliers pistolets, de petits jeunes gens dont je n'aurais pas voulu pour des soldats du train, et qui, parce qu'ils mettent des pattes de mouche sur du papier blanc, ont l'air de mépriser un vieux capitaine des dragons de la Garde Impériale, retraité chef de bataillon, entré dans toutes les capitales de l'Europe avec Napoléon …[44]
>
> [As for the writers, they are odd customers, little youngsters, whom I would not have wanted as soldiers in the transport corps and who, because they put their scribbly handwriting on white paper, look like they despise an old captain of dragoons of the Imperial Guard, who retired as a battalion commander and entered every capital city of Europe with Napoleon …]

The veteran soldier denounces the arrogance of young journalists, while considering that these civilians would have made mediocre soldiers in Napoleon's army. The latter are proud of their writing skills, which contrast with the old man's unpolished outspokenness. However, as Giroudeau explains, the smartest ones do not write:

> [Il y a] enfin le rédacteur qui ne rédige rien et qui n'est pas le plus bête, il ne fait pas de fautes celui-là, il se donne pour un écrivain, il appartient au journal, il nous paye à dîner, il flâne dans les théâtres, il entretient une actrice, il est très heureux.[45]
>
> [Finally, there is the writer who writes nothing and who is by no means the stupidest one, he's the one that makes no mistakes, he claims to be a proper writer, he belongs to the newspaper, he buys us dinner, he wanders about in theatres, he keeps an actress and he is very happy.]

It is assumed from his speech that Giroudeau considers military service to be more important than a writing career, as journalists manipulate information to flatter the dominant opinion and to promote their own cause. The rest of the story contradicts neither his critical analysis of journalism nor his claim about the ineptitude of young men. Young journalists are depicted as lazy and only interested in fast profit. Lousteau, for instance, gives Lucien a cynical lesson about the literary institution, where quickly

44 Balzac, *Illusions perdues*, 191.
45 Ibid.

written articles and mediocre books are preferred to literary masterpieces and where exploiting other people's work is a safer way to make a fortune.[46]

Whilst Balzac's text oscillates between contempt for old soldiers, sympathy for their hardship and admiration for their wit, it is less ambivalent with regard to Napoleon and the excess of ambition caused by his example. This is illustrated in the first part of *Illusions perdues*. Lucien Chardon (he is not yet allowed to bear his mother's name, *de Rubempré*) dreams of becoming a poet and is supported in this ambition by his mistress, Louise de Bargeton. Expressing concerns regarding her role in the promotion of his literary career, he is nevertheless soon convinced that he will succeed thanks to his talent rather than his mistress. The narrator then comments: 'L'exemple de Napoléon, si fatal au Dix-neuvième Siècle par les prétentions qu'il inspire à tant de gens médiocres, apparut à Lucien qui jeta ses calculs au vent en se les reprochant' [Napoleon's example, so fatal to the nineteenth century owing to the aspirations that it gives to so many mediocre people, revealed itself to Lucien, who threw his schemes to the wind and felt ashamed of them].[47] The main clause and the relative clause indicate that Lucien prefers Napoleon's glorious path over scheming. The Emperor's path is that of genius, as Lucien's free indirect speech indicates: 'Son génie luirait tôt ou tard comme celui de tant d'hommes, ses prédécesseurs, qui avaient dompté la société' [Sooner or later, his genius would shine as it had done with so many men, his predecessors, who had tamed society].[48] However, the adjectival phrase ('si fatal…') claims that this path is unsuitable for mediocre people and implies that Lucien could be such a person. Lucien's fate is foretold in this sentence: he is condemned to lose his illusions because he is not the Napoleonic genius that he thought he was. Moreau de la Sarthe stated that ambition was incompatible with women's weaker constitution.[49] Accordingly, Balzac's text shows that ambition in an epicene young man can have dangerous, even 'fatal' consequences.

46 Ibid. 201.

47 Balzac, *Illusions perdues*, 52.

48 Ibid.

49 Moreau de la Sarthe, *Histoire naturelle*, I, 712 and II, 397.

Vautrin makes a similar observation when meeting Lucien: 'C'est le défaut des Français dans votre époque. Ils ont été gâtés tous par l'exemple de Napoléon' [This is the flaw of the Frenchmen of your time. They have all been spoiled by Napoleon's example].[50] It is suggested, through both the omniscient narrator and one of the main characters of *La Comédie Humaine*, that Lucien, as a man of his generation, has been fashioned by a Napoleonic ambition that tallies with the typical masculine ideals of strength and power. The text shows that his generation, overwhelmed by the Napoleonic legend, is struggling to build a new masculine ideal that would replace Hulot and Giroudeau's obsolete virile ideals.

The Weak Man and the Aspiring Artist

Young men in July Monarchy narratives are the opposite of Napoleonic soldiers. Rather than being strong, brave and honourable, they appear weak, cowardly, self-interested and erratic. These flaws recall nineteenth-century physicians' belief that women were inconstant owing to their physical sensitivity and their generally weaker constitution. In other words, like eunuchs, hermaphrodites and homosexuals in scientific texts, these fictional and insecure young men contribute to the deprecation of manhood since they are unable to conform physically and, above all, psychologically to virile standards. Arguably, their selfishness, passivity and inability to focus on specific problems impede their success and condemn them to mediocrity, social failure and even death. Their predicaments in the fictional world convey the anxiety of the young bourgeois and aristocratic generation towards the structures of July Monarchy society and its doubts regarding its ability to renew its elite.

Certainly, male protagonists of July Monarchy narratives have no monopoly on weakness and insecurity, but the representation of these flaws undergoes a shift throughout the first half of the nineteenth century. Male characters in early nineteenth-century novels are also defined as changeable

50 Balzac, *Illusions perdues*, 510.

and discontented. For instance, Chateaubriand's René stresses his inconstancy and his thirst for solitude; Madame de Duras' Olivier lacks *caractère* and energy; and Stendhal's Octave is melancholic and misanthropic.[51] In *The Male Malady*, Margaret Waller analyses how, in what she calls *mal du siècle* novels (from Chateaubriand's *René* [1802] to Sand's first *Lélia* [1833]), male protagonists adopt a feminine, apparently powerless, position as a means of self-aggrandizement and, ultimately, re-empowerment.[52] She considers *La Confession d'un enfant du siècle* (along with Sainte-Beuve's *Volupté* [1834] and Balzac's *Le Lys dans la vallée* [1836]) an example of a second wave of *mal du siècle* novels. Whilst some of the features of the early *mal du siècle* novels remain (first-person narration, a weak hero and complex relations with the father), they distinguish themselves by the fact that the heroine, although objectified, resists the protagonist's desire and, consequently, does not partake in his self-glorification.[53] Moreover, although depicted as feminine and sentimental, the early *mal du siècle* male protagonists conserve highly virile traits, as Stacie Allan shows when examining Staël's and Duras's novels. For instance, both Oswald in *Corinne* (1807) and Olivier in the eponymous novel (1822) are military men who distinguished themselves on the battlefield.[54]

What also sets male characters of early nineteenth-century novels apart is that their feebleness is explained. In early *mal du siècle* novels, former soldiers experience difficulties acclimatizing to civilian life and fulfilling their national duties, that is, choose a wife and father children.[55] Incidentally, as was seen previously, this difficult transition is also the case for older soldiers in July Monarchy novels, as if the latter perpetuated the reflection that was addressed ten or twenty years earlier. Moreover, Restoration *mal*

51 François-René de Chateaubriand, *Atala. René. Les aventures du dernier Abencérage*, Jean-Claude Berchet, ed. (Paris: Garnier-Flammarion, 1996), 169 and 177–79; Madame de Duras, *Ourika. Édouard. Olivier ou le Secret*, Marie-Bénédicte Diethelm, ed. (Paris: Gallimard, 'Folio Classique', 2007), 238; Stendhal, *Armance*, Armand Hoog, ed. (Paris: Gallimard, 'Folio Classique', 1975), 50–51, 104 and 133.

52 Waller, *The Male Malady*, esp. Chapter One: 'Rereading the Mal du Siècle', 8–28.

53 Ibid., 178–83.

54 Allan, *Writing the Self*, 93–95 and 98–99. See also Waller, *The Male Malady*, 64–65.

55 Allan, *Writing the Self*, 99–105.

du siècle Romantics use the language of infertility and celibacy to assume *queer* identities threatening the democraphic future of French society, as Andrew Counter has highlighted.[56] Importantly, many male protagonists hide a secret in which their *mal du siècle* takes root. For René, it is his sister's incestuous love for him; for Duras's Olivier and Stendhal's Octave, their sexual impotence.[57] While some July Monarchy plots contain the germ of a secret, the logic of concealment and revelation is not taken as far as in early nineteenth-century novels. For instance, d'Albert's alleged homosexuality can be seen as a secret, but the early confession and the general complexity of the plot and of the narration shift attention away from it: it is ineffable but not secret.

While Chateaubriand, Staël and Constant took great pains to redeem the flaws of their heroes (notably by the means of confession), July Monarchy writers paint an uncompromising portrait of their protagonists. Their original sin, the one from which all their other flaws stem, is their resemblance to women or feminized figures. When Lucien's maid prostitutes herself so that he might gather enough money to pay for his journey back to Angoulême – an example of his indifference to others – the narrator claims that he is similar to 'toutes ces natures à demi féminines' [all those half feminine natures], highlighting his symbolic androgyny.[58] Virtually all Balzac's characters who cross Lucien's path compare him to a woman. Even Daniel d'Arthez, one of his few real friends, calls him a 'femmelette' [sissy] in a letter to Lucien's sister, Ève Séchard, because he likes 'paraître' [to be noticed].[59] The use of 'femmelette' is doubly insulting, as it compares him to a woman and implies weakness as well as a lack of energy.[60] Ève is also aware of the vanity of her brother, since she tells their

56 Andrew Counter, *The Amorous Restoration: Love, Sex, and Politics in Early Nineteenth-Century France* (Oxford: Oxford University Press, 2016), esp. 96.

57 On Restoration impotent protagonists, see Waller, *The Male Malady*; Schehr, *Parts of an Andrology*, 85–87; Gutermann, 'Le désir et l'entrave'. Counter (*The Amorous Restoration*, 142–73) considers that the *Olivier*/*Armance*' novels are *about* and *made of* gossips on sexuality in general.

58 Balzac, *Illusions perdues*, 381. See also 'cet homme à moitié femme' [this man half woman] (Balzac, *Splendeurs et misères des courtisanes*, 137).

59 Balzac, *Illusions perdues*, 406.

60 'Femmelette', in *Dictionnaire de l'Académie française*, I, 741.

mother: 'dans un poète il y a [...] une jolie femme de la pire espèce' [in a poet there is [...] a pretty woman of the worst kind] (more on *jolie* later).[61] Daniel and Ève's lucidity, when analysing Lucien's personality and absence of morals, is shared by Vautrin, who is usually Lucien's most faithful ally, but nevertheless considers the poet to be 'une femme manquée' [a would-be woman].[62] Distorting the French expression *garçon manqué* [tomboy], Vautrin's words suggest failure in Lucien's character. In fact, Lucien's allies imply that his feminized nature is inferior both to that of a man (which goes without saying in the nineteenth-century mindset), but also to that of a woman. This negative perception of an effete character as incomplete recalls that conveyed through some of the protagonists analysed in the previous chapter.

The comparison with women is a pretext to explore Lucien's 'naturally' feminine flaws in depth. His inconsistency and egocentricity are widely recognized. D'Arthez describes Lucien as lazy, easily influenced by Lousteau (who remains unnamed) and oblivious to his friends' sacrifices.[63] Lucien himself, in a rare moment of lucidity which recalls early *mal du siècle* protagonists who relished confessing their faults, points out his flaws in his farewell letter to his sister: 'ma cervelle est intermittente' [my brain power is intermittent].[64] For the narrator of *Splendeurs et misères des courtisanes*, Lucien is one of 'ces hommes faibles et avides, pleins de tendresse dans le cœur et de lâcheté dans le caractère' [those weak and greedy men, full of tenderness in their hearts and cowardice in their characters].[65] As we have seen, Vautrin shares this general opinion when talking to Camusot. The latter paraphrases the convict's speech as follows: 'Le poète est faible; il est femme' [The poet is weak; he is a woman] – and Vautrin is not blinded by his love when identifying the young man's flaws.[66] Interestingly, various characters' contempt for Lucien is contagious, as even scholarly readers

61 Balzac, *Illusions perdues*, 471.

62 Balzac, *Splendeurs et misères des courtisanes*, 677.

63 Balzac, *Illusions perdues*, 407–08.

64 Ibid., 500. On early *mal du siècle* protagonists, see Waller, *The Male Malady*, 52 and 76, for instance.

65 Balzac, *Splendeurs et misères des courtisanes*, 226. See also 137.

66 Ibid., 501.

do not remain objective when discussing him. The English writer A. S. Byatt, for instance, confessed at the beginning of an article that she 'had intensely disliked Lucien', listed many examples of the poet's selfishness and concluded: 'He is not only weak and narcissistic. He is fatally stupid.'[67] Lucien's ultimate sin, it seems, is that his indecisiveness and pusillanimity contribute to his untimely, avoidable death.

Whilst negative opinions are not as vehemently stated in *La Confession d'un enfant du siècle*, it appears that Octave's weakness is presented as a corruption of his former nature, an intermediate step between goodness and malice, as he observes: 'Tu as commencé par être bon; tu deviens faible, et tu seras méchant' [You started with being good; you are becoming weak and you will be wicked].[68] Maxime Foerster argues that Octave's weak masculinity is rooted in his rejection of the traditional norms of heterosexual love based on masculine domination and debauchery.[69] While this analysis is certainly accurate for the Octave of the first part of the novel, it overlooks the fact that the young man ends up embracing the heterosexual script by oppressing his mistress Brigitte. What is clear is that Octave and the other young protagonists have relinquished typical masculine characteristics (activity, strength, reason and willpower) but are not able to replace them with positive alternatives. After a successful start, Lucien cannot distinguish himself professionally, whereas Octave fails to build a stable relationship with Brigitte based on love and trust.

While passivity and weakness are also typical of male protagonists of the early nineteenth century, July Monarchy writers are concerned with inserting their characters and their flaws into a social milieu that is more realistically depicted.[70] Lucien's story, for instance, is deeply inscribed in the social context of his time. Like many young men of his generation, he is ambitious.[71] His ambition mainly relates to his desire for socio-economic

67 A. S. Byatt, 'The Death of Lucien de Rubempré', in Franco Moretti, ed., *The Novel*, II: *Forms and Themes* (Princeton and Oxford: Princeton University Press, 2006), 389–408 (389 and 403).

68 Musset, *La Confession d'un enfant du siècle*, 336.

69 Foerster, *The Politics of Love*, 105–06.

70 See Waller, *The Male Malady*, 180.

71 Balzac, *Illusions perdues*, 24, 82, 117, 306 and 359; Balzac, *Splendeurs et misères des courtisanes*, 99 and 491.

recognition. Indeed, Lucien Chardon recoils from the idea of following the bourgeois path of his father. He is ashamed of his father's common origins and yearns to erase *Chardon* from his birth certificate and bear the name of his aristocratic mother: *de Rubempré*. Later, he submits himself to the convict Vautrin in order to fulfil his dreams of wealth and nobility, acting as a kept man and using his mistresses to satisfy his interests. By denouncing the dangers of ambition in young men, Balzac's novels echo a specific social context that seems to be mostly absent from the early *mal du siècle* novels.

Lucien's obsession with his name should be understood in a socio-historical context that maintained the privileges of the Ancien Régime. In a society in which key positions were held by elders, as vestiges of the Empire or even the Ancien Régime, educated young men could not satisfy their ambitious desire for prestigious careers. Ralph Kingston observes that the July Monarchy administration knew that 'educated young men from good families were living in poverty while languishing as copy clerks'.[72] Scholars have shown that work is a pivotal component in the foundation of masculine identity and, as a result, the lack of employment prospects and insecurity in the workplace are thought to play a significant role in the twentieth-century 'crisis' of masculinity.[73] In a nineteenth-century French bourgeois society that promoted hard work, the lack of good employment prospects was also likely to be emasculating.

Accordingly, in both July Monarchy fiction and society, a career in art proved to be an appropriate alternative to post-Napoleonic young men who were unable to express virility through strength, power and self-control. Gutermann recalls that, in opposition to the bourgeois model, young men often dreamt 'au "devenir artiste", au "génie" et ainsi au "grand homme", dont la figure vient se substituer dans l'imaginaire social à celle du héros de guerre' [of 'becoming an artist', of 'genius' and thus of the 'great man', whose figure is replacing that of the war hero in the social imagination].[74]

72 Ralph Kingston, *Bureaucrats and Bourgeois Society: Office Politics and Individual Credit in France 1789–1848* (Houndmills, Basingstoke and New York: Palgrave Macmillan, 2012), 131.

73 Brittan, *Masculinity and Power*, 25; Badinter, *XY*, 31–32 and 135; Beynon, *Masculinities and Culture*, 14 and 86–88; Edwards, *Cultures of Masculinity*, 8–9.

74 Gutermann, 'Le désir et l'entrave', 63–64.

As Sandra Gilbert and Susan Gubar have pointed out, literature (and allegedly art in general) was perceived as a male activity in the nineteenth century and authorship was judged to be a procedure similar to paternity.[75] Young men's maxim seemed to be: I write (or paint, or sculpt), therefore I am a man. Many male characters in our corpus accordingly choose an artistic career. Lucien, d'Albert and Sténio are poets; Octave is a painter; and Wenceslas Steinbock and Sarrasine are sculptors. This strategy of masculinization is not only fictional; Catherine Nesci has shown how Eugène Delacroix became a *grand homme* thanks to his painting, *La Barque de Dante* (1822), whereas his brothers distinguished themselves by choosing a military career. She concludes:

> Pour Eugène [...], la voie militaire pour atteindre le statut de grand homme se trouva barrée après l'effondrement de l'Empire. L'art et la panthéonisation (du vivant de l'artiste) prirent donc le relais de l'héroïsme martial et de la grandeur politique; l'artiste, pourtant d'une santé précaire, triompha des guerriers.[76]
>
> [For Eugène [...], the military path to reach the status of a great man was blocked after the Empire collapsed. Art and deification (during the artist's lifetime) therefore took over from martial heroism and political greatness; the artist, although sickly, prevailed over the warriors.]

If the medium of art can be as powerful a device for virilization and self-aggrandizement as military glory, as Eugène Delacroix's case shows, it also appears to be a means of expressing uncertainty and anxiety in two different ways. Firstly, artists express their disillusionment through the arts. This is the approach of a first-person narrator, such as Octave in *La Confession d'un enfant du siècle*, whose reflections relate to his inability to find his place in society and his unease in the environment in which he lives. These thoughts may echo those of the writer, and by extension, those of his contemporaries. Secondly, young male artists also express their anxiety through their failure to produce art. Octave introduces himself as a painter but never describes himself as actively practising any of his artistic

75 Gilbert and Gubar, *The Madwoman in the Attic*, 3–11.

76 Catherine Nesci, 'Splendeurs et misères du "grand homme": De la catabase médiévale aux masculinités modernes (Dante, Delacroix, Balzac)', in *Masculinités en révolution*, 135–54 (137).

interests in the story, a characteristic that is shared by d'Albert and, to a lesser extent, by Wenceslas.[77] Barbéris has observed this attitude in several characters of *La Comédie humaine* and has pointed to their 'incapacité matérielle et psychologique à travailler' [material and psychological incapacity to work].[78]

Octave explains his incapacity to concentrate on a specific activity as follows: 'Je savais par cœur une grande quantité de choses, mais rien par ordre, de façon que j'avais la tête à la fois vide et gonflée, comme une éponge' [I knew a great deal of things by heart, but nothing in order, so that my head was at once empty and swollen, like a sponge].[79] Octave's squidgy head recalls Lucien's reference to his 'cervelle intermittente'. Like the feminine Lucien, Octave is easily distracted and changeable. He studied the law and medicine and worked at a bank for a short period of time but lacks the masculine stability that would help him take up the careers that his father has suggested.[80] As Grimsley argues, Octave's inability to choose reveals his immaturity.[81] The young man justifies this incompetence by claiming: 'je serai un homme, mais non une espèce d'homme particulière' [I will be a man, but not a particular kind of man].[82] Foerster points out that, in the following sentence (removed from the 1840 edition), Octave takes off his clothes to assert his lack of professional ambition along with his will to be simply a man.[83] His refusal to define himself as a man belonging to a specific corporation in order to embody a human ideal could appear as a philosophical – hence virile – statement. Yet, it paradoxically hinders his masculinity because, in his case, generality becomes the absence of identity and values.

Young men's attraction to an artistic occupation appears to be partly determined by how failure was perceived in the realm of the arts. Artistic

77 Musset, *La Confession d'un enfant du siècle*, 95. On this characteristic in d'Albert, see Laforgue, *L'éros romantique*, 207–08 and 216–19.

78 Barbéris, *Le monde de Balzac*, 191.

79 Musset, *La Confession d'un enfant du siècle*, 95.

80 Ibid., 92.

81 Grimsley, 'Romantic Emotion', 128.

82 Musset, *La Confession d'un enfant du siècle*, 93.

83 Foerster, *The Politics of Love*, 104.

failure was widely accepted in Romantic culture because it was inscribed in 'the image of the artist in conflict with society, a rebel whose originality is measured by the misunderstanding to which he falls victim, or by the scandal which he provokes'.[84] Artistic unpopularity was not a failure in itself. It was even glorified given that it obliquely indicated success and revealed the artist's uncompromised genius: such flattering depiction allowed him to sublimate his difficult position.[85] This posture also enabled him to delay indefinitely the questioning of his talent, as the answer would be given in an undetermined future, probably after his death, and would not be his concern. In Octave's case, adopting the posture of the misunderstood artist allows him to avoid producing any work of art, whether a masterpiece or a work of little artistic value, and, thus, to avoid failing in the strict sense of the term.

Lucien, who dreams of becoming an 'enfant sublime' [sublime child], as he is ironically called by Sixte du Châtelet and Blondet, seems, at first glance, more committed to the arts.[86] The second part of *Illusions perdues* depicts him as determined to pursue his literary career by following first the arduous path of literature offered by D'Arthez and his Cénacle, then the easier path of journalism suggested by Lousteau and the editor Finot, thanks to which he enjoys immediate (but short-lived) success. For Lucien, art is little more than a rung on the social ladder. As Per Nykrog and Annie Jourdan have both noted, he only commits himself to a literary career in the hope of accessing glory – hence the choice of the fast-tracked path of journalism – and, above all, joining the upper classes.[87] The use of *de*

84 Pierre Bourdieu, 'The Link between Literary and Artistic Struggles', Elinor Dorday, trans., in *Artistic Relations*, 30–39 (31).

85 O'Donovan, 'Avatars of the Artist', 231; Paul Smith, 'Literature and Art', in *French Studies*, 61/1 (January 2007), 1–13 (3).

86 Balzac, *Illusions perdues*, 40 and 250.

87 Per Nykrog, '*Illusions perdues* dans ses grandes lignes: Stratégies et tactiques romanesques', in Françoise Van Rossum-Guyon, ed., *Balzac: Illusions perdues: 'l'œuvre capitale dans l'œuvre'*, *CRIN*, 18 (1988), 34–46 (36); Annie Jourdan, 'Le triomphe de Lucien de Rubempré: Faux poète, vrai dandy', in *Balzac: Illusions perdues*, 62–77 (67).

Rubempré as a penname shows that he seeks the renown and the wealth that his birth has denied him.

Accordingly, throughout *Illusions perdues*, Lucien is a passive poet. His poetic production occurs prior to his encounter with Madame de Bargeton, that is, before the beginning of the story.[88] The novel's third-person narrator and the anonymous author of an article that celebrates the return of Lucien to Angoulême after his failure in Paris share a common irony when mentioning Lucien.[89] Not only do they call him a poet when he has renounced poetry, but they also call him the author or the poet of *Les Marguerites* [*The Daisies*] when his collection of sonnets is yet to be published.[90] The frequency of this denomination increases in the third part of *Illusions perdues* when it becomes certain that *Les Marguerites* will not be published. Its repetition ironically emphasizes Lucien's failure in the literary realm.

Splendeurs et misères des courtisanes confirms the fragility of Lucien's artistic ambition and the limited importance that he gave to his literary career, as he has renounced it by the time he officially becomes *de Rubempré*:

> Lucien [...] abandonna si bien toute pensée de gloire littéraire qu'il fut insensible aux succès de son roman, republié sous son vrai titre de *L'Archer de Charles IX*, et au bruit que fit son recueil de sonnets intitulés *Les Marguerites* vendu par Dauriat en une seule semaine. 'C'est un succès posthume', répondit-il en riant à mademoiselle des Touches qui le complimentait.[91]
>
> [Lucien [...] gave up all thought of literary glory so completely that he was indifferent to the success of his novel, republished under its real title of *Charles IX's Bowman,* and his collection of sonnets entitled *The Daisies* which sold out by Dauriat in only one week. 'It is a posthumous success', he replied with a laugh to Mademoiselle des Touches, who was complimenting him.]

Lucien's detachment vis-à-vis his literary success invalidates the plot of *Illusions perdues* by showing that the poet's main actions (writing and seeking a publisher, for instance), which would serve to affirm his social

88 Jourdan, 'Le triomphe de Lucien de Rubempré', 67.

89 Balzac, *Illusions perdues*, 467–68 and 483.

90 Ibid., 218, 261, 384, 423, 470 and 476; Balzac, *Splendeurs et misères des courtisanes*, 106.

91 Balzac, *Splendeurs et misères des courtisanes*, 113–14.

and masculine standing, are devoid of meaning in the sequel. Worse, Lucien's association with Vautrin is here perceived as suicidal. By portraying himself as dead, the young man implies that Vautrin's rescuing him did not stop his suicide, but only delayed it. The poet Lucien dies at the end of *Illusions perdues* to give way to the passive accomplice of a scam. He ends his life by hanging himself in his prison cell (after considering suicide several times during the plot) because he cannot bear to be dishonoured by his confessions revealing his close relationship with a criminal. Lucien's incentive to kill himself is his pride, which differentiates him from other suicidal Romantic figures of the late eighteenth and early nineteenth centuries, such as Goethe's Werther, Foscolo's Jacopo Ortis and Alfred de Vigny's Chatterton, who are motivated by impossible love or unattainable poetic aspirations.[92] Lucien is ashamed, not of his actions, but of their potential reception by his circle, notably because they impede his social ascension. His obsession with suicide is presented as a continuous failure, since not only can suicide itself be interpreted as a failure, but the repetition of failed suicide attempts illustrates Lucien's inability to commit to any life or death decision.[93]

Failing in their artistic ambitions, Lucien and Octave thus attempt to prove their virility by adopting a customary masculine posture: fighting and, more specifically, duelling, to defend their honour, after they have been offended by a physical or verbal provocation.[94] This practice stemmed from military and aristocratic codes of honour but was progressively adopted by the bourgeoisie in the wake of the revolutionary and imperial wars. Duelling

92 Waller (*The Male Malady*, 15–16) points out that, in contrast, few *mal du siècle* heroes commit suicide.

93 Alex Lascar states: 'l'un des faits les plus marquants, les plus notoires de *La Comédie humaine* est […] le suicide de Rubempré à la Conciergerie' [one of the most striking and best-known facts in *La Comédie humaine* is […] Rubempré's suicide at the Conciergerie prison]: 'Le suicide dans *La Comédie humaine*', *L'Année balzacienne*, 3rd ser., 17 (2016), 129–50 (130). Lucien is not the only man of our corpus to think about suicide. Wenceslas also tries to kill himself and is saved by Lisbeth Fischer. Octave considers committing suicide as well. Sténio also attempts to commit suicide, then kills himself at the end of *Lélia*. For more on young men's suicide in *La Comédie humaine*, see Barbéris, *Le monde de Balzac*, 537–50.

94 Sohn, *'Sois un homme!'*, 122–23.

constituted a civilized framework for violence that differentiated it from fighting with knives or bare hands, which characterized the working classes in the view of members of the ruling classes.[95] Octave prefers duelling over fighting precisely for these reasons: 'je résistai [...] à l'envie que j'avais de le [mon adversaire] frapper ou de l'insulter, ces sortes de violence étant toujours hideuses et inutiles, du moment que la loi permet le combat en règle' [I resisted [...] the temptation to hit or insult my opponent, these kinds of violence being always hideous and pointless, as long as the Law permits lawful fights].[96] Both Lucien and Octave fight a duel with a friend. The former fights with the member of the Cénacle Michel Chrestien, who spat at him because of his cruel article against D'Arthez (in return, Lucien slapped him so that Chrestien would not be the only offender). The latter fights with a friend who played 'footsie' with Octave's mistress. In both cases, the duel is presented as a half-success in affirming masculinity. Octave and Lucien lose the fights and are injured in the process but compensate for their failure by showing unexpectedly brave behaviour. These events are nonetheless isolated and do not eclipse the overall impression of passivity that are conveyed throughout the plots.

In short, the arts in the novels of the 1830s and 1840s, at first glance, help young men to compensate for their 'feminine' weakness, which prevents them from exercising traditional careers in society, and to develop a kind of alternative masculinity to manly and bourgeois identities. However, the artistic path fails to overcome the protagonists' generational weakness. Lucien renounces his faltering artistic ambitions and chooses the easy and illegal path of fraud to satisfy his dreams of social promotion, whereas Octave abandons his artistic posturing to seek refuge in a dysfunctional relationship apparently devoid of links with social and artistic

95 François Guillet, 'Le duel et la défense de l'honneur viril', in *Le triomphe de la virilité*, 83–124 (98–100, 104–05 and 112–14). On the gentrification of the duel, see Nye, *Masculinity and Male Codes of Honor*, 133 and 145. Historians generally agree that duelling was revived during the first half of the nineteenth century: see Nye, *Masculinity and Male Codes of Honor*, 135; Sohn, *'Sois un homme!'*, 122; Guillet, 'Le duel', 96 and 117.

96 Musset, *La Confession d'un enfant du siècle*, 84.

issues. Social disengagement and vain ambition appear to be these young men's only options.

The Beautiful Man

Estranged from the virile and strong model of the Napoleonic soldier because of his weakness, Lucien de Rubempré also diverges from this manly ideal because of his effeminate beauty. While noticeable, Lucien's beauty is by no means exceptional. This type of beautiful man appears elsewhere in the works of Balzac and other writers of the July Monarchy period, notably in the corpus of this study. Calyste du Guénic in *Béatrix*, Wenceslas Steinbock in *La Cousine Bette* and Sténio in *Lélia* share the same epicene, Greek and indolent beauty. However, no representation of physical perfection is as thorough as that of Lucien.

Lucien's beauty corresponds to the eighteenth-century neoclassical aesthetics discussed in the previous chapter that considered beauty to be the privilege of men. However, his exquisite features do not resemble the combination of physical and moral virile characteristics that Winckelmann highlighted in Greek sculptures such as that of Laocoön. On the contrary, Lucien unambiguously embodies the aesthetic model highlighted by Solomon-Godeau and Fend: that of a sensual and passive ephebe who unites feminine traits such as softness and roundness in a male body. In short, Lucien's beauty is fashioned in the same mould as that of Théodore/Madeleine de Maupin. Like Gautier's heroine, the young poet is a dignified impersonator of Hermaphroditus. The only difference with her is that he is not a female cross-dresser, but rather a man who is keen to leave his mark on a men's world: the world of publishing.

However, while referring to a past aesthetic model, *Illusions perdues* and *Splendeurs et misères des courtisanes* are novels rooted in the ideology of the July Monarchy. Indeed, they are grounded in the Romantic assumption that male beauty cannot exist as a positive attribute.[97] Balzac's reference to

97 Schehr, *Parts of an Andrology*, 83.

neoclassical beauty is thus different from Gautier's approach. Male beauty is no longer used to rehabilitate a potentially homosexual character, especially in his own eyes. In fact, it has almost the opposite function: it is used to discredit him as foppish. I therefore examine here how Balzac combines both neoclassical and Romantic perspectives to promote Lucien's beauty through meticulous references to the previous, neoclassical aesthetic tradition and to disparage his attractiveness as anachronistic and potentially weakening.

The young poet's portrait uses Greek mythology abundantly. He is compared to Greek gods and ancient heroes endowed with feminine physical qualities, such as delicacy, indolence and sensuality: Bacchus, Apollo, Antinous and a figure of Girodet, which might be Endymion – *Sarrasine*'s readers must have been familiar with its homoerotic implications.[98] Particularly detailed is Lucien's initial portrait in *Illusions perdues*. It is constructed in opposition to his friend David Séchard, a kind-hearted, virile man, whose physical unattractiveness recalls the mythological figures of Vulcan and Silenus.[99] Here is how Lucien is portrayed:

> Lucien se tenait dans la pose gracieuse trouvée par les sculpteurs pour le Bacchus indien. Son visage avait la distinction des lignes de la beauté antique: c'était un front et un nez grecs, la blancheur veloutée des femmes, des yeux noirs tant ils étaient bleus, des yeux pleins d'amour, et dont le blanc le disputait en fraîcheur à celui d'un enfant. Ces beaux yeux étaient surmontés de sourcils comme tracés par un pinceau chinois et bordés de longs cils châtains. Le long des joues brillait un duvet soyeux dont la couleur s'harmonisait à celle d'une blonde chevelure naturellement bouclée. Une suavité divine respirait dans ses tempes d'un blanc doré. Une incomparable noblesse était empreinte dans son menton court, relevé sans brusquerie. Le sourire des anges tristes errait sur ses lèvres de corail rehaussées par de belles dents. Il avait les mains de l'homme bien né, des mains élégantes, à un signe desquelles les hommes devaient obéir et que les femmes aiment à baiser. Lucien était mince et de taille moyenne. À voir ses pieds, un homme aurait été d'autant plus tenté de le prendre pour une jeune fille déguisée, que, semblable à la plupart des hommes fins, pour ne pas dire astucieux, il avait les hanches conformées comme celles d'une femme.[100]

98 Balzac, *Illusions perdues*, 23, 124 and 227. On Endymion, see Nesci, 'Splendeurs et misères du "grand homme"', 152–53.

99 Balzac, *Illusions perdues*, 23.

100 Ibid., 23–24.

> [Lucien was standing in the graceful pose that sculptors found for the Indian Bacchus. His face had the distinction of lines of ancient beauty; he had a Greek forehead and nose, the velvet whiteness of women, eyes so blue that they looked black, eyes so full of love and whose whiteness was as fresh as that of a child. Those beautiful eyes were topped by eyebrows that looked like they had been traced with a Chinese paintbrush and edged with long chestnut eyelashes. Along the cheeks shone a silky fuzz whose colour harmoniously matched that of his naturally curly blond hair. A divine suaveness exuded from his golden white temples. An unparalleled nobleness was stamped on his short chin, raised without abruptness. The smile of sad angels wandered on his coral-like lips, whose colour was enhanced by beautiful teeth. He had the hands of a high-born man, elegant hands, which men must obey and that women like to kiss. Lucien was slim and of average height. When seeing his feet, a man would have been even more tempted to take him for a girl in disguise since, like most astute men, not to say shrewd men, he had hips shaped like those of a woman.]

Lucien's portrait primarily aims to transform him, like hermaphroditic characters, into a masterpiece. A resemblance to sculptures and paintings is underlined and developed through an abundant network of substantives and adjectives denoting artistic perfection, for instance in the choice of colours characterizing his features (marble, white and gold), or in the reference to a paintbrush. Reviving the neoclassical tradition of beautiful ephebes, Lucien's portrait patently refers to Greek sculpture and emphasizes his grace and harmony.

While nonetheless disguised, there is an erotic dimension to this portrayal. The comparison with the god of wine Dionysus-Bacchus, for instance, is ambivalent. As Marie Delcourt has claimed, Dionysus is not only an effeminate ephebe in the Hellenistic culture, but also an androgynous god associated with cross-dressing rites in more ancient traditions.[101] He is therefore particularly suitable for describing Lucien's epicene beauty and his symbolic transvestism (see 'un homme aurait été d'autant plus tenté de le prendre pour une jeune fille déguisée', which recalls Madeleine de Maupin). The reference to Bacchus also suggests a connection with another effeminate and homoerotic Greek ephebe: Endymion, who, in Girodet's painting, lies on a panther skin, a Dionysian symbol.[102] This connection

101 Delcourt, *Hermaphrodite*, 39–42.
102 Fend, *Les limites de la masculinité*, 97.

is also indicated by Lucien's golden and curly hair, similar to that of the young man painted by Girodet. Incidentally, curls (which also appear in the portrayals of homosexual convicts, as well as those of Théodore and d'Albert) recall the principle of the undulating line defining feminine and androgynous bodies, as examined in the previous chapter.

Moreover, Lucien's effeminacy is conveyed in his portrait through the qualities of smoothness, suaveness and slimness, as well as through the description of his facial features. His femininity becomes more explicit towards the end of the portrait, as the prose draws the reader towards his girlish feet and womanly hips. A wide pelvis was a feature of the statues of Bacchus and Apollo, embodiments of ideal juvenile beauty according to Winckelmann. The art historian described 'the most beautiful statues' of the god of wine as follows: 'always with delicate, round limbs, and the full expanded hips of the female sex'.[103] This wide pelvis was also a noticeable trait of homosexuals and eunuchs according to Cabanis.[104]

Finally, the portrait emphasizes Lucien's nobility. His aristocratic origin is physically visible in his hands and his chin, but it is also reflected in his attitude and his 'natural' capacity to be obeyed. In his physiognomy at least, Lucien has been able to achieve his dream of social promotion that French society denies him. However, while contributing to the physical aggrandizement of the character, referring to Lucien's noble – maternal – ascendance is a convoluted way of asserting once again his femininity. Aristocracy had been perceived as effeminate since the eighteenth century.[105] In brief, the network of references to describe Lucien's appearance certainly highlights his physical perfection, which is comparable to artworks and ancient gods. His feminine nature, however, conveyed directly by comparing him to a woman, or indirectly by references to effete figures and aristocracy, undermines the positive connotation of beauty and questions Lucien's manliness. This early portrait suggests that Lucien is no match for the virile Parisian society.

103 Winckelmann, *History of the Ancient Art*, 93. See also Fend, *Les limites de la masculinité*, 50–51.

104 Cabanis, *Rapports*, I, 388–89.

105 See notably Rauch, *Crise de l'identité masculine*, 28–30; Clark, 'The Rhetoric of Masculine Citizenship', 12.

This long description of Lucien's physical perfection is apparently not enough; his attractiveness needs to be stated repeatedly. The poet's beauty is mentioned countless times throughout *Illusions perdues* and *Splendeurs et misères des courtisanes*. As Franco Moretti argues, with reference to *Illusions perdues*, beauty is 'simply affirmed and reiterated'.[106] Lucien's beauty is repeatedly brought up and commented on by the omniscient narrator, as well as by Lucien's friends and enemies alike. Strikingly, these new references bring no new information to the initial portrait and even appear gratuitous. For instance, the narrator refers to Lucien's beauty when recounting a discussion about an editorial disagreement: 'Dauriat [plongea] un regard assassin sur *le beau Lucien* qui le regarda d'un air agréable' [Dauriat looked daggers at *the beautiful Lucien* who was considering him pleasantly].[107] The aesthetic allusion seems irrelevant in this context. Yet, it catalyses other characters' reactions towards, and perceptions of, him. In this example, all Dauriat's animosity appears to be concentrated on the poet's beauty, whereas it is in fact directed against his naivety, provincialism and excessive ambition. By focusing a great deal on Lucien's physical and feminine attractivity, *Illusions perdues* marginalizes the more masculine facets of his character, such as his wit.[108] Lucien is reduced to his beauty to such an extent that his few qualities seem to have vanished in *Splendeurs et misères des courtisanes* to yield ground to a beautiful puppet in the hands of Vautrin.

Not only do Lucien's ephebic looks conceal his assets, they also distort the principle of the Swiss physiognomist Johann Kaspar Lavater. He claimed that physiognomy reflected the connection between the soul and

106 Franco Moretti specifies that 'beauty is never described', because he considers that beauty is apprehended only through comparisons and metaphors: 'Homo Palpitans: Balzac's Novels and Urban Personality', in *Signs Taken for Wonders: Essays in the Sociology of Literary Forms*, Susan Fischer, David Forgacs and David Miller, trans. (London and New York: Verso, 1988 [1983]), 109–29 (113). See also Barthes's claim (*S/Z*, 40): 'La beauté (contrairement à la laideur) ne peut vraiment s'expliquer: elle se dit, s'affirme, se répète en chaque partie du corps mais ne se décrit pas' [Beauty (contrary to ugliness) cannot really be explained: it declares itself, affirms itself and repeats itself in each part of the body, but cannot be described].

107 Balzac, *Illusions perdues*, 219 (my emphasis).

108 Ibid., 255, 284 and 327.

the body. A beautiful man was thus a moral man.[109] *Illusions perdues* does not give in to this theory. The physical opposition between Lucien and his friend David Séchard is coupled with a moral opposition. Despite his unattractiveness, David appears to be a good man: intelligent, hard-working and a devoted husband, father, son and friend – maybe too devoted given that he is swindled by his father and unintentionally ruined by Lucien. Indeed, what Balzac's novel suggests is that Lucien's effeminate external appearance is not condemned for itself, but rather because it mirrors his inner nature. His feminine beauty discredits him as a manly, powerful and virtuous character, whilst supporting his feminine vanity, inconstancy and foolishness.[110]

Lucien's beauty is even regarded as deceptive by numerous characters. The dangers of beauty in young men are best expressed by the secret agent Peyrade to his daughter Lydie:

> La beauté chez les hommes n'est pas toujours le signe de la bonté. Les jeunes gens doués d'un extérieur agréable ne rencontrent aucune difficulté au début de la vie, ils ne déploient alors aucun talent, ils sont corrompus par les avances que leur fait le monde, et il faut leur payer plus tard les intérêts de leurs qualités![111]
>
> [Beauty in men is not always a sign of goodness. Young men endowed with good looks encounter no difficulty at the beginning of their lives, so they exert no talent, they are corrupted by the advances that society makes to them, and they must later pay interest on their qualities!]

Peyrade is wary of Lucien – and rightly so, since Lydie, who is attracted to the young man, will be abducted and raped on Vautrin's orders. Balzac's novels both confirm and challenge Peyrade's opinion. Lucien's physical perfection means that he benefits from substantial love and support from female and male characters alike. He enjoys the indulgence of his mother, sister and friend when he makes mistakes, the patronage of his mistress Madame de Bargeton, the help of his lovers Coralie and Esther to earn a living, the artistic assistance of his friends from the Cénacle and the

109 Mosse, *The Image of Man*, 25–27.

110 Balzac, *Illusions perdues*, 83–84 and 357.

111 Balzac, *Splendeurs et misères des courtisanes*, 188.

protection of Vautrin. This contributes to his idleness, since he prefers journalism over literature, arranged marriage over real love and prostitution (of his girlfriend and arguably of himself) over hard work. As the aid that is bestowed on him does not last – his relatives become increasingly aware of his shortcomings, his mistress abandons him and his girlfriends die – Lucien does not succeed in his endeavours. This experience of failure is even more difficult for him because, following Peyrade's belief, as an attractive man, he had not been prepared for it.

Lucien's beauty is perhaps the most striking of the protagonist's characteristics, and it overshadows other aspects of his personality. It is grounded in neoclassical aesthetics, which promoted male and yet effeminate beauty, embodied in mythological figures such as Apollo, Bacchus and Endymion. Accordingly, the young poet's beauty makes him an anachronistic and thus maladjusted character during the nineteenth century. Hardly a quality, Lucien's feminine beauty reflects his effete inner nature, causes his masculinity to be questioned and highlights his inability to act honourably.

The Homosexual Protégé

If a neoclassical beauty such as Lucien's would not have raised suspicion in the eighteenth century, the same does not apply in Romantic times. Whereas neoclassical paintings were not interpreted as homoerotic in their day, Fend indicates that art critics increasingly felt uneasy towards the end of the Empire and the beginning of the Restoration when confronted with the effeminacy and sensuality of Girodet's *Endymion*, whose homoerotic elements became perceptible.[112] Potts also argues that the Apollo Belvedere, who used to be regarded as an embodiment of ideal male beauty, became 'a slightly outdated image' in the early nineteenth century.[113] There is thus no reason to believe that beauty such as Lucien's was not interpreted as homoerotic by nineteenth-century readers.

112 Fend, *Les limites de la masculinité*, 141–49.

113 Potts, *Flesh and the Ideal*, 125.

Balzac's novels certainly create ambiguity regarding the relations between Lucien and his mentor, but they also develop a critical discourse based on the stereotypes traditionally associated with homosexuality in order to subvert them. As Lawrence Schehr has argued: 'Dans un monde ostensiblement hétérosexuel, Balzac rend les positions d'adresse, de sexualité, et de genre tout à fait *queer*' [In a world that is conspicuously heterosexual, Balzac makes the positions of speech, sexuality and gender completely *queer*].[114] In using irony, Balzac demonstrates that sexual categories are heteronormative constructions that need to be deconstructed in order to give a full account of *La Comédie humaine*'s Parisian society.[115] In this section, I explore how the Balzacian narrative system deconstructs homosexuality in *Illusions perdues* and *Splendeurs et misères des courtisanes*.

When Vautrin meets Lucien, who is willing to commit suicide, at the end of *Illusions perdues*, he offers him a pact: 'si tu veux signer le pacte, me donner une seule preuve d'obéissance, elle est grande, je la veux! eh! bien la diligence de Bordeaux portera quinze mille francs à ta sœur…' [If you want to sign the pact and give me a single piece of proof of obedience – a great one, I want it! – well! the Bordeaux stagecoach will carry fifteen thousand francs to your sister…].[116] This proof of obedience, in exchange for protection and wealth, has raised considerable controversy among readers when discussing Vautrin and Lucien's sexuality. For some critics, there is no doubt that Lucien is homosexual and that the pact is a strictly sexual one, whereas the homoerotic nature of Lucien and Vautrin's relationship is highly dubious for others.[117] Michael Lucey considers that those who

114 Lawrence R. Schehr, 'Homme-diégèse', in Robert Mahieu and Franc Schuerewegen, eds, *Balzac ou la tentation de l'impossible* (Paris: SEDES, 1998), 135–41 (141; emphasis in original).

115 Ibid., 138.

116 Balzac, *Illusions perdues*, 520. For more on the pact in *Illusions perdues*, see Mireille Labouret, 'Méphistophélès et l'Androgyne: Les figures du pacte dans *Illusions perdues*', *L'Année balzacienne*, 2nd ser., 17 (1996), 211–30; André Vanoncini, 'Le pacte: Structures et évolutions d'un motif balzacien', *L'Année balzacienne*, 3rd ser., 3 (2002), 279–92 (284); Chamard-Bergeron, 'Vous croyez aux amis', 305–11.

117 The debate is summarized by Philippe Berthier, 'Balzac du côté de Sodome', *L'Année balzacienne*, 1st ser. (1979), 147–77 (153–54); Michael Lucey, *The Misfit of the Family: Balzac and the Social Forms of Sexuality* (Durham, NC, and

openly mention Lucien's homosexuality are too adamant and he rightly highlights that homosexuality is omnipresent *around* Lucien without ever designating him directly.[118] His analysis nonetheless overlooks the strategy of avoidance that is adopted by Balzac who deliberately chooses to be vague about this relationship.

The interest in Lucien and Vautrin's relationship does not lie in the knowledge of its exact nature (in other terms, whether or not they have a sexual relationship), but rather in the self-censorship adopted by the narrator to discuss their bond. This in turn leads to discrediting Lucien as a young man with no willpower. This narrative censorship shares similarities with that employed by nineteenth-century scientists in their discourses about homosexuality. The resemblances between scientific and literary texts are based on the construction of homosexuality as a sign of social non-conformity, or even abnormality. However, Balzac, I suggest, does not simply borrow the rhetorical tools of the scientific community to condemn homosexuality *per se*; rather, he reassesses the conduct that is commonly linked to homosexuality. Furthermore, he thoroughly reflects on the ability of language to influence our perception of interpersonal relationships.

Chapter 1 has shown that male homosexuality was associated with effeminacy in scientific writings. This was particularly clear in Ambroise Tardieu's excerpt in which he described homosexuals' alleged obsession with clothes, jewels, perfumes and make-up. This representation of homosexuals echoes not only the emphasis on feminine beauty in Lucien de Rubempré's portrait, but also d'Albert's inclination for fashion and the numerous hours that he dedicates to his *toilette*, like the men described by Hubert Lauvergne and Étienne Esquirol. D'Albert's coquetry in *Mademoiselle de Maupin* is a striking example of this phenomenon:

London: Duke University Press, 2003), 186–88. Whilst acknowledging Vautrin's homosexuality, Barbéris rejects the homoerotic nature of Lucien and Vautrin's couple (*Le monde de Balzac*, 431–37). Likewise, D. A. Miller refers to 'Vautrin's Platonized homosexuality': 'Balzac's Illusions Lost and Found', *Yale French Studies*, 67 (1984), 164–81 (175).

118 Lucey, *The Misfit of the Family*, 186.

> – Il m'est revenu que beaucoup d'entre eux avaient amèrement critiqué ma façon de me mettre, et avaient dit que je m'habillais d'une manière trop efféminée: que mes cheveux étaient bouclés et lustrés avec plus de soin qu'il ne convenait; que cela, joint à ma figure imberbe, me donnait un air damoiseau on ne peut plus ridicule; que j'affectais pour mes vêtements des étoffes riches et brillantes qui sentaient leur théâtre, et que je ressemblais plus à un comédien qu'à un homme: – toutes les banalités qu'on dit pour se donner le droit d'être sale et de porter des habits pauvres et mal coupés.[119]
>
> [I have heard that many of them had bitterly criticized my attire and had said that I dressed too effeminately: that my hair was curled and pomaded more carefully than it should have been; that all this, combined with my beardless face, gave me the most ridiculous dandy looks; that I dressed loudly with rich, shiny fabrics that stank of theatre and that I looked more like a comedian than a man: – all the trite remarks that people say to give themselves the right to be unkempt and wear poor and ill-fitting clothes.]

Notably, one of the ways in which d'Albert's physical femininity is revealed is through the fact that he is beardless. A beard and a moustache were generally considered to be visible signs of manhood and their absence could suggest homosexuality.[120] His effeminacy is also presented as a choice, given that he deliberately accentuates his feminine looks by dressing like a woman, according to his rivals. Like the homosexuals described in scientific writings, d'Albert wears colourful fabrics, shiny accessories and curls his hair. Interestingly, his feminine appearance is deemed to be that of an actor, which recalls the principle of performativity already analysed in the previous chapter. Artificiality is even more important to him since, unlike Lucien, d'Albert is unattractive.[121] Faking effete beauty through cosmetic and sartorial tricks may be his way of acquiring some of Lucien's sexual power. In contrast, virility is seen here to be devoid of affectation, excusing both unclean clothes and poor taste in attire.

As scientific treatises and this extract from *Mademoiselle de Maupin* show, overt effeminacy was commonly linked to homosexuality or at least homoerotic behaviour during the early nineteenth century. *Illusions perdues*

119 Gautier, *Mademoiselle de Maupin*, 115.

120 Sohn, *'Sois un homme!'*, 27–28; Corbin, 'Introduction', 8; Revenin, 'Homosexualité et virilité', 398.

121 Gautier, *Mademoiselle de Maupin*, 224.

offers even more striking resemblances with the scientific texts examined. In the novel, d'Arthez writes to Ève Séchard to dismiss her concerns about Lucien's possible criminality: 'Rassurez-vous, Lucien n'ira jamais jusqu'au crime, il n'en aurait pas la force' [Do not worry, Lucien will never go as far as to commit a crime, he would not have the strength], whereas the physician of the *bagne* of Toulon Lauvergne affirms in *Les Forçats* that effete convicts 'ne tueront pas comme assassins, puisqu'une femme n'en a pas la force physique!' [will never kill like murderers, since a woman does not have the physical strength to do so!].[122] Three points must be made about these two comments. Firstly, the common intertextual thread linking Balzac's and Lauvergne's texts is the impossibility for effete men to commit murder. Such an act would require physical strength, which is not supposed to exist in either women or men sharing characteristics with women. The resemblance between the two expressions above points to the common belief that effeminate convicts, women, feebleness and the inability to kill could be grouped under the same banner. Secondly, not being a murderer in the making appears to be a redeeming quality in neither case. The lasting impression is that both Balzac's novel and Lauvergne's treatise blame their character or subject of study for not being virile enough to commit an action that is highly reprehensible on legal and moral grounds. Both texts go so far in their criticism of physical and *de facto* social weakness that they end up promoting criminal acts. This is essential to understand how the criminal Vautrin benefits from such a prominent status in Balzac's novel. Finally, the similarity not only between ideas but also in the wording suggests that one text may have influenced the other. *Les Forçats* and the third part of *Illusions perdues* were both published at the beginning of the 1840s (1841 and 1843 respectively). Although no evidence has been found that Balzac read Lauvergne, the possibility cannot be excluded.[123] It remains true that the two texts refer to analogous ideas regarding effeminacy and potential homosexuality, in all likelihood shared by many contemporaries.

122 Balzac, *Illusions perdues*, 406–07; Lauvergne, *Les Forçats*, 299.

123 Berthier ('Balzac du côté de Sodome', 148–49) mentions only Vidocq and Benjamin Appert as Balzac's sources about penitentiary homosexuality.

Balzac's use of scientists' strategy of *avoidance-discourse* goes even further than the scientific treatises by not referring to male homosexuality where such a reference would be expected. Balzac's in-depth description of the galleries of the Palais-Royal emphasizes the fact that these galleries are a place for female prostitution.[124] However, the description does not mention, or even allude to, the fact that they had also been an important space for male prostitution and homosexual subculture since the end of the eighteenth century.[125] Although this absence of homosexual reference hardly constitutes incontestable evidence of Balzac's reluctance to discuss homosexuality, such omission is nonetheless surprising. It is likely that Balzac knew about the existence of male prostitutes since their presence raised many complaints from shopkeepers and locals at the time.[126] Its omission only suggests that self-censorship was practised by Balzac where issues regarding homosexuality were concerned.

Another strategy used by the narrator of *Splendeurs et misères des courtisanes* bears resemblance to the discourse of scientists, but with different effects. Some physicians chose to write the most explicit extracts about homosexuality in Latin (notably those describing anal penetration).[127] This strategy, which Nigel Smith terms *disguise-discourse*, intends 'to protect the sensibility and the morality of readers, [...] to flaunt knowledge of the secret and thus assert power'.[128] Balzac's narrator also chooses a language

124 Balzac, *Illusions perdues*, 209–14. See also Balzac, *Splendeurs et misères des courtisanes*, 59–60.

125 Sibalis, 'The Palais-Royal', 117–29 (esp. 121–23 for the July Monarchy period); Peniston, *Pederasts and Others*, 131; Murat, *La loi du genre*, 41 and 49.

126 Sibalis, 'The Palais-Royal', 121–22.

127 Aron and Kempf (*Le pénis et la démoralisation de l'Occident*, 57), as well as Smith ('Silence', 88), give the example of a Dr Moll using Latin in his texts. *Disguise-discourse* is not limited to anal intercourse. When having to describe what happens to a woman's body during sexual intercourse with a man, Moreau de la Sarthe (*Histoire naturelle*, II, 162) quotes a text by Haller called 'Quae feminis in coitus accident' ['The things which may happen to women during intercourse'] and written in Latin, 'dans une langue où la décence ne s'oppose point à la liberté et à l'exactitude de l'expression' [in a language where decency is not opposed to freedom and accuracy of expression]. See also Rosario, *The Erotic imagination*, 95–96.

128 Smith, 'Silence', 87–88.

that is not understood by everyone: prison slang. In this case, the choice is justified by the fact that some characters use this language. The convicts, recognizing in Carlos Herrera their fellow criminal Jacques Collin, alias Vautrin, ask him whether he came to the Conciergerie to free his *tante*, Théodore Calvi, who has been sentenced to death. In order to explain the meaning of *tante*, the narrator tells an allegedly true story:

> Pour donner une vague idée du personnage que les reclus, les argousins et les surveillants appellent *une tante*, il suffira de rapporter ce mot magnifique du directeur d'une des maisons centrales au feu lord Durham, qui visita toutes les prisons pendant son séjour à Paris.
>
> […]
>
> 'Je ne mène pas là Votre Seigneurie, dit-il, car c'est le quartier des *tantes* …
>
> – *Hao!* fit lord Durham, et qu'est-ce?
> – C'est le troisième sexe, milord.'[129]

> [To give a vague idea of the character that the recluses, the coppers and the guards call *an aunt*, it is enough to recall these magnificent words by the warden of one of the central prisons to the late Lord Durham, who visited all the prisons during his stay in Paris.
>
> […]
>
> 'I am not taking Your Lordship there, he said, because it is the *aunts*' quarters …
>
> – *Hao!* Lord Durham said, and what is that?
> – It is the third sex, my lord.']

This anecdote explains the significance of the slang word *tante*, a male homosexual prisoner, by using a phrase as enigmatic as the original expression. The 'troisième sexe' [third sex] entry in Courouve's dictionary of male homosexuality shows that this expression was not common in Balzac's time.[130] Like the Latin texts, the Balzacian narrator hides explicit sexual reference. However, this concealment is the reason why the anecdote arouses the curiosity of the readers.[131] Those who had also read

129 Balzac, *Splendeurs et misères des courtisanes*, 599 (emphasis in original).
130 Courouve, 'Troisième sexe', in *Vocabulaire de l'homosexualité masculine*, 212–16.
131 Smith, 'Silence', 88.

Mademoiselle de Maupin would nevertheless have been acquainted with the sexually ambiguous connotations of 'troisième sexe' and would therefore join the happy few who were familiar with, or at least able to decipher, non-heteronormative references. Andrew Counter points out a similar phenomenon regarding (hetero)sexual encounters in his study of French Restoration novels. He analyses how ellipses in these contexts create a sense of complicity between the reader and the text. His conclusion is well suited to Balzac's novels: ellipses 'shape and create political communities based on a mutual sense of unspoken knowledge and interpretative competence'.[132]

We must take a closer look at Balzac's relation to his readers. Readership was mainly female or feminized during the July Monarchy, or at least perceived as such.[133] Schehr, however, identifies the implied reader of *Le Père Goriot* – the first novel in which Vautrin appears – as male and heterosexual.[134] Similarly, Judith Lyon-Caen's analysis of the letters sent to Balzac by readers reveals that his readership was not exclusively female: it included readers of both sexes and from different social milieux.[135] How to explain these apparent contradictions concerning real and assumed readership? It appears that Balzac's – implicitly male and heterosexual – narrator is addressing a male narratee, but that this narratee is feminized and thus submitted, through the narrator, to Balzac's will.

Accordingly, Balzac's strategy of self-censorship does not simply aim to protect his readers' sensibility and morality, but also assert his power and confine the convict to the exotic role of social and sexual pervert by using slang.[136] More importantly, Balzac fosters a relation of superiority and seduction with the readers of his novels. He creates an erotic auctorial

132 Counter, *The Amorous Restoration*, 18.

133 Lise Queffelec, 'Le lecteur du roman comme lectrice: Stratégies romanesques et stratégies critiques sous la Monarchie de Juillet', *Romantisme*, 53 (1986), 9–21 (10–11 and 17). See also Lyon-Caen, *La lecture et la vie*, 90.

134 Schehr, 'L'homme-diégèse', 138.

135 Lyon-Caen, *La lecture et la vie*, esp. 136–37.

136 On the latter function, see Berthier, 'Balzac du côté de Sodome', 149.

bond, and even a homosocial one, with his potential male readers.[137] The main purpose of this bond, I have argued, is to generate collusion with those able to grasp the meaning of the story about Lord Durham and the *tantes*. This conniving and seductive bond between two men, one coded as powerful and the other as submissive, recalls a fictional relationship: that between Vautrin and his protégé Lucien.

Neither openly blamed nor praised, Vautrin and Lucien's bond generates paradoxical discourses throughout *Splendeurs et misères des courtisanes*. On the one hand, the Balzacian narrator adheres to the morals of his time by avoiding being explicit about the nature of the relationship between Lucien and Vautrin.[138] For example, he uses the word *ami* [friend] rather than *amant* [lover] to 'translate' the slang word *tante*.[139] On the other hand, in order to deconstruct (hetero)sexual categories, the novel reveals the hypocrisy of French society concerning homoerotic relations by depicting the censorship that is practised. The author has some of his protagonists employ his strategy of *avoidance-discourse*. They are disinclined to designate explicitly the nature of the links that may unite Vautrin and Lucien. Twice in conversation some characters are about to let slip a compromising word but choose another expression, one that is more socially acceptable, to describe the relations between the two men.

137 Rivers ('*Inintelligibles pour une femme honnête*', 18–19) highlights the importance of a homosocial bond between a male author and a male reader for *Mademoiselle de Maupin*.

138 In his introduction to the volume of *Romantisme* dedicated to 'Sodome et Gomorrhe', Bordas ('Introduction', 8) points to the practice of allusion in the discourse about homosexuality. Berthier ('Balzac du côté de Sodome', 153) mentions the fact that Vautrin loves men 'dans les trous du discours'[in the gaps of the discourse]; Schehr ('L'homme-diégèse', 136) highlights 'l'impossibilité ontologique de dire l'homosexualité *a priori*' [the ontological impossibility of talking about homosexuality *a priori*]; Lucey (*The Misfit of the Family*, 220) refers to 'the procedures of semi-censorship, euphemism, and innuendo' used to discuss same-sex relationships; Diana Knight uses Eve Kosofsky Sedgwick's *Epistemology of the Closet* to refer to the notion of secrecy that is intrinsic to discourses about homosexuality: 'Skeletons in the Closet: Homosocial Secrets in Balzac's *La Comédie Humaine*', *French Studies*, 57/2 (April 2003), 167–80 (esp. 172).

139 Balzac, *Splendeurs et misères des courtisanes*, 601 and 606.

Firstly, Madame Camusot, the wife of the investigating magistrate, visits Lucien's ex-mistress, Diane de Maufrigneuse, in order to ensure her husband's promotion. She reveals to the duchess that it is likely that Vautrin 'a mis en lieu sûr les lettres les plus compromettantes des maîtresses de son … – Son ami, dit vivement la duchesse' [has put away in a safe place the most compromising letters from the mistresses of his … – His friend, the Duchess said quickly].[140] Madame Camusot, out of decency, hesitates to qualify the bond between the poet and the convict in front of the duchess. Diane de Maufrigneuse is, however, able to understand what her interlocutor has implied, as the sharpness of her reply indicates. It is inadmissible for her to hear a friend of hers (and above all an ex-lover) being called the lover of a convict.

Secondly, very soon after the meeting between the two women, Camusot and Monsieur de Grandville, the general Prosecutor, discuss Vautrin's case. Camusot explains: 'Il [Vautrin] est en ce moment auprès de votre condamné à mort [Théodore Calvi], qui fut jadis au bagne pour lui ce que Lucien était à Paris … son protégé!' [Right now, Vautrin is with your condemned man [Théodore Calvi], who used to be at the *bagne* what Lucien was in Paris … his protégé!].[141] The magistrate practises self-censorship in order not to offend Grandville by insinuating that an aristocrat has the same morals as a convict. In contrast, the relationship between Vautrin and Calvi is made relatively explicit in the text, as it is designated as 'l'intimité de Jacques Collin avec Théodore Calvi' [intimacy between Jacques Collin and Théodore Calvi], probably because the latter belongs to the underworld.[142] The fact remains that, in both cases, speakers seem

140 Ibid., 650–51.

141 Ibid., 669. There are numerous other examples of euphemism in the novel that characterize Lucien and Vautrin's relationship: Lucien is 'le compagnon intime d'un forçat évadé' [the intimate companion of an escaped convict] for the narrator (427), 'l'ami, l'élève d'un forçat évadé' [the friend, the disciple of an escaped convict] for Prosecutor Grandville (521); Vautrin is 'l'ami intime, le conseiller de Monsieur Lucien de Rubempré' [the intimate friend, the adviser of Monsieur Lucien de Rubempré] for the police (440) and Madame Camusot mentions the 'mariage de ce forçat avec Lucien de Rubempré' [marriage of this convict with Lucien de Rubempré] (554). See also the warden's ironical answer to the doctor's observation about Vautrin's paternity (569–70).

142 Ibid., 671.

to renounce designating the relationship between Lucien and Vautrin as homoerotic. In having his characters adopt this strategy, Balzac certainly imitates contemporary society. He also suggests that, whilst characters are eager to consider Vautrin and Lucien (as well as Calvi) as lovers, homosexuality is voluntarily overlooked, because it is deemed to be a threat to the new bourgeois social order.[143]

Regarding this discursive strategy, Lucey states: 'Given that everyone in the justice system knows this word [*tante*], what it means, and how to use it, it is quite interesting to note how close people come to applying it to Lucien, without ever quite daring to do so.'[144] He shows that the text carefully stresses the family tie between Vautrin and his aunt Jacqueline Collin in order to suggest by contrast that the word *tante* does not always refer to a relative but may also have *queer* connotations.[145] Lucey, however, probably overestimates the importance of this word: certainly, Grandville, Camusot and even his wife, who is involved in the judge's work out of self-interest, would be familiar with prison slang, but it is doubtful that Madame de Maufrigneuse, who comes from the high aristocracy, would have heard of such a word.

Lucey's otherwise remarkable analysis of sexual relationships in Balzac also neglects the context of Balzac's writing, namely that the avoidance strategy that Balzac gives to his characters was in fact typical of the manner in which homosexuality was discussed, notably in scientific discourse, during the first half of the nineteenth century. Moreover, the word *protégé* itself, used by Camusot, is not devoid of ambiguity. In *La Cousine Bette*, Crevel, who courts Adeline Hulot, asks her: 'savez-vous que votre monstre d'homme a *protégé* Jenny Cadine, à l'âge de treize ans?' [do you know that your monster of a husband *protected* Jenny Cadine at the age of thirteen?].[146] The emphasis on *protégé* implies that Hulot has helped Jenny to launch her acting career in exchange for her favours. *Protégé*, in the world

143 See Lucey's (*The Misfit of the Family*, 209–18 and 220–22) analysis of the collusion between Vautrin and Prosecutor Grandville, as an embodiment of aristocracy, against the new bourgeois legal bureaucracy incarnated by Camusot.

144 Ibid., 219.

145 Ibid, 218–19.

146 Balzac, *La Cousine Bette*, 34 (emphasis in original).

of *La Comédie humaine*, is thus a socially acceptable synonym for *lover*, whose homoerotic connotations are comprehensible, even for upper-class interlocutors, when talking about two men. The hypocrisy of *La Comédie humaine*'s upper-class society consists in tolerating questionable relationships, whether adulterous or paedophile, as long as they are heterosexual, whereas homosexual relationships, even when they are consensual and legal (or at least not illegal), are forcefully denied by erasing any word that could explicitly designate them.

The text thus shifts back and forth between two universes and their own jargons. In the underworld, relationships between two men are acceptable and designated accordingly, but this universe and its language are probably unknown to the readers who can only establish their understanding on the clues given by the text. Conversely, in the upper-class world, these relations are officially unacceptable, and, consequently, this universe has no appropriate vocabulary at its disposal. Balzac's text, nonetheless, shows that vocabulary from the upper classes is not guileless. Indeed, its narrator states: '[l]e grand monde a son argot' [high society has its own slang].[147] Through understatement and comparison with heterosexual relationships, this aristocratic vocabulary manages to develop a level of explicitness that does not quite match the penitentiary vocabulary, but that is still comprehensible for most protagonists. Beyond rejecting the physical bond between Lucien and Vautrin, the aristocracy also reject, through their reluctance to use explicit vocabulary, the symbolic connotations of homosexuality, notably the 'flaws' with which it is associated. After all, it is Lucien, rather than Vautrin, who is mainly targeted. Accordingly, the criticism shifts from same-sex relationships *per se* to the person who allegedly practises them or, from a Foucauldian perspective, from sodomitic action to effete identity. Lucien appears to be blamed less for his intimate association with a criminal than for the incentives behind this collusion: his social ascension to compensate for his 'natural' and social weakness.

What the novel demonstrates by alluding to homosexuality is that someone as insecure and emotionally unstable as Lucien (as he is torn between Esther, Clotilde de Grandlieu, Madame de Maufrigneuse, Madame

147 Balzac, *Splendeurs et misères des courtisanes*, 657.

de Sérizy and Vautrin) is doomed to failure. He is unable to be a substitute for the Law since his intrinsic weaknesses have undermined his ability to exercise authority in a patriarchal society. Lucien fails in his professional endeavours, legal confrontation with Camusot and matrimonial ambition with Clotilde de Grandlieu. He also lacks steady political convictions, as he moves from a liberal newspaper to a royalist one to the great displeasure of his fellow journalists. The date chosen for his suicide (15 May 1830), shortly before the promulgation of the conservative Ordinances of Saint-Cloud and therefore before the political and social commitment of the Parisians during the July Revolution, indicates his social and political incapacity. This shows that the revitalization of the social order can come neither from Lucien, Octave or any other young man, nor from the patriarchal elites, who were either senile, like the Napoleonic soldiers, or portrayed as clinging to the privileges of the Ancien Régime like Prosecutor Grandville.

The only character who embodies a way forward for this post-imperial society is a character who 'n'aime pas les femmes' [does not like women], mocks Peyrade's lust for women ('Ah! papa Peyrade, vous aimez encore assez les femmes pour? … Mais vous m'attrapez; moi, j'ai soixante ans et je m'en prive très bien…' [Ah! Papa Peyrade, you still like women enough to? … but you catch me out; I am sixty years old and I easily live without them]) and complains about '[l]es hommes assez bêtes pour aimer une femme' [the men stupid enough to love a woman]; a protagonist who is as openly homosexual as it was possible to be in 1840s literary fiction – according to Diana Knight, he is 'the only explicitly self-identified homosexual in the whole of *La Comédie humaine*'.[148] This character is Vautrin.

The Reconciled Man

The leap from Lucien to Vautrin is performed by the convict himself, who invites the reader to assimilate the two characters. He deliberately

148 Respectively Balzac, *Le Père Goriot*, 233; Balzac, *Splendeurs et misères des courtisanes*, 317 and 627; Knight, 'Skeletons in the Closet', 171.

creates a blurred identity when expressing his desire to become Lucien ('je me ferai vous' [I will become you]; 'Ce beau jeune homme, c'est moi!' [This handsome young man, it is me!]).[149] He bestows on Lucien the financially successful social life of which both were deprived due to the modest means of their families. The criminal perceives the poet as an example of perfection owing to his beauty, youth and aristocratic origins. However, the convict's apparently innate adaptability enables him to surpass Lucien. This quality is notably based on his ability to reconcile both his masculine and feminine sides; it therefore allows him to tolerate the changes in society and to assert his social role.

Although there is neither a defence nor a valorization of homosexuality within the novels, through Vautrin Balzac envisages the possibility for a man to assume his homosexuality. As a member of the underworld, as well as an escaped convict and a Machiavellian figure,[150] Vautrin is the villain of *La Comédie humaine* par excellence. And yet, despite his villainy – and perhaps because of it –, the character is overall admired rather than vilified. The narrator praises the way the criminal conducts his business and manipulates his entourage, whether it is Lucien, Esther, Camusot or Grandville. For instance, with regard to convincing the young woman to prostitute herself to Nucingen, the narrator states: 'En révélant *la supériorité* de ce corrupteur [Vautrin], *cette finesse* indique le procédé par lequel il avait soumis Lucien' [In revealing this corruptor [Vautrin]'s *superiority*, *this cunning* indicates the method by which he had subdued Lucien].[151] Lucien, on the other hand, does not receive the same narratorial benevolence.[152] The protector-protégé couple that Vautrin and Lucien incarnate is characterized

149 Balzac, *Illusions perdues*, 515 and 519. See also Balzac, *Splendeurs et misères des courtisanes*, 132, 133 and 564.

150 Balzac, *Splendeurs et misères des courtisanes*, 410, 411 and 650.

151 Ibid., 265 (my emphasis). The titles of chapters of the third part of the novel also reveal this admiration: for example, 'XXIX Admirable invention de Jacques Collin' [Jacques Collin's admirable invention] and 'XXXV Audace du prévenu' [The defendant's audacity].

152 This scorn is shared by many readers. Compare, for example, Byatt's description of Lucien, quoted earlier, with that of Vautrin: 'Vautrin has the grandeur of Milton's Satan and also his motivation': 'The Death of Lucien de Rubempré', 392.

by a relationship of opposition. Whilst Lucien's alleged homosexuality is notably conveyed through his feminization and is therefore criticized as a sign of weakness and insecurity, Vautrin's virility is never questioned. He is clearly depicted as incarnating masculine values.

The process through which Vautrin's virilization operates uses the same narrative strategies as that of Lucien's effeminacy, namely reference to ancient mythology. Rather than suggesting the beauty and grace of ephebes, references to Greek and Roman heroes here recall the image of strength, which is typical of the manly ideal. In *Splendeurs et misères des courtisanes*, Vautrin is undressed in front of Camusot. The narrator comments on the scene as follows: 'on put admirer un torse velu d'une puissance cyclopéenne. C'était l'Hercule Farnèse de Naples sans sa colossale exagération' [all could admire a hairy torso of Cyclopean power. It was the Farnese Hercules without his colossal exaggeration].[153] The mythological hero Hercules is a patent symbol of both manly strength and virile attractiveness. The criminal is old and ugly (his face is covered with scars); yet, his body is eroticized. Magistrates and witnesses constantly expect him to remove his clothes, officially to check the mark of the *bagne* on his shoulder.[154] When Madame Poiret, who used to live in the same boarding house as Vautrin at the time of *Le Père Goriot*, asks to see his bare chest in order to recognize him, she provokes the hilarity of Camusot and the clerk (and of Vautrin himself), whereas they do not question their own fascination for his naked body. Vautrin's manly attributes (his Herculean shoulders and hirsute torso) can clearly seduce both sexes.

The third section of *Splendeurs et misères des courtisanes*, 'Où mènent les mauvais chemins' [Where wicked ways lead], and the respective cross-examinations of Vautrin and his accomplice reveal the fundamental opposition of character between the master and his disciple. Vautrin is so resourceful and assertive during the interrogation that the magistrate cannot help but question his own certainties. For instance, when Vautrin makes up a lie to explain his presence at the late Esther's home, the narrator concludes: 'tout en [dans sa tirade] était si vraisemblable [...] que le juge en

153 Ibid., 479. See also 70, 411, 499, 552 and 562; Balzac, *Illusions perdues*, 516.
154 Balzac, *Splendeurs et misères des courtisanes*, 484–85.

fut ébranlé' [everything in his speech was so plausible [...] that the judge was shaken].[155] The reader accordingly feels a sense of complicity with the convict. No such feeling is bestowed on Lucien. When the young man meets Camusot, he appears defeated even before the cross-examination begins.[156] He shows weakness when harassed by the judge and admits not only the real identity of the so-called Carlos Herrera but also the suspect provenance of his money. The text lays heavy emphasis on the difference between the poet and the man of action (an obvious allusion to Lucien and Vautrin): 'l'un se livre au sentiment pour le reproduire en images vives, il ne juge qu'après; tandis que l'autre sent et juge à la fois' [one abandons himself to feeling in order to reproduce it in lively images, he only judges afterwards, while the other feels and judges at once].[157] Lucien, out of lack of ingenuity and fear that 'une plus infâme vérité' [a more odious truth] – his sexual relations with Vautrin? – be unveiled, undoes all the work that Vautrin, thanks to his leonine courage and acumen, carried out with Camusot.[158] Lucien blames the magistrate's subtlety, but an intelligent man such as he should have known better.

Through Lucien, the text denounces the inability of young men to use the resources with which they have been endowed. Lucien is an educated man, but he is unable to use his mind in practical situations. As Richard Terdiman indicates – and as Balzac illustrates in *La Comédie humaine* – the French educational system did not prepare young people for the realities of life, notably to face the massive economic and demographic transformations of the first half of the nineteenth century.[159] Conversely, Vautrin does not benefit from the same educational background – although we learn that he studied at the College of the Oratorian Fathers.[160] Yet, his experience has taught him to be audacious and to use common sense. Vautrin's masculinity is not only an expression of strength, action and violence, but also of resourcefulness, cunning and capacity to cope with adversity.

155 Ibid., 475. See also 487.
156 Ibid., 502.
157 Ibid., 509.
158 Ibid.
159 Terdiman, 'Structures of Initiation', 204.
160 Balzac, *Splendeurs et misères des courtisanes*, 482. See also 713.

In his farewell letter, Lucien acknowledges his failure, since he did not measure up to the ambitions of his protector ('vous avez voulu faire [de moi] un personnage plus grand que je ne pouvais l'être' [you wanted to make me a greater character than I could be]).[161] Despite his ingratitude, he also admits, albeit reluctantly, the superiority of his mentor. Lucien's style is ambivalent, probably because the false priest, to whom the letter is addressed, is not its only recipient. The poet assumes that it will be read by any representative of the justice system, Camusot included. The ambivalence of Lucien's letter also lies in the expression of mixed feelings towards his protector. It oscillates between the formal *vous* and the affectionate *tu*, between the relatively endearing term 'Mon cher abbé' [my dear Abbot], the colder 'monsieur l'abbé' [Abbot] and the frankly critical 'grandiose statue du mal et de la corruption' [spectacular statue of evil and corruption]. Whilst fascinated by Vautrin, Lucien also appears resentful and even contemptuous, as the letter ends with these words: 'Ne me regrettez pas: mon mépris pour vous était égal à mon admiration' [Don't feel regret for me: my contempt for you was equal to my admiration]. Like Vautrin, Lucien endorses the emotional filiation between them ('fils spirituel' [spiritual son]; 'paternelle tendresse' [fatherly tenderness]) but denies any real paternal links ('celui que vous aviez adopté' [he whom you had adopted]). This may be in reaction to Camusot's previous revelation that Vautrin claimed to be his father. Lucien was shocked by the usurpation of identity, perhaps because this lie implied the incestuous nature of their relationship or the convict's feelings towards the poet if their bond is platonic.[162]

After deploring his ingratitude towards his protector and his inability to fulfil Vautrin's ambitious projects, the tone of the letter changes, as the different appellatives have already shown. Lucien's portrayal of his mentor reveals his fascination for Vautrin's evil side and his awe in the face of his 'poésie du mal' [poetry of evil].[163] The poet reminds Vautrin of his claim when they met in *Illusions perdues* that he belonged to the posterity of both Cain and Abel: good to his friends and cruel to his enemies.[164] Lucien likens

161 Ibid., 530. The letter (530–32) is reproduced with variations in 570–72.

162 Ibid., 507–08.

163 Ibid., 531.

164 Balzac, *Illusions perdues*, 516.

the so-called abbot to a demon and a lion (a common comparison in the novel, but here containing a degree of cruelty that is absent elsewhere), before comparing Vautrin to Napoleon and other grandiose historical figures:[165]

> Il y a la postérité de Caïn et celle d'Abel [...]. Caïn, dans le grand drame de l'Humanité, c'est l'opposition. [...] Parmi les démons de cette filiation, il s'en trouve, de temps en temps, de terribles, à organisations vastes, qui résument toutes les forces humaines [...]. Ces gens-là sont dangereux dans la Société comme des lions le seraient en pleine Normandie [...]. Quand Dieu le veut, ces êtres mystérieux sont Moïse, Attila, Charlemagne, Mahomet ou Napoléon; mais, quand il laisse rouiller au fond de l'océan d'une génération ces instruments gigantesques, ils ne sont plus que Pugatcheff, Robespierre, Louvel et l'abbé Carlos Herrera.[166]
>
> [There is Cain's posterity and then there is Abel's [...]. Cain, in the great drama of Humankind, is the opposition. [...] Among the demons of this line, one can find, from time to time, terrible ones, with vast organizations, who sum up all human strength [...]. Those people are dangerous in Society like lions would be in the heart of Normandy [...]. When God wants it, those mysterious beings are Moses, Attila, Charlemagne, Muhammed or Napoleon; but, when he lets those gigantic instruments rust at the bottom of the ocean of a generation, they are only Pugachev, Robespierre, Louvel and Abbot Carlos Herrera.]

Vautrin is aligned with one of those mysterious and dangerous men meant for great things, like Napoleon; but, unlike the Emperor, he is seen as being abandoned by God. Lucien emphasizes not only Vautrin's ability to fascinate, but also the danger that he represents and the fear that he arouses. This comparison with Napoleon links Vautrin to patriarchal traditions, whilst the difference between them underlines the subversion of patriarchal leadership in the convict.[167] Vautrin is a proteiform character who combines his malice, inherited from Cain, with his Abel side,

165 Jeannine Guichardet has analysed the character of Vautrin as the greatest 'doublure' of Napoleon in *La Comédie humaine*: 'Doublures historiques en scène parisienne', *L'Année balzacienne*, 2nd ser., 5 (1984), 307–25 (322–25).

166 Balzac, *Splendeurs et misères des courtisanes*, 531.

167 Vautrin is not devoid of patriarchal aspirations: 'J'ai besoin de deux cent mille francs, parce que je veux deux cents nègres, afin de satisfaire mon goût pour la vie patriarcale' [I need two hundred thousand francs for I want two hundred negroes in order to satisfy my taste for patriarchal life] (Balzac, *Le Père Goriot*, 168).

defined by empathy and generosity. In other words, rather than being characterized by either uncompromised manliness or excessive effeminacy, Vautrin combines the qualities that are supposed to be typical of each side: tenderness and aggressiveness. Like hermaphroditic characters such as Madeleine de Maupin, his success lies in his ability to reconcile contrary natures.

Yet, Vautrin's feelings for Lucien and, more generally, for beautiful young men (Eugène de Rastignac in *Le Père Goriot* and Théodore Calvi in the final part of *Splendeurs et misères des courtisanes* – who, incidentally shares Lucien's beauty and blondness) are often seen as problematic and are reduced to anti-authoritarian behaviour.[168] Vautrin's homosexuality would simply be a metaphor of his transgression of legal, social and moral codes. Certainly, this dimension is present in Balzac's works and the polemical nature of the criminal and the poet's relationship is essential in the dynamic of the novels. However, this should not conceal the fact that the depiction of Vautrin's same-sex love mostly attempts to redefine social relations and, specifically, parental relations. Indeed, Vautrin and Lucien's relationship subverts homosocial bonds because it combines homoeroticism and parental affection. The qualities displayed by the criminal, namely tenderness and protectiveness, are salient in what Élisabeth Badinter has named *l'homme réconcilié* [the reconciled man], 'le *gentle man* [*sic*] qui sait allier solidité et sensibilité' [the *gentle man* who can combine solidity and sensitivity].[169] Vautrin is a reconciled man, not only because he reconciles opposite characteristics, but also because this reconciliation is with himself. It is a means of being at peace with himself by reinventing his identity.[170]

Badinter has called for a paternal revolution, which would place more value on the role of the father in parenting, as long as he can express his femininity when nurturing his offspring.[171] Seen in this light, would

168 Balzac, *Splendeurs et misères des courtisanes*, 666. Berthier, 'Balzac du côté de Sodome', 170; Gerald H. Storzer, 'The Homosexual Paradigm in Balzac, Gide, and Genet', in George Stambolian and Elaine Marks, eds, *Homosexualities and French Literature: Cultural Contexts/Critical Texts*, pref. Richard Howard (Ithaca and London: Cornell University Press, 1990 [1979]), 186–209 (186).

169 Badinter, *XY*, 239. On *l'homme réconcilié*, see 239–74.

170 See 'Réconcilier', in *Dictionnaire de l'Académie française*, II, 584.

171 Badinter, *XY*, 247–71.

Vautrin be the ideal modern father? *Splendeurs et misères des courtisanes* seems to suggest that he would. Like Badinter's ideal father, Vautrin is able to combine paternal and maternal love, by mobilizing his femininity, as well as nurturing and stimulating his child.[172] Indeed, when he learns of Lucien's death, he tells the doctor and the warden of the prison: 'Vous n'êtes pères [...] que d'une manière; ... je suis mère, aussi!' [You are fathers [...] only in one way; ... I am a mother too!].[173] Moreover, his pathetic speech to Prosecutor Grandville, in which he expresses his affection with tender terms such as 'ce cher enfant' [this dear child] and 'le petit' [the little one], certainly aims at gaining his interlocutor's sympathy by showing love and suffering for Lucien.[174] However, it should not be dismissed as a mere rhetorical strategy, because it sincerely demonstrates the extent of his affection, in opposition to his image of the rough master of the underworld. Vautrin defines his role as maternal and behaves like a devoted mother to Lucien. He displays 'empressement maternel' [maternal attentiveness] towards him, tucks him in and shares his secrets.[175] Vautrin summarizes his sentiment to the Prosecutor as follows: 'Ah! jamais une bonne mère n'a tendrement aimé son fils unique comme j'aimais cet ange' [Ah! Never did a good mother tenderly love her only son like I loved that angel!].[176]

Furthermore, Badinter defines the new man as emerging during a time of social uncertainty. He thus resembles the former convict of the 1830s and 1840s who can transcend his criminal status and become a loving father – or a fatherly lover. Certainly, Vautrin does not become a good man thanks to Lucien (he will keep exercising his Machiavellian mind until the end of the novel), but his love modifies readers' interpretation of the character. They are invited to admire Vautrin by an oxymoron that highlights the paradoxical nature of the criminal: 'n'est-il pas *monstrueusement beau* par son attachement digne de la race canine envers celui dont il fait son ami?'

172 Ibid., 258–65.

173 Balzac, *Splendeurs et misères des courtisanes*, 569.

174 Ibid., 677–78. Lucey, *The Misfit of the Family*, 216–18. See also Berthier, 'Balzac du côté de Sodome', 173.

175 Balzac, *Illusions perdues*, 504. See also Balzac, *Splendeurs et misères des courtisanes*, 573 and 677.

176 Balzac, *Splendeurs et misères des courtisanes*, 677.

[is he not *monstrously attractive* in his attachment worthy of the canine race towards the one whom he made his friend?].[177] His dog-like, faithful love reminds us of the love of his former neighbour Goriot for his ungrateful daughters, as this model of fatherhood was also likened to a dog ('son sentiment irréfléchi élevait [Goriot] jusqu'au sublime de la nature canine' [his thoughtless feeling raised Goriot to the sublime of canine nature]).[178] Vautrin thus becomes another 'Christ de la Paternité' [Christ of Fatherhood].[179] Geneviève Vinciguerra has analysed this love of a father for his children, a love which is so passionate that it verges on madness. She has shown that it is grounded in a demiurgical function – the father creates his children – and in his ability to live through them.[180] And what is Vautrin if not Lucien's maker? The convict tells Lucien during their first encounter : 'Je veux aimer ma créature, la façonner, la pétrir à mon usage, afin de l'aimer comme un père aime son enfant. […] ce marquis de Rubempré, je l'ai créé et mis au monde aristocratique' [I want to love my creature, fashion it, mould it to my use, in order to love it like a father loves his child. […] This marquess de Rubempré, I created him and brought him into the artistocratic world].[181] He also claims that he 'revivait dans le corps élégant de Lucien dont l'âme était devenue la sienne' [lived again in Lucien's elegant body, whose soul had become his].[182] Beyond its homoerotic connotations, Vautrin's love for Lucien is paternal and maternal, as well as possessive and controlling.

Philippe Berthier is thus right to point out the intensity of the relationships between men in prison in Balzac's novel.[183] Vautrin's relationships with his protégés, inside or outside prison, are more intense than those of other protectors. The same generosity is directed towards Calvi, both Lucien's predecessor and successor in Vautrin's heart. The Head of Security

177 Ibid., 563 (my emphasis).

178 Balzac, *Le Père Goriot*, 178.

179 Ibid., 282.

180 Geneviève Vinciguerra, 'Trois drames de la paternité frustrée: Goriot, Vautrin, Balzac', *Littératures*, 17 (1970), 3–14 (4–5).

181 Balzac, *Illusions perdues*, 519.

182 Balzac, *Splendeurs et misères des courtisanes*, 133.

183 Berthier, 'Balzac du côté de Sodome', 150.

Bibi-Lupin, also a former convict, claims that Jacques Collin used to make 'de bien belles *patarasses*' [beautiful *patarasses*] for Calvi.[184] These are, as the text explains, pads of tow and cloth to protect the convict's flesh from the rings of the chain. This anecdote contradicts D. A. Miller's claim that Vautrin is only sexually involved with Calvi.[185] It would be simpler if the convict merely sought sexual intercourse with his protégés. Yet, the subversiveness of his love lies in its ambivalence: it reproduces ideal parental love, whilst concurrently expecting favours in return.

In the final part of *Splendeurs et misères des courtisanes*, the former convict Vautrin is integrated into the social order. He becomes the new Head of Security, the representative of the *Nom-du-Père*, by making a bargain with Prosecutor Grandville. He will give Grandville the compromising letters that the Duchesse de Maufrigneuse, the Comtesse de Sérizy and Clotilde de Grandlieu wrote to Lucien if Calvi is not guillotined. The narrator concludes the novel as follows: 'Après avoir exercé ses fonctions pendant environ quinze ans, Jacques Collin s'est retiré vers 1845' [After having exercised his functions for about fifteen years, Jacques Collin retired around 1845].[186] Vautrin has thus been appointed Head of Security just before the July Revolution and has exercised his career during most of Louis-Philippe's reign. For Berthier, the fact that a criminal – and an openly homosexual one – is the representative of the Law shows that the Law has become meaningless during this period.[187] However, as Lucey has observed, it is the context of the Restoration (Charles X is about to promulgate the Ordinances of Saint-Cloud) that allows the King and the aristocratic prosecutor to hire a criminal.[188] Such an appointment would not have been possible during the July Monarchy, in which the collusion between Law and Aristocracy was not as strong. Vautrin takes advantage of the flaws of an Ancien Régime legal system, which seems to be more lenient towards homosexuality, to pave his way through the more homophobic but also less corruptible bourgeois system.

184 Balzac, *Splendeurs et misères des courtisanes*, 610 (emphasis in original).

185 Miller, 'Balzac's Illusions Lost and Found', 177.

186 Balzac, *Splendeurs et misères des courtisanes*, 726.

187 Berthier, 'Balzac du côté de Sodome', 175.

188 Lucey, *The Misfit of the Family*, 221.

The conclusion of *Splendeurs et misères des courtisanes* implies a peaceful career for Vautrin and a gentrification of his character. He becomes the living embodiment of the new bourgeois France, as his Herculean appearance predisposed him to be a champion of French values. Before the feminine Marianne, it was the virile mythological hero Hercules who was for a time the symbol of the French Republic.[189] It is nonetheless incorrect to assert, as Gerald Storzer does, that Vautrin loses his (homo)sexual identity in order to be integrated into society.[190] On the contrary, his rescuing Calvi suggests that he does not renounce his relationship with young men to access his new functions. Grandville even tells him that he can have the young Corsican work for him.[191] At the end of the novel, while Vautrin's love for men is not openly stated, his abhorrence for women is established as a work ethic. He reiterates his contempt for men, especially men of power, who succumb to women's charms, and claims in an openly sexist speech:

> Oh! combien de force acquiert un homme quand il s'est soustrait, comme moi, à cette tyrannie d'enfant, à ces probités renversées par la passion, à ces méchancetés candides, à ces ruses de Sauvage! La femme, avec son génie de bourreau, ses talents pour la torture, est et sera toujours la perte de l'homme.[192]
>
> [Oh! What strength a man acquires when he has escaped, like me, that childish tyranny, that integrity overturned by passion, that candid wickedness, that Savage cunning! Woman, with her executioner-like genius, her talents for torture, is and always will be man's downfall.]

He then concludes: 'Je régnerai toujours sur ce monde, qui, depuis vingt-cinq ans, m'obéit …' [I will always rule over this world, which has been

189 Potts, *Flesh and the Ideal*, 135; Solomon-Godeau, *Male Trouble*, 203; Rauch, *Crise de l'identité masculine*, 31; Clark, 'The Rhetoric of Masculine Citizenship', 10; Fend, *Les limites de la masculinité*, 104; Maira and Roulin, 'Constructions littéraires de la masculinité', 9.

190 Storzer, 'The Homosexual Paradigm', 189–90. In contrast, Saint-Paulien considers that it is not Vautrin but society that has demeaned itself to Vautrin's level (*Napoléon*, 268).

191 Balzac, *Splendeurs et misères des courtisanes*, 725–26.

192 Ibid., 724–25.

obeying me for twenty-five years], indicating his authority over criminals, as well as Camusot, Grandville and the likes of them.[193]

It is therefore erroneous to consider that his appointment illustrates the decline of the character, as Vinciguerra suggests when she claims that he 'rentre dans le rang' [toes the line] or when André Vanoncini calls him a 'banal défenseur du contrat social' [a banal defender of the social contract].[194] Vautrin reassures Jacqueline Collin about his new functions: 'Je ferai toujours trembler tout notre monde!' [I shall always make everyone in our world tremble] and plans to take revenge on Corentin, the secret agent whom he holds responsible for Lucien's death.[195] Furthermore, Vautrin is mentioned in *La Cousine Bette* during a conversation between Victorin Hulot and the aunt of the ex-convict.[196] This reference reveals that he has kept his subversive nature. Certainly, the former criminal and convict becomes indirectly involved with the Minister for War, a former Napoleonic soldier. Certainly, he defends the familial order as well as the interests of the State under the threat of Lisbeth Fischer and Valérie Marneffe.[197] However, in order to do so, he is eager to organize the murder of the latter and Crével. Vautrin's distinctive identity – namely his homosexuality and criminality – is integrated into a new order, while simultaneously subverting it. 'La dernière incarnation de Vautrin' [Vautrin's last incarnation] (the title of the final part of *Splendeurs et misères des courtisanes*) is the incarnation of a reconciled man who, thanks to his decisiveness, resourcefulness and influence, is able to contribute to the renewal of the July Monarchy bourgeois society by playing an active social role and even exercising authority as a high-ranking police officer. Vautrin reinvents his patriarchal attributes by instilling maternal traits in them and subverts homosocial bonds between men by asserting his love for young protégés.

In brief, the reconciled man can combine masculine and feminine qualities and act as both a paternal and a maternal figure without challenging his virility. This association of antithetical qualities moves away

193 Ibid., 725.

194 Vinciguerra, 'Trois drames de la paternité frustré', 13; Vanoncini, 'Le pacte', 291.

195 Balzac, *Splendeurs et misères des courtisanes*, 721.

196 Balzac, *La Cousine Bette*, 431–32.

197 Baron Hulot has embezzled government money to fund Valérie's lifestyle.

from traditionally disparaging perceptions of homosexuality, such as those characterizing Lucien. Above all, it allows such a man to consolidate his position in society by subverting the opposition between patriarchy and male insecurity and placing his non-normative sexual identity at the core of his masculinity.

Let us briefly return to *La Confession d'un enfant du siècle* and its range of protagonists. Can an equivalent character to Vautrin be found in Musset's novel? Certainly, Balzac's character is unique, but Henri Smith, Brigitte's lover, represents the positive alter ego of Octave, as he seems devoid of Octave's generational torments. Octave himself also appears as a reconciled man in the final chapter of the novel. He is perceived as 'l'enfant qui devient homme' [the child who becomes a man], whereas he previously described himself as a child who thought of himself as a man after unsuccessfully attempting to kill a sleeping Brigitte.[198] He is in a process of maturation, which has not yet been achieved, but which promises a better future. This change is emphasized by a change of narrator. The final chapter of the novel is told by a third-person narrator, which not only constitutes a rupture with the past, but also undermines the egocentric traits that Octave had thus far exhibited in showing indifference towards Brigitte's activities and mocking her piety.[199] He is able to act as an adult, make responsible choices and sacrifice his own happiness to the person whom he loves. After finding a letter from Brigitte to Henri in which she confessed her love to the latter, Octave decides to renounce his mistress, to whom he is noxious, and to act as her brother, a role that implies a tenderness and protectiveness similar to that of a father. He 'gives' her to Smith, described as 'brave, bon et honnête' [brave, good and honest].[200] The optimism of Octave's reconciliation with himself and with his former mistress is confirmed by the beautiful weather during their farewell.[201]

This bittersweet ending constrasts with the impression of stagnation that prevails in the novel as a commentary on contemporary society. The signs of insecurity and failure in July Monarchy narratives represent the

198 Musset, *La Confession d'un enfant du siècle*, 348. See also 345.
199 Ibid., 232 and 245.
200 Ibid., 349. See also 291 and 293.
201 Ibid., 347 and 351.

difficulty of young men to conform to the masculine side of the sexual dichotomy. This masculine malaise is notably due to the discrepancy between the ideal image of virility embodied by the Napoleonic soldier and the scarcity of means available to young men to attain this manly model. A strong correlation is therefore established between male protagonists and children, as well as women, based on similar alleged weaknesses that determine young men's inability to assume a virile position in society. Balzac's treatment of homosexuality not only illustrates the censorship that affects homosexual behaviour, but also the symbolic use of homosexuality as a symptom of young men's maladjustment in the new bourgeois society that expects productive men. Yet, paradoxically, homosexuality in Balzac's novels is presented as a transcending value that allows the reconciled man to move on to the next stage: not only is he freed from social rules, but he can even create his own rules, as does Vautrin in becoming the Head of Security. Similarly, the new Octave of the end of *La Confession d'un enfant du siècle* sets out his own rules in 'giving' Brigitte to Henri.

This transfer from Lucien to Vautrin and, to a lesser extent, from Octave to the new Octave/Smith, in other words from poorly appreciated characters to more popular figures, indicates that characters themselves are less culpable than their malaise.[202] As Barbéris notes, men such as Lucien are not the only ones responsible for their failures.[203] The system that creates a gap between ambition and possible achievements renders young men insecure and inclined to run away from their responsibilities. As such, the novels analysed here do not invite a return to the patriarchal and manly model, since its representatives are as blameworthy as the fragile young men and are often ridiculed. These texts suggest an alternative model, represented by Vautrin. The ex-convict contrasts with both the traditional

202 Byatt's antipathy towards Lucien has already been mentioned. Although not as explicit, Michelle Coquillat also shows that she disapproves of Octave's conduct and justifications for his actions. She notably states: 'il soupire pour la forme, mais il est aussi très satisfait: il se donne le beau rôle, et son attendrissement est sur lui-même' [he sighs *pro forma*, but he is also very satisfied: he makes himself look good, and he feels sorry for himself]: *La poétique du mâle* (Paris: Gallimard, 1982), 334. See also Bellet, 'La confession politique', 133.

203 Barbéris, *Le monde de Balzac*, 395.

representations of manliness and the victims of the patriarchal system, by reconciling so-called masculine values (power and reason) and feminine qualities (tenderness and care). This combination of opposite natures enables Vautrin to exercise his masculine power, whilst openly displaying his homoerotic and parental affection for young men. More generally, it reveals that confident masculinity can only exist through the recognition of the virtues of femininity. The connections between masculinity and femininity are further explored in the next chapter, which analyses the literary figure of the masculine woman.

CHAPTER 4

The Butch

In the first chapter, I referred to what French nineteenth-century scientists, and especially Virey, described as masculine women: women who infringed sexual norms by displaying 'masculine' physical and psychological features. Such a medical type also appeared in contemporary fiction in varied forms: the intellectual Lélia in Sand's eponymous novel (first version: 1833), the writer and composer Félicité des Touches in Balzac's *Béatrix* (1839–45) and the old maid Lisbeth Fischer in *La Cousine Bette* (1846) by the same author. These masculine women are thus ambiguously gendered because they display characteristics that are usually attributed to men. As in other narratives previously studied, the question of gendered identity is therefore of vital importance to all the novels analysed in this chapter.

This identity question is asked in dichotomous terms that recall those in hermaphroditic narratives. The androgynous dimension of a character such as *Béatrix*'s Félicité des Touches is manifest, both for critics and characters in the plots.[1] Félicité appears in numerous novels of *La Comédie humaine*. In *Illusions perdues*, the narrator calls her 'cette illustre hermaphrodite' [this renowned hermaphrodite].[2] In *Béatrix*, Félicité is a mature

1 Owen N. Heathcote, '"Cet être amphibie qui n'est ni homme ni femme": Marginalizing Gender and Gendering the Marginal in Balzac's Camille Maupin', *Nottingham French Studies*, 41/2 (Autumn 2002), 37–46; Françoise Van Rossum-Guyon, *Balzac et la littérature réfléchie: Discours et autoreprésentations* (Montreal and Paris: Paragraphes, Département d'Études françaises de l'Université de Montréal, 2002), 81; Kyoto Murata, 'Sous le signe des monstruosités: La Camille Maupin de Balzac et George Sand', *George Sand Studies*, 26 (2007), 43–54 (44); André Vanoncini, 'La femme-livre dans *Béatrix* de Balzac', in Peter Schnyder, ed., *L'homme-livre: Des hommes et des livres, de l'Antiquité au XXe siècle* (Paris: Orizons, chez L'Harmattan, 2007), 153–65 (156).

2 Balzac, *Illusions perdues*, 368.

writer who is loved by the young Calyste du Guénic. Despite loving him back, she rejects him because she knows that she does not suit his conservative Breton family. When her friend Béatrix de Rochefide visits her, she encourages Calyste to fall in love with Béatrix, who soon shares his feelings. Abbot Grimont, the conservative priest of the equally old-fashioned Breton town of Guérande, designates her as 'cet être amphibie qui n'est ni homme ni femme' [this amphibian being that is neither man nor woman],[3] which situates her in a gender no man's land that denigrates her identity and even her humanity – *amphibie* is a term that is mostly used to refer to animals (and sometimes plants) that can equally live on land and under water.[4]

Félicité's penname is Camille Maupin, which combines the names of two other hermaphroditic characters (Camille in *Fragoletta* and Madeleine de Maupin). It is generally perceived as a masculine name ('le sieur Camille Maupin' [sir Camille Maupin], 'un nom d'homme' [a man's name], 'au pauvre Camille' [to the unfortunate Camille]), but the narrator and some characters also use it as a female name to refer to Félicité (for instance 'la célèbre Camille Maupin' [the famous Camille Maupin]).[5] As with *hermaphrodite*, used with the feminine demonstrative adjective *cette* in *Illusions perdues* despite its masculine grammatical gender, Balzac does not hesitate to distort French grammar in *Béatrix*. The narrator even combines masculine and feminine forms in the same phrase: 'la célèbre Camille Maupin *lui-même*' [the famous Camille Maupin *himself*], as if to assert concomitantly Félicité's masculine, feminine and hermaphroditic identities.[6] This means that masculine women like Félicité are not deprived of femininity because their gendered identity is complex. Although counterbalanced by masculine physical and behavioural characteristics, their femininity, I argue in this chapter, finds other social and emotional means to be reintegrated into their gendered identity. Romantic plots and maternal feelings remain

3 Honoré de Balzac, *Béatrix*, Michel Lichtlé, ed. (Paris: Librairie Générale Française, 'Le Livre de Poche Classiques', 2012), 119–20.

4 See 'Amphibie', in *Dictionnaire de l'Académie française*, I, 69.

5 Balzac, *Béatrix*, 119, 129, 343 and 236. See also 224 and 288.

6 Ibid., 242 (emphasis in original).

common narrative tropes in these three novels, from which even masculine female protagonists cannot escape.

Before moving on to the analysis of these narratives, one point must be clarified immediately: female masculinity does not equate to female homosexuality. The two notions are often conflated in academic writings. Even though the gender theorist Jack Halberstam admits in *Female Masculinity* that female masculinity is not limited to lesbianism and can be applied to heterosexual women, he nevertheless focuses almost exclusively on the cases of lesbians in his study.[7] Similarly, whilst reproaching Halberstam for his exclusivity, Jean Bobby Noble does not distance himself from this either in a book on the same topic. His analysis remains confined to 'deviant' identities (butches, drag kings and trans men) to the exclusion of straight, masculine ciswomen.[8] Incidentally, these studies partially reproduce the biased opinion of sexologists from the end of the nineteenth century and the beginning of the twentieth century, for example, Krafft-Ebing and Havelock Ellis. The latter considered masculinity to be proportional to lesbianism in homosexual women: the most 'degenerate' lesbian was the most masculine.[9] In all these cases, the masculinity of heterosexual women is *de facto* neglected or even denied. In contrast, while I do not reject the bond between female masculinity and lesbianism, as characters such as Lélia and Lisbeth have a relationship with other women that can be referred to as homoerotic, I do not consider lesbianism to be essential to female masculinity and I do not reduce the uniqueness of these concepts by substituting one for another.

7 Halberstam, *Female Masculinity*, 268.

8 Jean Bobby Noble, *Masculinities without Men: Female Masculinity in Twentieth-Century Fictions* (Vancouver and Toronto: University of British Columbia Press, 2004), xii.

9 For more on Krafft-Ebing and Havelock Ellis's categories of deviant sexualities and lesbians, see Esther Newton, 'The Mythic Mannish Woman: Radclyffe Hall and the New Woman', *Signs*, 9/4 (Summer 1984), 557–75 (565–68); Lillian Faderman, *Surpassing the Love of Men: Romantic Friendship and Love between Women from the Renaissance to the Present* (London: The Women's Press, 1985), 239–53; Carroll Smith-Rosenberg, *Disorderly Conducts: Visions of Genders in Victorian America* (Oxford and New York: Oxford University Press, 1985), 268–72 and 275–80.

Masculine women in French fiction of the July Monarchy have thus far raised little interest. This chapter therefore intends to rectify this lack of interest by highlighting how female masculinity in Balzac's and Sand's novels is used to defy gender prescriptions and advocate, to some extent, women's empowerment in depicting independent female agents. Firstly, it shows that Félicité, Lisbeth and Lélia's difference compared to conventionally feminine women emphasizes their rejection of typically feminine behaviour and their ability to conquer the reputedly masculine realms of willpower and intellect. It also analyses masculine women's refusal of heteronormative sexuality, as well as social prescriptions that are supposed to define a woman's role in a patriarchal society, namely marriage and maternity. Importantly, this analysis follows the conclusions of Carroll Smith-Rosenberg's study of the New Woman, a 'single, highly educated, economically autonomous' American masculine woman, although such a woman lived in a later period (1870–1936). Her study shows that female university graduates and social reformers were accused of being mannish lesbians and seducers of innocent girls, in order to contest their fight for women's rights.[10] Accordingly, I argue that French masculine women were also considered to be threats to patriarchy and had subsequently to be discredited, domesticated or excluded from July Monarchy society. Nevertheless, a pivotal question in Balzac's and Sand's novels is that of masculine women's intellectual, moral and influential superiority. They often appear to be superior to both submissive women and insecure young men, and even to exercise agency in a context of gender inequity.

Portraits of Masculine Women: Monumentalization and Alienation

We know how important descriptions are in Balzac's novels. Physical depictions of characters are codified so that the reader can comprehend their

10 Smith-Rosenberg, *Disorderly Conducts*, 245 and 280–82.

personalities. As Christopher Prendergast reminds us: 'descriptive materials exist not simply as neutral informants, but have a distinct semantic function as integral, signifying components of the general thematic structure of the novel'.[11] For instance, the analysis of Lucien's portrait has revealed that his effeminate beauty and his aristocratic allure suggest the passivity and ambiguous sexuality that are constitutive of his character and therefore his inability to measure up to the manly Parisian society.

Similarly, the portraits of Lisbeth Fischer, Félicité des Touches and, in Sand's novel, Lélia, determine their intrinsic masculinity, although they do not conform to the mannish models that can be found in scientific texts. This section shows that masculine women in these stories do not resemble men physically, but that their female masculinity appears in subtle details, rather than being emphatically stated. I also study how literary strategies of aggrandizement situate masculine women outside gender conventions.

Narrative techniques used for conveying information about the characters' masculine physique vary depending on the writer. Félicité's and Lisbeth's portraits both belong to the Balzacian realist tradition; the long and detailed descriptions (Félicité's is an excellent example of this type) contrast with the absence of such a portrayal in *Lélia*. In George Sand's novel, the main protagonist is a clever woman suffering from melancholia. She is loved and desired by two men: the young poet Sténio, whose love she is incapable of reciprocating despite her feelings for him, and Magnus, a mad Catholic priest. As Béatrice Didier observes, little is known about the main protagonists of *Lélia* because their physical appearance is schematic; indications are parsimoniously given to readers so that they can fashion their own portrait of the heroine.[12] Naomi Schor calls 'idealist effect' Sand's refusal to provide a realist-style portrayal that would create the illusion of the real (*à la* Balzac).[13] The poet Sténio best articulates this effect when stating: 'Qu'est-ce donc que Lélia? Une ombre, un rêve, une

11 Christopher A. Prendergast, 'Towards a Re-assessment of *Béatrix*', *Essays in French Literature*, 9 (1972), 46–62 (50).

12 Béatrice Didier, 'Le corps féminin dans *Lélia*', *Revue d'Histoire Littéraire de la France*, 76/4 (July–August 1976), 634–51 (638).

13 Naomi Schor, *George Sand and Idealism* (New York: Columbia University Press, 1993), 46.

idée tout au plus' [What is Lélia? A shadow, a dream, an idea at most].[14] *Lélia* is an allegorical novel. As Waller accurately points out, the characters are conceived more as ideas than autonomous beings.[15]

However, despite differences in narrative technique, the female protagonists in Sand's and Balzac's novels share common characteristics, such as their age. All of them are mature women, in their forties or at least older than their lovers; although Lélia's age is not revealed, Sténio unkindly notes some silver hairs and wrinkles.[16] The age difference, we will see later, consolidates masculine women's status as maternal and mentoring figures to their lovers.

Whilst masculine women are depicted in relations with young men, they are also constructed in opposition to more traditional female protagonists: Lélia's sister Pulchérie, Félicité's rival Béatrix de Rochefide, Lisbeth's cousin Adeline Hulot, her niece Hortense Hulot and her friend Valérie Marneffe. These feminine women follow restrictive roles since they are either devoted wives and mothers (Adeline and Hortense) or egoistic seductresses, divided into vain aristocrats (Béatrix), middle-class courtesans (Valérie) and prostitutes (Pulchérie). An obvious physical dissimilarity that conveys their gender difference is the colour of their hair. All the masculine characters are dark-haired – and have a dark fuzz on their arms or their upper lip –whereas feminine women are blonde.[17] Lélia's hair can be correlated with George Sand's own dark hair, although the writer denied any autobiographical connection between herself and the heroine of her novel.[18] Likewise, Balzac found inspiration in the writer of *Lélia* to create Félicité.[19] The contrast between blonde women and brunettes is described

14 George Sand, *Lélia*, Pierre Reboul, ed. (Paris: Gallimard, 'Folio Classique', 2003 [Garnier, 1960]), 47.

15 Waller, *The Male Malady*, 137. On allegory in *Lélia*, see also Schor, *George Sand and Idealism*, 55–68

16 Balzac, *Béatrix*, 122 and 144; Balzac, *La Cousine Bette*, 162; Sand, *Lélia*, 289.

17 Dark hair: Balzac, *Béatrix*, 145 and 147–48; Balzac, *La Cousine Bette*, 54; Sand, *Lélia*, 13, 45 and 157–58. Blonde hair: Balzac, *Béatrix*, 175; Balzac, *La Cousine Bette*, 48, 53 and 266; Sand, *Lélia*, 157.

18 Isabelle Hoog Naginski, *George Sand: Writing for Her Life* (New Brunswick and London: Rutgers University Press, 1991), 114–15.

19 On the connections between Sand and Félicité, see Geneviève Delattre, 'L'imagination balzacienne au travail: La lecture créatrice', *Cahiers de l'Association*

most vividly in *Béatrix* as an exemplification of the two women's rivalry.[20] Blonde women are regularly depicted as more feminine than brunettes. Félicité tells her suitor, the young Calyste du Guénic: 'Les blondes sont plus femmes que nous, nous ressemblons trop aux hommes, nous autres brunes françaises' [Blonde women are more womanly than us; we French brunettes look too much like men].[21] In Balzac's novels, the association between fairness and femininity is depicted as coming from Eve, the original woman: Béatrix, Adeline and Valérie are all assimilated to Eve.[22]

In contrast, having dark hair and dark skin was a sign of masculinity according to nineteenth-century norms. It must be recalled that complexion was deemed to be a secondary sexual characteristic in scientific writings. Virey reminded his readers: 'Communément, la petite fille est [...] plus blonde que le petit garçon; [...] son teint est moins vif ou plus blanc' [Usually, the little girl is [...] fairer than the little boy; her complexion is less bright, or whiter], and: 'La femme a communément [...] une peau blanche et délicate' [Woman usually has [...] white and delicate skin].[23] The adverb *communément* suggests that dark-haired and dark-skinned women did not belong to the female norm, which absurdly excludes most women. With the simple physical detail of hair colour, Sand and Balzac assert their heroines' singularity and their masculinity without renouncing their female identity and their beauty.

Whilst Lélia and Félicité's hair colour distinguishes them from more feminine female characters, they do not conform to the image of modern

---

internationale des études françaises, 15 (1963), 395–406 (402); Isabelle de Courtivron, 'Weak Men and Fatal Women: The Sand Image', in *Homosexualities and French Literature*, 210–27 (211); Catherine Barry, 'Camille Maupin: Io to Balzac's Prometheus?', *Nineteenth-Century French Studies*, 20/1–2 (Fall–Winter 1991–92), 44–52 (45); Aline Mura, *'Béatrix' ou la logique des contraires* (Paris: Champion, 1997), 175–76; Van Rossum-Guyon, *Balzac et la littérature réfléchie*, 84; Isabelle Hoog Naginski, '*Lélia*, ou l'héroïne impossible', *Études littéraires*, 35/2–3 (2003), 87–106 (101–02); Murata, 'Sous le signe des monstruosités', 43–45; Vanoncini, 'La femme-livre', 153–54.

20 Balzac, *Béatrix*, 213 and 289.

21 Ibid., 177.

22 Ibid., 175; Balzac, *La Cousine Bette*, 48, 225, 233 and 242.

23 Virey, *De la femme*, 69–70 and 172.

'butch' lesbians. Whilst a mannish woman was deemed to be repulsive according to Virey, Sand's and Balzac's protagonists remain physically attractive so that they can seduce lovers and be involved in heterosexual romantic plots. This ideological confinement of women to the sentimental sphere is illustrated by Sténio when he claims 'là où il n'y a pas d'amour, il n'y a pas de femme' [where there is no love, there is no woman].[24]

Lisbeth Fischer has a distinctive place in the corpus of this chapter, since she is the only masculine woman who is overtly depicted as a negative character. Humiliated by the condescending generosity of the Hulots and having her 'lover' taken away from her by her niece Hortense, she joins forces with her neighbour Valérie to take revenge on the Hulots. Her concise portrayal is constructed in opposition to that of her beautiful and aristocratic cousin Adeline[25] and highlights the unattractiveness of the spinster:

> Lisbeth, de cinq ans moins âgée que madame Hulot, et néanmoins fille de l'aîné des Fischer, était loin d'être belle comme sa cousine; aussi avait-elle été prodigieusement jalouse d'Adeline. […] Paysanne des Vosges, dans toute l'extension du mot, maigre, brune, les cheveux d'un noir luisant, les sourcils épais et réunis par un bouquet, les bras longs et forts, les pieds épais, quelques verrues dans sa face longue et *simiesque*, tel est le portrait concis de cette vierge.[26]

> [Lisbeth, who was five years younger than Madame Hulot, although the daughter of the eldest Fischer, was far from being as beautiful as her cousin; she had therefore been tremendously jealous of Adeline. […] A peasant girl from the Vosges in every sense of the word, thin, dark, with shiny black hair, thick eyebrows meeting together in a bunch, long and strong arms, wide feet, a few warts on her long, *simian* face, this is the quick portrait of this virgin.]

Looking older than her older cousin, Lisbeth is viewed as a peasant's daughter, a working-class woman and, above all, an old maid unable to attract men's favour. As the ultimate spinster of *La Comédie humaine*, she is dry, skinny, warty and apish.[27] Such an unflattering portrait is not only constructed in opposition to her cousin's evident femininity, but also

24 Sand, *Lélia*, 47. I will return to this trope in 'The Sexuality of Masculine Women'.

25 Balzac, *La Cousine Bette*, 47–48.

26 Ibid., 54 (emphasis in original).

27 For instance, ibid., 25, 61, 126 and 165.

to that of Valérie and Wenceslas Steinbock, the young Polish aristocrat whom she calls her lover and of whom she takes care like a mother.[28] The latter is portrayed as a soft and childish figure, notably due to his Slavonic origins.

Lisbeth's portrait reveals that alienation is characteristic of the portrayal of masculine women. Alienation is supported by references to animality (as in the description of Lisbeth's face), notably to big cats.[29] In the following extract, Lisbeth has just learned from Valérie that Wenceslas is about to marry her cousin Hortense:

> La physionomie de la Lorraine était devenue terrible. Ses yeux noirs et pénétrants avaient la fixité de ceux des tigres. Sa figure ressemblait à celle que nous supposons aux pythonisses [...]. Elle avait glissé sa main crochue entre son bonnet et ses cheveux pour les empoigner et soutenir sa tête, devenue trop lourde; elle brûlait! La fumée de l'incendie qui la ravageait semblait passer par ses rides comme par autant de crevasses labourées par une éruption volcanique. Ce fut un spectacle sublime.[30]
>
> [The appearance of the Lorraine had become terrible. Her piercing black eyes had the fixity of those of tigers. Her face resembled that of a pythoness [...]. She had slipped her claw-like hand between her bonnet and her hair to grasp the latter and support her head, which had become too heavy; she was burning hot! The smoke of the fire that was ravaging her seemed to go through her wrinkles as it would through cracks made by a volcanic eruption. It was a sublime spectacle.]

Lisbeth appears as an otherworldly creature ('pythonisses'), exhausted by emotion, but still frightening. She is compared to a tiger, that is, a ferocious and exotic animal. The peak of the extract is, however, its ending. The old maid seems to turn into a volcano that is about to explode with fury, a scary spectacle that fascinates the narrator.

Moreover, Lisbeth is often assimilated to a savage, either through use of the terms 'sauvage' [savage] and 'sauvagerie' [savagery], or through

28 Valérie: ibid., 198; Steinbock: 89–90, 163 and 270–71.

29 Sand, *Lélia*, 147; Balzac, *Béatrix*, 149 and 291; Balzac, *La Cousine Bette*, 199. For more on the comparison between Lisbeth and animals: Pascale Krumm, 'La Bette noire de Balzac', *Australian Journal of French Studies*, 28/3 (September–December 1991), 254–63 (257–61).

30 Balzac, *La Cousine Bette*, 136–37.

references to reputedly primitive people, whether undetermined, Mohicans or Corsicans.[31] These mentions highlight her instinctive, animalistic and yet scheming nature, which makes her highly dangerous. Interestingly, both Lisbeth and Félicité are perceived as Mediterranean – Corsican for Lisbeth and Italian and Spanish for Félicité.[32] However, the two masculine women have respectively Germanic and Celtic roots: Lisbeth is from Lorraine and Félicité from Brittany. It appears that Balzac altered their origins to indulge in the stereotype about the hot blood of Mediterranean people. The qualifications 'Lorraine' and 'paysanne des Vosges' [peasant woman from the Vosges] are nevertheless frequently employed to describe Lisbeth. Conversely, they are not used for her cousin Adeline, a native of the same region, as if she were able to eradicate her rural origins.[33] Whilst the famous Camille Maupin is regarded with suspicion by the Bretons as a Parisian, she in fact belongs to the wild landscape of Brittany, as the walk that she undertakes with Béatrix and Calyste to Le Croisic shows.[34] Both women's savagery accordingly appears as a by-product of their provincial nature, by definition 'natural' and 'savage' in opposition to Parisian, cultural and civilized identity.

Depicting ancient civilizations constitutes another means of conveying masculine women's estrangement. Even though the references to neoclassical arts and Greek mythology are less pronounced in their portraits than in those of effeminate male protagonists, such as Lucien de Rubempré and Calyste du Guénic,[35] they are still constitutive of masculine women's description. Félicité's vast forehead, for instance, is compared to that of a 'Diane chasseresse' [hunting Diana], a Greek reference that underlines Félicité's masculinity given that it relates to the masculine activity of this goddess, even though Diana also fulfils the feminine function of

31 For instance, ibid., 57–58, 61 and 144–45.

32 Balzac, *Béatrix*, 145.

33 A notable exception: Balzac, *La Cousine Bette*, 500. Claudia Moscovici mistakenly believes that Adeline Fischer is noble by birth: 'Gendered Spheres in Balzac's *La Cousine Bette*', *International Journal of Politics, Culture and Society*, 10/3 (1997), 469–96 (476).

34 Balzac, *Béatrix*, 298.

35 Ibid., 125–26.

aiding childbirth.[36] Along with dark eyes, a prominent forehead was a sign of thoughtfulness.[37] This physical trait is also a characteristic of Claude Vignon, a talented male writer in *La Comédie humaine*.[38]

The longest reference to Greece is also the most explicit allusion to Félicité's female masculinity. The narrator admires her lower back and states:

> La chute des reins est magnifique, et rappelle plus le Bacchus que la Vénus Callipyge. Là, se voit la nuance qui sépare de leur sexe presque toutes les femmes célèbres, elles ont là comme une vague similitude avec l'homme, elles n'ont ni la souplesse, ni l'abandon des femmes que la nature a destinées à la maternité; leur démarche ne se brise pas en un mouvement doux. Cette observation est comme bilatérale, elle a sa contrepartie chez les hommes dont les hanches sont presque semblables à celles des femmes quand ils sont fins, astucieux, faux et lâches.[39]
>
> [The lower back is magnificent and recalls more Bacchus than Venus Callipyge. Therein is visible the nuance that separates almost all famous women from their sex, they have there a certain vague similarity with men, they have neither the suppleness nor the abandon of women that nature destined to motherhood; their gait is not broken by a soft move. This observation is almost two-sided, it has its counterpart in men whose hips are almost similar to those of women when they are astute, shrewd, false and cowardly.]

The comparison with Bacchus is reminiscent of Lucien's description, but it serves a different purpose. Whereas the comparison intended to highlight Lucien's effeminacy in *Illusions perdues*, as Bacchus is an effete god, here it stresses Félicité's masculinity, since Bacchus is a male god in opposition to the female and highly feminine Callipygian Venus. This characteristic is a pretext for the narrator to generalize about those women who differ from others because of their solid gait. This observation can be linked to those of Cabanis and Esquirol who pointed out the typical walk of homosexuals. If the gait of so-called effeminate men is judged to be feminine, as was also indicated in *Mademoiselle de Maupin*,

36 Ibid., 146.

37 Michael D. Garval, *A Dream of Stone: Fame, Vision, and Monumentality in Nineteenth-Century French Literary Culture* (Newark, DE: University of Delaware Press, 2004), 117 and 121. See also Didier, 'Le corps féminin dans *Lélia*', 639.

38 Balzac, *Béatrix*, 187 and 205–06.

39 Ibid., 148.

it appears logical that the walk of masculine women should be depicted as manly. The allusion to the poet of *Illusions perdues* is clear when the narrator mentions men with female hips who are shrewd, as well as hypocritical and cowardly – moral weaknesses that have been highlighted in Lucien with similar words.

Alternatively, cultural references to Egypt are far more prevalent when defining the physical characteristics of masculine women. Napoleon's campaign in Egypt (1798–1801) and its concomitant scientific expedition had raised nineteenth-century France's interest in this ancient civilization and its alleged mysteries.[40] Along with Cleopatra who is depicted as a short authoritarian brunette, Félicité is compared to Isis.[41] One of the main goddesses of the Egyptian pantheon, Isis is known for her intelligence and her skills in witchcraft, and she is interpreted as mysterious, ideally feminine and yet androgynous: 'Ce visage, plus rond qu'ovale, ressemble à celui de quelque belle Isis des bas-reliefs éginétiques. Vous diriez la pureté des têtes de sphinx, polies par le feu des déserts, caressées par la flamme du soleil égyptien' [This face, more round than oval, resembles that of a beautiful Isis in bas-reliefs from Egina. It looks like the purity of sphinxes' heads, polished by the fire of deserts, caressed by the flame of the Egyptian sun].[42] Egypt sustains an enigmatic atmosphere that presents Félicité herself as an enigma.[43] Moreover, by alluding to a powerful Egyptian queen and an astute Egyptian goddess who deceived the sun-god Ra to steal his powers, the narrator portrays Félicité as a *femme fatale*, that is, an unattainable woman who potentially threatens male authority.[44] This comparison with beautiful seductresses highlights her femininity, while asserting her

40 See for instance Michel Brix, 'Nerval et le rêve égyptien', *Romantisme*, 120 (2003), 37–46 (39).

41 Balzac, *Béatrix*, 149.

42 Ibid., 145. On Isis, see Brix, 'Nerval et le rêve égyptien', 39–40; Vanoncini, 'La femme-livre', 157.

43 Prendergast, 'Towards a Re-assessment of *Béatrix*', 52–53. This important trope will be developed in the final section of this chapter.

44 On *femmes fatales*, see Mario Praz, *The Romantic Agony*, Angus Davidson, trans., foreword Frank Kermode, 2nd edn (Oxford: Oxford University Press, 1970 [1933]), 213–16.

pretension to power and, as will be seen, depicting her as a potential threat to masculinity.

The reference to bas-reliefs and sphinxes also highlights Félicité's metaphorical transformation into a statue. This process of petrification has been analysed by Harkness as objectifying female protagonists in order to restrict their autonomy and abandon them to male desire.[45] In Sand's novel, Lélia is turned into a stone or a marble monument by the male protagonists. Sténio, for instance, when talking about her to the Irish priest Magnus, states: 'Ce granit, incessamment lavé par l'eau qui s'échappe du glacier, n'est pas plus froid que la main de Lélia' [This granite, continuously washed by the water that flows from the glacier, is not as cold as Lélia's hand].[46] The reference to stone paradoxically renders corporeal an otherwise allegorical character. Lélia's dual nature is expressed by Magnus as 'corps et âme' [body and soul].[47] The junction between her physicality and her spirituality is – as was the case for Félicité – her forehead: this wide, bright and polished physical feature is uncommonly described in the novel in order to convey her sublimity.[48] Although objectified and idealized by male protagonists, Lélia is nonetheless able to reverse her own petrification and to petrify her admirers as a modern Medusa. Magnus confesses: 'le regard méchant et froid de Lélia [...] me pétrifiait' [Lélia's wicked and cold gaze

45 Harkness, 'Resisting Petrification', esp. 166–69. Lélia's petrification is at the core of Harkness's article, but he also notes this process in *Béatrix* about Béatrix de Rochefide (171). See also Harkness, *Men of their Words*, 127–29; Jad Hatem, 'Scission du cœur et désordre du corps dans *Lélia* de Sand', *Romantisme*, 91 (1996), 19–34 (32–33).

46 Sand, *Lélia*, 83. See also 221, for instance.

47 Ibid., 81. On the opposition between body and mind, see Didier, 'Le corps féminin dans *Lélia*', 636; Nancy Rogers, 'Psychosexual Identity and the Erotic Imagination in the Early Novels of George Sand', *Studies in the Literary Imagination*, 12/2 (Fall 1979), 19–35 (22); Hatem, 'Scission du cœur', esp. 20; Evelyne Ender, '"Une femme qui rêve n'est pas tout à fait une femme": Lélia en rupture d'identité', *Nineteenth-Century French Studies*, 29/3–4 (Spring–Summer 2001), 226–46. On the corporeality of the first Lélia, see Isabelle Hoog Naginski, 'Les deux *Lélia*: Une réécriture exemplaire', *Revue des sciences humaines*, 226 (April–June 1992), 65–84 (81 and 83).

48 Sand, *Lélia*, 13, 46, 60, 70 and 164.

[…] petrified me].[49] This ability to escape reification, displayed by Lélia and other masculine women, creates anxiety.

The process of monumentalization in *Béatrix* mostly originates in the narrator's portrayal of Félicité ('le granit de la statue égyptienne adouci par le temps' [the granite of Egyptian statue softened by time], 'femme colossale' [colossal woman]), rather than being the product of male protagonists' attempts to define and therefore limit the masculine woman.[50] It firstly expresses her innate mystery, as Françoise Van Rossum-Guyon observes: 'Elle [Félicité] est décrite de façon à prendre des proportions "colossales" dignes de la statuaire antique. Les comparaisons avec les héroïnes de la Fable et les divinités anciennes […] confèrent au personnage une dimension surhumaine, mystérieuse, mythique et quasiment divine' [Félicité is described so as to take 'colossal' proportions worthy of ancient statuary. Comparisons with heroines of Fables and ancient divinities […] bestow a superhuman, mysterious, mythical and almost divine dimension on the character].[51] Félicité's transformation into a statue also suggests her grandeur. She is not just any sculpture: she seems to have been fashioned by the great Greek sculptor Phidias.[52] This monumentalization mirrors a sense of admiration for the writer Camille Maupin whose transformation into a living monument aims to recognize her literary greatness during her lifetime: in short, she is a masterpiece producing masterpieces. This follows the nineteenth-century tendency analysed by Michael Garval that consisted in identifying writers and their works as monuments.[53] In other words, it highlights her exceptionality and distances her from the popular masses of Guérande by depicting her as a powerful and cold statuary goddess. This process nevertheless runs the risk of objectifying the exceptional woman and alienating her from other women given that the latter are denied her power.

49 Ibid., 86.

50 Balzac, *Béatrix*, 146–47 and 148.

51 Van Rossum-Guyon, *Balzac et la littérature réfléchie*, 84.

52 Balzac, *Béatrix*, 147.

53 She concretizes the nineteenth-century mindset according to which '[t]he truly great writer accedes naturally to statufication': Garval, *A Dream of Stone*, 59. See also 34–61.

The same process of petrification is used in *La Cousine Bette* when Valérie Marneffe deploys her fashion skills to improve the appearance of her friend. Thanks to the use of cosmetics and elegant clothes, Lisbeth's unkempt looks are highlighted as an asset. The narrator observes:

> Bette, comme une Vierge de Cranach et de Van Eyck, comme une Vierge Byzantine, sorties de leurs cadres, gardait la froideur, la correction de ces figures mystérieuses, cousines germaines des Isis et des divinités mises en gaine par les sculpteurs égyptiens. C'était du granit, du basalte, du porphyre qui marchait.[54]
>
> [Like a Virgin painted by Cranach and Van Eyck, like a Byzantine Virgin, coming out of their frames, Bette kept the coldness, the rigid bearing of these mysterious figures, first cousins of Isis and the divinities put in a sheath by Egyptian sculptors. She was walking granite, basalt and porphyry.]

The explicit comparison with Nordic and Byzantine virgins is flattering for Lisbeth, as it bestows the positive attribute of ideal beauty upon her. It also indirectly connects her to Egyptian statues. As the virginal figures of these Flemish and Germanic painters are usually fair, it is unlikely that the resemblance between the latter and Lisbeth should relate to their complexion. What links them, if not their common coldness? As for Félicité, the Egyptian references underscore Lisbeth's mystery. As for Lélia, the emphasis on stone underlines her coldness and inflexibility, thus contradicting the more positive references to her artistic qualities.[55] Although the topic of the ugly old woman who tries to appear beautiful with the aid of finery may serve comical purposes, the reference to petrification here cancels out any preposterous implications to show the spinster as an estranged and menacing individual.

Certainly, the physical characteristics of Félicité, Lélia and Lisbeth do not suggest that they are mannish women and, apart from Lisbeth, do not prevent them from being the love interests of younger men. Nevertheless, the references to savagery, the emphasis on the dark hair of the protagonists, their exoticization and petrification render these women unconventional.

54 Balzac, *La Cousine Bette*, 199.

55 The emphasis on stone, however, is associated with fire in the extract expressing Lisbeth's jealousy, already quoted (Balzac, *Bette*, 137).

These characteristics position them as enigmatic, inaccessible and even threatening figures, and thus highlight the writers' unease towards their masculine creations.

The Intellectual and Independent Woman

Although the masculinity of masculine women is not immediately manifest in their physical portraits, it becomes more apparent in the way their mentality and activities are depicted. Balzac and Sand follow the path of scientists by deeming psychological make-up, habits and activities to be relevant when determining the masculinity of masculine women. I will briefly explore the interests of fictional masculine characters before looking in a more detailed fashion at their intellectual activities and livelihood. In drawing a parallel with women's conditions during the July Monarchy, I show how these novels, consciously or not, advocate women's autonomy.

Let us briefly examine our three protagonists' areas of interest. Firstly, two of them, Lélia and Félicité, are fond of smoking. The latter smokes cigarettes, cigars and hookah, whilst Lélia takes snuff.[56] For a nineteenth-century reader, smoking was coded as masculine, since only men smoked; emancipated women were exceptions.[57] Lélia and Félicité also ride horses and the latter even hunts.[58] Furthermore, both women cross-dress, although, like Fragoletta in Latouche's novel, these outfits are put on for exceptional circumstances, such as a masquerade ball and a walk on the cliffs.[59] Diana Holmes points out the freedom of movement that male attire allowed in comparison with the restraint of the ample skirts and delicate shoes

56 For instance, Balzac, *Béatrix*, 120, 171–72, 255–56, 262 and 287; Sand, *Lélia*, 63.

57 Sohn, *'Sois un homme!'*, 34–36. For more on female dandies or *lionnes*, see Miranda Gill, *Eccentricity and the Cultural Imagination in Nineteenth-Century Paris* (Oxford: Oxford University Press, 2009), 87 and 89.

58 Sand, *Lélia*, 311; Balzac, *Béatrix*, 162 and 209.

59 Sand, *Lélia*, 45; Balzac, *Béatrix*, 298.

of female clothes, which echoed social coercion and justified masculine women's desire to be rid of them.[60] Finally, unlike stereotypical women, Félicité, Lélia and Lisbeth are generally indifferent to fashion and appearance.[61] The difference between these protagonists is that Lisbeth is regarded as unattractive, whereas Lélia and Félicité are considered beautiful but simply do not care about the attention that their beauty attracts. However, Béatrix accuses Félicité of showing women's pettiness and coquetry in their rivalry for Calyste's love.[62]

Beyond these trivial characteristics, Lélia, Lisbeth and Félicité distinguish themselves by displaying 'des qualités d'homme' [male qualities].[63] They are depicted as being strong, active, bold and hard-working.[64] Their physical strength is merely a metaphor for their moral integrity. Félicité, for instance, shows determination in the way that she successfully leads her life and career, whilst Lisbeth works in the *passementerie* [trimmings] trade and teaches herself how to read and write at a mature age.[65] Their strong character is perceptible in the influence that they all exercise over men. Lisbeth's 'volonté puissante' [powerful will], for example, enables her to dominate the weak Wenceslas.[66]

More importantly, these three characters are perceived as very clever, at a time when reason was deemed to be a male prerogative. *Béatrix* emphatically states Félicité's intelligence. Beyond her remarkable career as a writer, her uncommon intellect, perspicacity and eloquence are stressed by the narrator.[67] Lisbeth is also shown to be a clever woman, although her intelligence is presented as less refined and more analogous to common sense than that of Félicité.[68] Unlike *Béatrix*'s narrator, that of *La Cousine*

60 Diana Holmes, *French Women's Writing: 1848–1994* (London and Atlantic Highlands: Athlone, 1996), 26.

61 Sand, *Lélia*, 137; Balzac, *Béatrix*, 143, 166 and 273; Balzac, *La Cousine Bette*, 61–62.

62 Balzac, *Béatrix*, 257.

63 Balzac, *La Cousine Bette*, 59.

64 For instance, Sand, *Lélia*, 103, 147 and 158; Balzac, *Béatrix*, 224 and 298; Balzac, *La Cousine Bette*, 90–92 and 140.

65 Balzac, *Béatrix*, 142; Balzac, *La Cousine Bette*, 55.

66 Balzac, *La Cousine Bette*, 90. See also 91 and 100.

67 Balzac, *Béatrix*, 149, 197 and 325.

68 Balzac, *La Cousine Bette*, 57–58.

Bette does not need to tell the reader that Lisbeth is intelligent; her machinations speak for themselves. As the narrator sums up: 'Lisbeth pensait, madame Marneffe agissait' [Lisbeth thought, Madame Marneffe acted].[69]

As for Lélia's intelligence, Sand adopts a similar strategy of showing rather than telling. In other words, the narrator spends little time stating that Lélia is a smart woman but allows her to speak and demonstrate her ability to reason by herself. Waller highlights her 'power of rhetoric'.[70] For instance, early on, Lélia denounces vanity as one of men's greatest and least acknowledged passions; despite their selfishness, men act with benevolence in order to aggrandize themselves. She goes on: 'Le christianisme lui-même, qui a produit ce qu'il y a de plus héroïque sur la terre, le christianisme, qu'a-t-il pour base? L'espoir des récompenses, un trône élevé dans le ciel' [Christianity itself, which produced the most heroic things on Earth, Christianity, what are its grounds? The hope of reward, a throne raised up to Heaven].[71] Her choice of example is particularly pertinent because Christianity holds an important place in the system of beliefs of the recipient of her speech, Sténio, and, allegedly, the reader. Through the simplicity of her question and her brief but powerful response, Lélia deconstructs charity, an essential Christian virtue, as deeply entrenched in human self-interest. Whilst such reasoning seems to rule in favour of her detractors who accuse her of atheism – more on this in the final part of this chapter – it mainly illustrates her ability to challenge common beliefs, elaborate complex thoughts and encourage her listeners to reason by themselves.

Not only are Félicité and Lélia remarkably clever, they are also represented as intellectual women, who (as was shown in Chapter 1) are criticized and pitied by the scientific community. In Félicité's case, Balzac establishes strong links between his heroine and the literary context of his time by addressing the issue of the *femme auteur* [woman writer]. The attempts made by intellectual women to become writers were regarded with mockery and even hatred by men in the nineteenth century. As Christine Planté has shown, *la femme auteur* was considered to be mannish because

69 Ibid., 204–05.

70 Waller, *The Male Malady*, 157.

71 Sand, *Lélia*, 29.

she exercised a profession and intellectual faculties that were reputed to be masculine.[72] In *De la femme*, Virey claimed: 'Celles qui se sont le plus distinguées dans cette carrière [la littérature] ont souvent mérité l'épithète *mascula* qu'Horace donnait à Sapho' [Women who most distinguished themselves in literature often earned the epithet *mascula* that Horace gave to Sappho].[73] The adjective *mascula* appears in Horace's *Epistles* (I, 19, verse 28: *temperat Archilochi Musam pede mascula Sappho*, that is: 'manly Sappho moderates with her foot Archilochus' Muse').[74] Olivier Thévenaz has indicated that this adjective has raised controversy since Antiquity. Sappho was considered masculine either because of her fame in the masculine art of poetry or because of her lesbian reputation.[75] Unsurprisingly, Félicité des Touches is compared with Sappho several times.[76]

Female writers were deemed to be a threat to the separation between feminine receptiveness and instinct, and masculine originality and intelligence, and were thus treated with hostility.[77] They questioned the social order because they proved that they were able to think for themselves and their minds were not therefore inferior to those of men.[78] Their refusal to conform to gendered functions and their reputed lack of feminine delicacy raised scientists' concern about their ability to fulfil their domestic duties:

> Par quel attrait peuvent-elles fixer le jeune homme qui cherche une compagne? Quels secours peuvent en attendre des parents infirmes ou vieux? Quelles douceurs répandront-elles sur la vie d'un mari? Les verra-t-on descendre du haut de leur génie, pour veiller à leurs enfants, à leur ménage? Tous ces rapports si délicats qui font le

72 Christine Planté, *La petite sœur de Balzac: Essai sur la femme auteur* (Paris: Seuil, 1989).

73 Virey, *De la femme*, 5.

74 Translation from Latin in: Anastasia-Erasmia Peponi, 'Fantasizing Lyric: Horace, *Epistles*, 1.19' in Michael Paschalis, ed., *Horace and Greek Lyric Poetry* (Rethymnon: University of Crete, 2002), 19–45 (40).

75 Olivier Thévenaz, 'Sappho au banquet d'Horace', *Camenae*, 12 (June 2012), 1–14 (11): <http://www.paris-sorbonne.fr/IMG/pdf/2-_O-_Thevenaz.pdf>, accessed 29 March 2017.

76 Balzac, *Béatrix*, 154 and 255.

77 Holmes, *French Women's Writing*, 30.

78 Planté, *La petite sœur de Balzac*, 57.

> charme et qui assurent le bonheur de la femme, n'existent plus alors: en voulant étendre son empire, elle le détruit.[79]
>
> [With what charm can they secure a young man seeking a partner? What assistance can infirm or old parents expect from them? What pleasures will they bestow on the life of a husband? Will they be seen descending from the height of their genius to look after their children and household? All these delicate relations that create the charm and ensure the happiness of women no longer exist: in wanting to expand their influence, they destroy it.]

The portrait of the ideal woman is realized in contrast to the intellectual woman's inability to display any zeal for familial life. As Holmes summarizes, '[t]he truly feminine woman, worthy of a degree of formal respect, was the chaste wife and mother confined within the home under male protection'.[80] Her functions were solely grounded in the domestic sphere and limited to her interactions with other relatives (husband, children and parents). She was not legally responsible for them, given that children were under their father's authority, but she had to ensure their happiness by her continuous care and affection. In *La Cousine Bette*, Adeline Hulot, devoted to her family to the point of abnegation – is she not compared to a martyr? – perfectly captures the essence of the ideal wife and mother who sacrifices her own happiness for that of her husband and children.[81]

This ideal woman was also depicted in arts and in politics. French women were victims of the disguised misogyny of nineteenth-century Romantic arts. The ideal image of women as innocent, consoling and nurturing figures was paradoxically depreciative because it also portrayed them as infantile, submissive and dedicated only to men's interests.[82] Furthermore, this 'disempowering idealization of women' was inherent to republican discourses.[83] Worshipping women as incarnations of the virtues of the nation (by means of allegories of France and the Republic, such as *Marianne*) enabled men to remove them from the political scene (as voters, for instance)

79 Cabanis, *Rapports*, I, 363.

80 Holmes, *French Women's Writing*, 9.

81 Balzac, *La Cousine Bette*, 45–46, 51, 287 and 430.

82 Claire Goldberg Moses, *French Feminism in the Nineteenth Century* (Albany: State University of New York Press, 1984), 17–18.

83 Holmes, *French Women's Writing*, 14.

since, as objects of men's adoration, they were not allowed to be subjects.[84] Holmes summarizes 'the problem of the woman writer' as 'how she makes her voice heard when what is asked of a woman is not art, or opinion, but only that "something else" which signifies the fulfilment of male needs'.[85]

In this context, Félicité des Touches constitutes both the paragon of the woman writer and an exception in the literary world. Balzac's narrator does not conceal his admiration for his character, whilst taking care to point out that, except for George Sand (who is here considered not to be the inspiration, but a younger rival that has overshadowed Camille Maupin's own success), no real woman writer is talented enough to measure herself against Félicité des Touches.[86] The narrator praises Camille Maupin but denies merit to her non-fictional colleagues: 'Elle [Camille Maupin] n'eut d'ailleurs rien de la femme auteur' [Camille Maupin had indeed nothing of the woman writer].[87] Camille Maupin's career is impressive by any female, male, fictional or real standards. A pioneer of romanticism, she has written plays influenced by Shakespeare and Lope de Vega and a sentimental novel in the manner of Constant's *Adolphe*. For these reasons, the narrator insists that 'elle joue un rôle dans l'histoire littéraire de notre époque' [she is playing a part in the literary history of our time].[88] Her intellectual faculties allow her to help her great-uncle in his research and even to write, at a young age, three of his books, allegedly relating to archaeology and history.[89] These two disciplines constituted high literary genres in the nineteenth century and were judged inaccessible to women because of their so-called lack of reason.[90] The fact that Félicité's uncle takes credit for her writing fictionalizes the phenomenon described in *Ni vues ni connues* [*Unseen, unknown*], published by the feminist goup Georgette Sand, as

84 Ibid., 14–15; Bertrand-Jennings, *Un autre mal du siècle*, 11; Allan, *Writing the Self*, 90. On the similar perception of women in poetry, see Planté, *La petite sœur de Balzac*, 190.

85 Holmes, *French Women's Writing*, 4.

86 Balzac, *Béatrix*, 154.

87 Ibid.

88 Ibid., 137.

89 Ibid., 139–40.

90 Planté, *La petite sœur de Balzac*, 242–43.

the *invisibilisation* [marginalization] of women in history, notably due to men claiming ownership of their works and discoveries.[91] She is also a talented musician and composer, a famous *salonnière* and an authority on the Parisian cultural scene.

As Louis-Philippe's reign was regarded as 'un véritable âge d'or de la femme auteur' [a true golden age for the woman writer],[92] Camille Maupin should simply constitute one representative, perhaps the most talented one – as is the prerogative of fiction –, of women writers. Yet, apart from George Sand and, to a lesser extent, Madame de Staël, intellectual women are almost non-existent in the world of *Béatrix*. This illustrates the fact that, regardless of many women publishing books, writing was still perceived as a masculine activity during the July Monarchy. Calyste's father, the reactionary Baron du Guénic, shows his incredulity when he hears that Félicité des Touches writes plays and books, and does not know any nineteenth-century female author. His awareness of women writers is limited to Mademoiselle de Scudéry and Madame de Sévigné, and he attributes these 'prodigies' to Louis XIV and his court.[93] Certainly, Baron du Guénic and the conservative community of Guérande are not literary experts, but Balzac's narrator shares their ignorance and misogyny where gifted women are concerned: 'en vingt siècles, à peine compte-t-on vingt grandes femmes' [in twenty centuries, scarcely twenty great women can be counted].[94] These characters and the narrator of *Béatrix* reflect the refusal of the critics of the July Monarchy to acknowledge works by female novelists and hence also their eradication from the French literary pantheon under the influence of Balzac and Stendhal who ensured the 'masculinization of a previously feminine form', as Margaret Cohen has shown.[95]

By advocating her right to pursue intellectual activities, as Félicité does, the masculine woman appears to be freed from the feminine imperatives of submission and discretion imposed by patriarchal norms. She therefore

91 Collectif Georgette Sand, *Ni vues ni connues: Panthéon, Histoire, mémoire: Où sont les femmes?* (Paris: Hugo Doc, 2017), 10–13.

92 Planté, *La petite sœur de Balzac*, 43.

93 Balzac, *Béatrix*, 130.

94 Ibid., 137.

95 Cohen, *The Sentimental Education*, 13.

achieves some of the goals of nineteenth-century feminists, notably with regard to education. Although not strictly speaking a feminist, George Sand was indignant at the discrepancy between male and female educations.[96] The educational opportunities for middle- and upper-classes girls (let alone working-class girls) were restricted in this period.[97] Educated in private schools or convents, girls were taught reading, spelling, history, geography and, above all, skills that were thought to be necessary to become a good homemaker (cooking and sewing), as well as recreational arts (painting, singing and playing the piano). This educational programme, devoid of disciplines such as Latin, sciences and mathematics, did not lead to the *baccalauréat* [secondary school certificate]. It conformed both to the domestic role that was expected of a married woman and to the belief in woman's 'natural' lack of rationality and her excessive sensitivity. The reform of education in France by Guizot, the Minister of Public Education, in 1833, compelled every commune to organize a primary school for boys.[98] However, it angered the writers of the feminist journal *La Tribune des femmes* [*Women's Platform*] because no action was taken for girls' education.[99]

Under these circumstances, it comes as no surprise that the education of fictional masculine women diverges from that of ordinary girls, as did that of many female writers of the nineteenth century.[100] As the narrator points out, Félicité has escaped 'l'éducation futile donnée aux femmes, [...] les enseignements maternels sur la toilette, sur la décence hypocrite, sur les grâces chasseresses du sexe' [the superficial education given to women, [...] a mother's lessons on dresses, hypocritical decency and the hunting graces of the fair sex].[101] She is an autodidact, as she 's'éleva toute seule, en garçon'

96 Kristina Wingård Vareille, *Socialité, sexualité et les impasses de l'histoire: L'évolution de la thématique sandienne d''Indiana' (1832) à 'Mauprat' (1837)* (Uppsala: Almqvist & Wiksell, 1987), 392–94. On Sand's reservations towards feminist movements, see 397–408.

97 Holmes, *French Women's Writing*, 11 and 13–14; Moses, *French Feminism*, 32–33.

98 For more on the law of 1833, see Broglie, *La Monarchie de Juillet*, 327.

99 For more on the reaction of feminists, see Moses, *French Feminism*, 83–84.

100 Alison Finch, *Women's Writing in Nineteenth-Century France* (Cambridge: Cambridge University Press, 2000), 9–16.

101 Balzac, *Béatrix*, 143.

[raised herself, as a boy].[102] This self-education is motivated by her passion for reading and her excellent memory. Félicité also studies music, not as a mere recreational art, but in order to compose operas.[103] Her situation, nonetheless, is exceptional because her instruction is regarded as astounding and her readings prodigious.[104] The narrator highlights the contrast between the expectations of such an education and the protagonist's reality. Excessive knowledge was believed to deprave girls, notably through the reading of 'licentious' novels, as suggested by the scientists Moreau de la Sarthe and Virey.[105] Félicité, nevertheless, remains chaste.[106] However, the narrator's conclusions are not overall positive, as her readings are considered to be a 'dépravation de l'intelligence' [depravity of intelligence].[107] He also questions the young Félicité's intelligence, as he suggests that it consists in accumulating knowledge rather than reasoning. More significantly, in accordance with scientific theories of the time, Félicité's body cannot endure her highly intellectual activities. She becomes ill, even though she recovers from her sickness, paradoxically, by undertaking another masculine activity: horse-riding.[108] In brief, the pages describing Félicité's education highlight its masculine component. They oscillate between admiration for this exceptional woman and concern for the danger that a so-called masculine upbringing can have on women to the point that the novel paradoxically defends manly physical activities in order to complement a masculine mind.

Although the description of Lélia's education is not as detailed as that of Félicité, they share some similarities. This is probably due to the fact that Balzac was inspired by Sand's description of her heroine's childhood to write the story of his own protagonist.[109] Lélia's confession to her sister Pulchérie

102 Ibid., 139.
103 Ibid., 140–41.
104 Ibid., 139.
105 Moreau de la Sarthe, *Histoire naturelle*, II, 238, 273 and 278; Virey, *De la femme*, 92, 98 and 106.
106 Balzac, *Béatrix*, 139.
107 Ibid.
108 Ibid., 140.
109 Nicole Mozet, *George Sand: Écrivain de romans* (Saint-Cyr-sur-Loire: Pirot, 1997), 75; Naginski, '*Lélia*', 102–03.

reveals her intellectual curiosity and her interest in science, spiritualism and poetry.[110] In these pages, the dangers of knowledge are emphasized given that, as for Félicité, study weakens Lélia's health. In Balzac's novel, concern about whether intellectual activities suit women's constitution is raised in echo with scientists' apprehensions. In Sand's novel, however, the criticism lies elsewhere. Women's study of poetry is not disparaged in itself but because it causes a discrepancy between their expectations and their inability to satisfy them owing to the mediocrity of their position in society as objects of a man's love.

Another salient characteristic of masculine women is their celibacy. As spinsters, Lélia, Lisbeth and Félicité represent social aberrations because they escape the social system drawn by the Napoleonic Code according to which women had the same legal status as minors and the insane and were submitted to the authority of their husband or their father[111]. Deprived of parental authority (or any sort of authority), a woman owed obedience to her husband, who owed her protection in return. As only the rights and duties of the married woman were defined, the unmarried woman did not appear to have the same legal status. It was unclear whether she could claim the same rights as a man because of her independence or whether she was determined only by her gender.[112] She was despised for her worthlessness and inability to attract suitors, and was perceived as a burden to her family, since no one was legally required to take care of her.[113]

As unmarried women, fictional masculine women are also financially independent. This puts Lisbeth and Félicité in an exceptional position for their time. According to the Civil Code, married women were not allowed to manage their wealth without the permission of their husbands.[114] Claire Goldberg Moses argues that gaining financial autonomy was almost

110 Sand, *Lélia*, 164–68.

111 *Code civil*, art. 212–26, 53–55.

112 On the debate about unmarried women's status, see Moscovici, 'Gendered Spheres', 469–70 and 478–79.

113 For more on women's choice of unmarried life, see Moses, *French Feminism*, 35; Jane Ussher, *Women's Madness: Misogyny or Mental Illness?* (New York: Harvester Wheatsheaf, 1991), 81.

114 *Code civil*, art. 217, 55.

impossible for working-class women, to which group the embroiderer Lisbeth Fischer belongs.[115] However, Félicité and Lisbeth owe their privileged situation not only to their celibacy, but mostly to their own skills. Félicité competently manages the fortune that she has inherited from her family and probably has an independent income generated by her writing.[116] Likewise, Lisbeth is an able worker in the *passementerie* industry who exercises supervisory responsibility within the factory. She cleverly assumes that the defeat of the Empire will have dire consequences for this industry. She thus rejects the partnership that her employers offer her just before their bankruptcy and becomes a simple independent worker with her own, albeit modest, income.[117]

The exceptional status of these masculine women raises doubts about their conventionally female nature, expressed both by other characters and by themselves. Félicité is the most emblematic case. She acknowledges her lack of femininity by saying 'Quelque peu femme que je sois' [However little of a woman as I may be].[118] Yet, she does so to show that she is not entirely deprived of feminine ability in the art of seduction and, therefore, that she is still a woman. Calyste goes further, since he denies Félicité's femininity, and even her femaleness, in a letter to Béatrix:

> vous m'avez démontré que Camille est un garçon: elle nage, elle chasse, elle monte à cheval, elle fume, elle boit, elle écrit, elle analyse un cœur et un livre, elle n'a pas la moindre faiblesse, elle marche dans sa force; elle n'a ni vos mouvements déliés, ni votre pas qui ressemble au vol d'un oiseau, ni votre voix d'amour, ni vos regards fins, ni votre allure gracieuse; elle est Camille Maupin, et pas autre chose; elle n'a rien de la femme [...].[119]
>
> [You have proved to me that Camille is a boy: she swims, hunts, rides a horse, smokes, drinks, writes, analyses a heart and a book, she has no weakness, she walks in strength; she has neither your fine moves nor your step that resembles the flight of a bird, nor your voice of love, nor your shrewd gaze, nor your graceful appearance; she is Camille Maupin and nothing else; she has nothing of a woman [...].]

115 Moses, *French Feminism*, 27.
116 Balzac, *Béatrix*, 141–42.
117 Balzac, *La Cousine Bette*, 55–56.
118 Balzac, *Béatrix*, 289.
119 Ibid., 278.

True, this portrait is written by a lover who wants to convince the woman whom he endeavours to seduce of his indifference to her rival, and it is constructed in opposition to Béatrix's own characteristics. Nevertheless, since Béatrix embodies the typical woman (owing to her fair hair, undulating gait, delicate voice and grace), the pronoun *vous* progressively designates, not only her, but all the persons of her sex. Calyste focuses on Félicité's physical and intellectual activities, as well as her lack of feminine characteristics, which Béatrix possesses. Like Virey, the young man chiefly appears to be obsessed with her non-feminine walk and assimilates her to a 'butch' woman. His virulent rejection of Félicité's seduction reveals the combination of fascination and fear that masculine women provoke.

In short, the figure of the masculine woman seems to outshine the traditional feminine woman – and perhaps the male protagonists – by her intellectual abilities and her independence. It is Camille Maupin, and not a man, for example, who incarnates the archetype of the writer in *La Comédie humaine.*[120] This allegedly flattering and positive portrait of masculine women is, however, thwarted by an intrinsic anxiety towards them that is reminiscent of scientists' concern for women who behave like men and enjoy masculine activities. In what follows, I analyse the depiction of masculine women's unconventional sexuality as a construct for expressing this fear, but also their autonomy.

The Sexuality of Masculine Women

Masculine women's relations with men reflect their non-conformism. Their sexuality represents a rupture with, and even a rejection of, traditional heteronormative sexuality. The female protagonists of *Lélia*, *La Cousine Bette* and *Béatrix* are bolder than many real-life feminists. To some extent, they put into practice the programme of Charles Fourier and the Saint-Simonian Prosper Enfantin who advocated women's sexual emancipation and free love. Indeed, the agenda of these philosophers

120 Van Rossum-Guyon, *Balzac et la littérature réfléchie*, 88–92.

remained mostly theoretical and encountered little popularity both among men – which is not surprising – and feminist women, because it did not favour the latter's social liberation and need for respect.[121] On the contrary, fictional masculine women take the path of sexual emancipation. The novels suggest, or even depict, alternative sexualities, including virginity, asexuality, sexuality with multiple partners and lesbianism. These sexualities escape male control because they are generally freely chosen by the female characters. They constitute a threat to men's supremacy in the private sphere and even raise the fear that the threat can be extended to the public sphere.

Lisbeth's virginity is an intrinsic part of her character because her role in the novel is defined by her status as the poor celibate relative of an apparently wealthy family. Lisbeth's complexity and position at the core of the plot nevertheless prevent her from being a stereotypical old maid.[122] Her virginity has no positive connotation, as references to the Virgin Mary are reserved for her married cousin Adeline.[123] Here, the novel adopts a similar perspective to that of nineteenth-century physicians. Chapter 1 has mentioned that prolonged chastity was frowned upon because it was believed to cause diseases such as anaemia, hysteria and cancers. In *La Cousine Bette*, the narrator adopts an even more severe perspective than that of physicians. Lisbeth's virginity is depicted as worse than a disease; it is a monstrosity:

> La Virginité, comme toutes les monstruosités, a des richesses spéciales, des grandeurs absorbantes. La vie, dont les forces sont économisées, a pris chez l'individu vierge une qualité de résistance et de durée incalculable. Le cerveau s'est enrichi dans l'ensemble de ses facultés réservées. Lorsque les gens chastes ont besoin de leur corps ou de leur âme, qu'ils recourent à l'action ou à la pensée, ils trouvent alors de l'acier dans leurs

121 On Fourier's viewpoint on women's sexuality, see Michael Clifford Spencer, *Charles Fourier* (Boston: Twayne Publishers, 1981), 91–95. On Enfantin, Saint-Simonians and free love, see Moses, *French Feminism*, 47–49 and 70–85; Antoine Picon, *Les saint-simoniens: Raison, imaginaire et utopie* (Paris: Belin, 2002), 93–94 and 133.

122 André Lorant, *Les Parents Pauvres d'Honoré de Balzac: 'La Cousine Bette'–'Le Cousin Pons': Étude historique et critique*, 2 vols (Genève: Droz, 1967), I, 46.

123 For instance, Balzac, *La Cousine Bette*, 48 and 493.

> muscles ou de la science infuse dans leur intelligence, une force diabolique ou la magie noire de la Volonté.[124]
>
> [Virginity, like all monstrosities, has peculiar treasures, absorbing greatness. Life, whose forces are spared, has developed in the virgin individual an incalculable quality of resistance and duration. The brain has been enriched by all its preserved faculties. When chaste people need their bodies or their souls in order to resort to either action or thought, they find steel in their muscles or infused science in their intelligence, a diabolical strength or the dark magic of the Will.]

However, the novel extracts essential qualities from this apparent aberration. As virgins do not exhaust their energy in sexual intercourse, they are thought to find within themselves strength that they can use for other activities. Shortly before this excerpt, the perceptive Valérie identifies an accomplice in the spinster, who is described as 'cette ardente fille, passionnée à vide' [this ardent spinster, passionate without fuel].[125] In other words, the narrator points out that all the ardour that Lisbeth feels and her great potential remain unused and therefore can be invested elsewhere: in plotting and scheming. This idea of virginity increasing one's potential can be found in Menville de Ponsan's text: 'Il est certain que la chasteté conservant la vigueur des fonctions vitales, et reportant dans tous les organes cette surabondance de vie qui se rencontre dans les parties génitales, doit augmenter l'énergie de toutes nos fonctions' [It is certain that, since chastity preserves the vigour of the vital functions and transfers the overabundance of life that is encountered in the genitalia to all the organs, it increases the energy of all our functions].[126] Accordingly, virgins can concentrate on other matters and their intellectual faculties are thus more developed than those of sexually active people, an idea that Balzac's narrator repeats several times: Lisbeth was previously portrayed as '[d]ouée d'une finesse devenue profonde, comme chez tous les gens voués à un célibat réel' [endowed with profound subtlety, like all those who are truly celibate].[127]

124 Ibid., 145.
125 Ibid., 144.
126 Menville de Ponsan, *Histoire médicale*, I, 262.
127 Balzac, *La Cousine Bette*, 58.

While chastity is thus first presented as having some benefits, this assertion is soon refuted. At the end of the quoted extract, a discursive twist points to the diabolical consequences of this 'monstrous' virginity. In other words, the narrator restores the viewpoint that was introduced at the very beginning. This positive digression is a means of enhancing the fundamental flaws of virginity. Physical failures that were correlated to it in scientific texts become moral ones in the fictional text, increased by the supposed qualities of chastity. The physical and, above all, intellectual strength that Lisbeth finds in her own virginity allows her to pursue her evil plan.

As virginity is an essential part of Lisbeth's identity, we need to investigate its causes. Apparently, her celibacy is not a deliberate choice. Firstly, it has been imposed by her family, who preferred to promote Adeline as a candidate for marriage. Secondly, the novel offers a simple, misogynistic and only partially true explanation: Lisbeth is a virgin because no man wanted her – because of her lack of dowry and beauty, along with her obstinate disposition. This is Adeline and Hortense's interpretation, convinced of their cousin's 'célibat perpétuel' [permanent celibacy].[128] The impossibility of Lisbeth getting married makes her act acrimoniously and jealously towards the beautiful Adeline and feeds her desire for retaliation.

However, Lisbeth's virginity is also of her own volition. She comes to Paris in the hope of finding a husband, and Baron Hulot finds her four suitors coming from commerce, administration and the army, while she is courted by an embroiderer.[129] She nevertheless rejects all of them. Her decision to remain unattached has different motivations. She wishes to attract a better-off, more prestigious husband, namely her cousin's brother-in-law, Marshal Hulot. From a medical perspective, Lisbeth, as an unmarried woman, is ill, and her quest for a husband is legitimate for health reasons. However, according to the logic of the novel, seeking a man of better birth than her own is presented as a usurpation of privileges. As the novel keeps reminding its reader, Lisbeth is a country woman and, in the Parisian world, almost a foreigner. Accordingly, remaining celibate is her way of conserving some influence over her close relations despite her

128 Ibid., 70.

129 Ibid., 60.

modest origins, a successful strategy throughout most of the novel. In brief, Lisbeth's virginity, whether or not she is responsible for it, is perceived as a social flaw that cannot find favour in the narrator. Paradoxically, both her celibacy and willingness to find a husband are equally disparaged. Despite its benefits, her virginity is considered monstrous.

Like Lisbeth's virginity, Lélia's complex sexuality deserves further attention. It is often read as frigidity. Pierre Reboul, the editor of *Lélia*, categorically states in his introduction: 'Ne mâchons pas les mots: Lélia est frigide' [Let's be honest: Lélia is frigid].[130] While the word is never used in Sand's text, references to coldness are nonetheless abundant.[131] Such references highlight Lélia's sometimes dubious femininity, in opposition to her warm and highly feminine sister Pulchérie, a famous courtesan. Sténio stresses the coldness of Lélia's body, usually as a metonymy for the supposed coldness of her disposition. For instance, he declares, with increasing resentment: 'Votre main est aussi froide que le marbre d'où vous sortez' [Your hand is as cold as the marble whence you came], 'Vous êtes froide, Lélia, oh! froide comme le marbre!' [You are cold, Lélia, oh! as cold as marble!] and 'c'est vous dont la froide haleine a flétri ma jeunesse dans sa fleur' [your cold breath wilted the prime of my life].[132]

Beyond these allegorical references to coldness, the text suggests that Lélia feels no pleasure during intercourse. When encountering her sister in the gardens of Prince dei Bambucci after years of separation, Lélia confesses her malady to Pulchérie and calls it '[l]a froideur de mes sens' [the coldness of my senses].[133] She also refers to a form of exhaustion of her body due to her erudite activities – this reminds us of doctors' warnings against women's intellectual activities in Chapter 1 – which may be a cause for her

130 Pierre Reboul, 'Introduction', in *Lélia*, iii–lxvii (xlviii).

131 Didier ('Le corps féminin dans *Lélia*', 641) suggests that listing the occurrences of the word 'froid' would be meaningful. See also Ender, 'Une femme qui rêve', 230; Harkness, 'Resisting Realist Petrification', 164. For more on the opposition between Lélia and Pulchérie, see Didier, 'Le corps féminin dans *Lélia*', 637; Rogers, 'Psychosexual Identity', 26–28; Vareille, *Socialité*, 199–200.

132 Sand, *Lélia*, 48, 216 and 234.

133 Ibid., 169. See also 315.

impotence (she uses the adjective *impuissant* several times).[134] However, this so-called frigidity must be further explained and relativized.

Before doing so, we need to understand what frigidity meant in George Sand's times. Virey's entry 'Frigidité et froideur' [Frigidity and coldness] in the *Dictionnaire des sciences médicales* defines frigidity as 'l'état d'un individu de l'un ou de l'autre sexe, mais principalement de l'homme, qui se montre impuissant ou incapable de génération, et même de coït' [the state of an individual of either sex, but mostly male, who is revealed to be impotent or incapable of procreation and even of coitus].[135] While frigidity is nowadays regarded as more common amongst women, Virey's article considers it to be a mostly male characteristic and, subsequently, develops the case of men at the expense of women.[136] Yet, paradoxically, the scientist specifies that there are more frigid women than frigid men.[137] This internal contradiction can be explained by the fact that, even though female frigidity was considered to be more frequent, it was regarded to be of little importance given that it did not prevent them from being fertile and fulfilling their marital duties, in contrast with men for whom frigidity equated to impotence, and consequently sterility.[138] Women's pleasure was not therefore a matter of concern within the scientific reflection. While Virey mentions the shame of the impotent husband unable to perform his duties, he neglects to describe the feelings of the frigid wife unable to enjoy hers.[139]

134 About her intellectual activities, ibid., 166. About her impotence, ibid., 163, 174 and 231.

135 Julien-Joseph Virey, 'Frigidité et froideur', in *Dictionnaire des sciences médicales*, XVII (1816), 11–28 (11).

136 *Le Petit Robert*, for instance, defines it as 'absence d'orgasme chez la femme' [absence of orgasm in a woman] and 'absence de désir sexuel et de satisfaction sexuelle (rare en parlant d'un homme et alors distinct de l'impuissance)' [absence of sexual desire and sexual satisfaction (rare when referring to a man and in this case different from impotence)] (1131). Naginski ('*Lélia*', 87) has rightly noted that frigidity was regarded as a male pathology. However, the scholar did not question Lélia's frigidity in her previous works (*George Sand*, 116 and 126; 'Les deux *Lélia*', 81).

137 Virey, 'Frigidité et froideur', 15.

138 Ibid., 15–16.

139 Ibid., 16–17.

The frigid woman is described as follows:

> La femme froide et stérile a [...] au plus haut degré le caractère de l'effémination comme ces femmelettes si blondes, si blanches, si délicates et énervées, presque sans gorge ou mamelles, n'étant presque ni réglées, ni pourvues de ces poils qui ombragent l'organe sexuel [...], et se faisant à peine entendre avec un petit filet de voix. [...] leur clitoris est presque introuvable et sans érection.[140]

> [The cold and sterile woman has [...] the highest degree of effeminacy like those wispy women who are so blonde, so pale, so delicate and highly nervous, almost flat-chested and without nipples, who almost never menstruate and are deprived of those hairs that shade the sexual organ [...] and who can barely be heard with their tiny voice. [...] their clitoris can barely be found and is not erect.]

This scientific description contrasts with literary discourses, since the portrayal of the frigid woman appears to be in contradiction with that of the masculine woman. Here, the frigid woman embodies an excess of feminine characteristics, such as fairness, paleness and fragility. All these characteristics are expected to exist in women, but they reach their peak in the frigid woman. While her description consists of the accentuation of femininity, that of the frigid man represents an inversion, not an emphasis, of masculinity, as he possesses the same effeminate characteristics as the frigid woman.[141] Indeed, Sténio, whose debauchery has rendered him impotent in the final part of the novel, corresponds to this portrait.[142]

Virey describes diverse causes of frigidity. They are either hereditary and caused by parental flaws (such as debauchery and incest) or accidental, such as the consumption of certain types of food, sexual inactivity, intense intellectual work, fatigue and mismatched spouses.[143] Whilst Virey does not explain whether these causes affect male or female bodies, it can be assumed that he is focusing almost exclusively on the former. For instance, he indicates that thyme and rue can cause frigidity in men but are aphrodisiacs for women.[144] However, he does not mention these plants when listing means of dissipating the symptoms of frigidity.

140 Ibid., 15.
141 Ibid., 14–15.
142 Sand, *Lélia*, 261.
143 Virey, 'Frigidité et froideur', 18–25.
144 Ibid., 20.

Freud later studied women's frigidity. Unlike Virey, he attributed it to their education. In his opinion, women associated sexuality with a proscribed act and thus refused, as a punishment, the pleasure that they might feel.[145] According to Freud, women's frigidity is not a physiological dysfunction but the consequence of women's own repressed personalities, although influenced by social convention. To put it simply, women hold the responsibility for their sexual displeasure.

Let us return to Lélia's absence of sexual pleasure. While Lélia is described – and describes herself – as cold in bed, she nevertheless has a sensual imagination, which contrasts with physical love.[146] Some of her dreams are recalled in bluntly erotic terms:

> Couchée sur la pierre des tombeaux, je m'abandonnais aux fureurs de mon imagination. Je rêvais les étreintes d'un démon inconnu; je sentais sa chaude haleine brûler ma poitrine et j'enfonçais mes ongles dans mes épaules, croyant y sentir l'empreinte de ses dents.[147]
>
> [Lying on gravestones, I gave in to the ardours of my imagination. I dreamt of the embraces of an unknown demon; I felt his hot breath burn my chest and I sank my fingernails into my shoulders, as if I felt the marks of his teeth.]

Her problem is that her sexual appetite is not satisfied by her lover: 'Je me sentais la poitrine dévorée d'un feu inextinguible et ses baisers n'y versaient aucun soulagement' [I felt as if my chest were devoured by an inextinguishable fire and his kisses poured no relief onto it].[148] She is therefore condemned to feel desire while her prostitute sister lives to 'jouir' [have (sexual) pleasure].[149]

As Lélia complains about the coldness of her senses and her excessive idealism, she moves imperceptibly towards a detailed tale of her sexual experience. Her first – and, sexually speaking, only – lover has had

145 Freud, 'On the Universal Tendency to Debasement in the Sphere of Love (Contributions to the Psychology of Love II)', in *The Freud Reader*, 394–400 (400).

146 Sand, *Lélia*, 166–67, 169, 178 and 207.

147 Ibid., 185.

148 Ibid., 174.

149 Ibid., 175.

a considerable influence over her sexuality. Yet, what Lélia describes is sexual and emotional abuse. She has seemingly been pressured by him into sexual intercourse.[150] She has been forced to yield to his advances ('je me dévouai en pâlissant et en fermant les yeux' [I sacrificed myself while growing pale and closing my eyes]) and reveals tense relations with him, as he shows anger and contempt towards her pain, which he dismisses as prudish hypocrisy.[151] Lélia also highlights his violence and even hints at having been raped. Contrasting her dream and reality, she states:

> À la place de cet être aérien […], je retrouvais l'homme, l'homme brutal et vorace comme une bête fauve et je m'enfuyais avec horreur. Mais il me poursuivait […] et il savourait son farouche plaisir sur le sein d'une femme évanouie et demi-morte.[152]
>
> [Instead of this aerial being […], I found man, brutal and voracious man, like a wild beast, and I ran away with horror. But he pursued me […] and he enjoyed his savage pleasure on the bosom of a fainted and half-dead woman.]

This traumatic sexual initiation has determined her subsequent relationships with men. She rejects any physical contact with Sténio, as if she fears being raped once again:

> [S]'abandonnant à un élan qui avait quelque chose de vorace et de fauve, il [Sténio] maîtrisa la femme à son tour, il l'étreignit dans ses bras, il colla sa bouche à cette bouche douce et molle dont le contact l'étonnait encore… Mais Lélia, le repoussant tout à coup, lui dit d'une voix sèche et dure:
>
> 'Laissez-moi, je ne vous aime plus!'[153]
>
> [Abandoning himself to an impulse that had something voracious and savage, Sténio overpowered the woman, he clasped her in his arms, he put his mouth on that soft and smooth mouth whose contact still surprised him… But Lélia, suddenly pushing him away, told him with a dry and hard voice:
>
> 'Leave me alone, I no longer love you!']

150 Ibid., 166.
151 Ibid., 167. See also 170–75.
152 Ibid., 175.
153 Ibid., 90. See also 217.

Unlike Virey's claim about frigid women's coldness, Lélia appears to be a sensual woman, even though male lovers are not the recipients of her sensuality. In other terms, and in opposition to a Freudian reading, Lélia's coldness is not of her own making; the responsibility lies with men who act brutally towards her. Eileen Boyd Sivert, Kristina Wingård Vareille, Margaret Waller and Isabelle Hoog Naginski have thus rightly seen in Lélia's story a denunciation of women's conditions in a patriarchal society where they hold a restrictive role and are psychologically and sexually abused.[154]

In short, Lélia's sexuality is stifled from the outset by her lover's violence. However, this abuse does not annihilate it. Her sexuality takes the less physical and more cerebral dimension of erotic dreams and therefore thwarts patriarchal conceptions of female pleasure.[155] As Lélia's pleasure does not depend on men's actions, it is invalidated because it does not tally with the sexual norm. Indeed, Phyllis Chesler suggests that women were expected not only to have orgasms but also to experience the right sort of orgasms in heteronormative relations.[156] Consequently, the label of frigidity appears as an attempt to deny this masculine woman the sexual satisfaction that she aspires to experience because of her rejection of traditional roles.

Félicité des Touches's sexuality also shocks her conservative community. The priest of Guérande sees her as a loose woman and implies that her house is a harem populated with male lovers.[157] Similarly, Calyste's mother indignantly points out: 'Elle a eu des aventures…' [she has had affairs…] and the abbot adds: 'Avec plusieurs hommes' [With several men].[158] Indeed, although unmarried, Félicité is not a virgin and has had several partners in

154 Eileen Boyd Sivert, 'Lélia and Feminism', *Yale French Studies*, 62 (1981), 45–66 (51); Vareille, *Socialité*, 145–46; Waller, *The Male Malady*, 144; Naginski, '*Lélia*', 90.

155 Arguably, the pleasure that Lélia feels in her dreams can be interpreted in terms of clitoral masturbation in opposition to vaginal intercourse with men (a pleasure that is despised by Freud as infantile): Freud, 'Three Essays on the Theory of Sexuality', in *The Freud Reader*, 239–93 (287–88).

156 Phyllis Chesler, *Women and Madness*, rev. and updated edn (Houndmills, Basingstoke and New York: Palgrave Macmillan, 2005), 105–06. See also Ussher, *Women's Madness*, 81.

157 Balzac, *Béatrix*, 119.

158 Ibid., 121.

the past: a first lover, an anonymous man who is considered to have 'made' Camille Maupin, the composer Conti, Calyste du Guénic and the journalist Claude Vignon. These multiple sexual experiences are also characteristic of masculine women's sexuality.

As was shown in Chapter 1, nineteenth-century scientists believed that female masculinity could be caused by an excessive quantity of sperm in the female body. Nymphomania, which was also called 'fureur utérine' [uterine fury], designated a disease caused by the exaltation of the sensitivity of the ovaries, as well as by the example of 'corrupt' people, such as servants, and the reading of so-called voluptuous texts. It consisted of expressing desire through lascivious discourses or gestures, provoking men and being violent to those who rejected their advances.[159] Women that were mostly affected by nymphomania were believed to resemble the Greek poetess Sappho, the cultural model of the masculine woman. They were short, tanned brunettes; they had large breasts, abundant menstruation and sensitive genitalia.[160] Incidentally, Félicité des Touches corresponds to the physical description of a nymphomaniac because, like Sappho, she is a short, dark-haired, 'Mediterranean' woman writer. Additionally, Virey argues that the constitution of many women writers was more erotic than that of other women.[161] His use of *érotique* is medical – he later mentions 'l'hystérie érotique' [erotic hysteria].[162] As Vernon Rosario shows, the medical meaning of *érotique* shifted at that time from 'melancholic' to 'erotomaniac' and even 'nymphomaniac'.[163] Moreau de la Sarthe's and Virey's medical definitions of a nymphomaniac refer less to a woman who has numerous sexual encounters than to one who is shameless in expressing her sexual desire.

In *Béatrix*, the protagonists' perception of Félicité is similar to the first meaning of nymphomaniac: she is condemned for her multiple lovers. Likewise, the narrator refers to her as a 'Don Juan femelle' [female Don

159 Moreau de la Sarthe, *Histoire naturelle*, II, 269–73; Virey, *De la femme*, 82 and 109–10.

160 Moreau de la Sarthe, *Histoire naturelle*, II, 271.

161 Virey, *De la femme*, 5–6.

162 Ibid., 109.

163 Rosario, *The Erotic Imagination*, 48–49.

Juan].[164] He nevertheless indirectly acknowledges the incoherence of his comparison, as he adds that she has neither debts nor conquests – two elements that are constitutive of Don Juan's identity. The comparison here stresses more her libertarian philosophical thought. Labelling her a nymphomaniac or a maneater is indeed unjustified. Her list of lovers is short, even though not all of them may be mentioned. Moreover, although Félicité is in love with Calyste and plans to marry the journalist Vignon, she does not have sex with either of them. The priest's reference to a harem and the narrator's exaggerated comparison with Don Juan reveal that Félicité's bad reputation is caused rather by her rejection of sexual conventions in openly expressing her desire, even in moderate proportions. Indeed, what shocks the male community is the fact that Félicité's fulfilling sexual life is not ruled by marriage and may not even be ruled by romantic feelings. Consequently, Félicité suffers from a common prejudice highlighted by Simone de Beauvoir: 'on confond avec entêtement femme libre et femme facile' [the free woman is obstinately mistaken for the easy woman].[165] The claim of her right to sexual freedom is discredited as wantonness.

Finally, masculine women's sexual autonomy is depicted, and sometimes discredited, as lesbianism. The viewpoint on lesbianism expressed in *La Cousine Bette* and *Lélia* is allusive and critical. It tallies with the brief studies of lesbianism in scientific writings of the time, which conceive of it as a dangerous sexual practice that must not be mentioned explicitly. In *La Cousine Bette*, sapphic relations are suggested between Lisbeth Fischer and Valérie Marneffe. The narrator uses a third party, the slanderous Parisians, to refer to rumours concerning their relationship: 'Lisbeth et Valérie offraient le touchant spectacle d'une de ces amitiés si vives et si peu probables entre femmes, que les Parisiens, toujours trop spirituels, les calomnient aussitôt' [Lisbeth and Valérie gave the moving spectacle of one of those friendships between women that are so strong and so unlikely that the Parisians, always too witty, immediately denigrate them].[166] In attributing these rumours to the wit of Parisians, the narrator seems to distance himself from them, but,

164 Balzac, *Béatrix*, 153.

165 Beauvoir, *Le deuxième sexe*, II, 446.

166 Balzac, *La Cousine Bette*, 198. See also 486–87.

merely by mentioning them, he raises doubts in the reader's mind and acknowledges the possibility that these slanders may be true. This paralipsis, a strategy that shares similarities with Nigel Smith's *avoidance-discourse*, helps to construct Lisbeth's homosexuality, while not taking the risk of making it an undeniable reality. But why does the narrator need to be so cautious when mentioning Lisbeth and Valérie's friendship? The two friends represent a double challenge to the traditional family unit. On the one hand, they claim to belong to it, in marrying (or being willing to marry) Marshal Hulot and Victorin Hulot's father-in-law Crevel respectively. On the other hand, as Lucey points out, by doing so, they would create a 'deviant' bond between them, who are outsiders, and respectable members of the bourgeoisie and the nobility.[167] The ambiguous relationship between Lisbeth and Valérie threatens familial order and is therefore severely punished; Valérie in particular meets a gruesome end.

In *Lélia*, the relationship between Lélia and her sister Pulchérie is unconventional because it suggests both lesbianism and incest. With her blonde hair, bright cheeks, round shoulders and predestined beauty (*pulcher* means 'beautiful' in Latin), Pulchérie embodies ideal femininity.[168] Pulchérie, the feminine sister, is attracted to Lélia because the latter is masculine. The courtesan told her sister when they were children: 'Regarde-toi, ma sœur: ne te trouves-tu pas belle? […] Tu ressembles à un homme' [Look at yourself, sister: don't you find yourself beautiful? […] You look like a man].[169] However, Pulchérie's love for her sister seems at first glance to be non-reciprocal. When the two sisters meet, they recall an episode of their childhood in which they fell asleep together on the grass, which raised 'une émotion pleine de charme et peut-être de honte' [a feeling full of charm and maybe of shame].[170] After dreaming of a dark-haired man, Pulchérie awoke and discovered the masculine beauty of her sleeping sister.

167 Lucey, *The Misfit of the Family*, 154–56. For more on lesbian relationships in *La Cousine Bette*, see James R. McGuire, 'The Feminine Conspiracy in Balzac's *La Cousine Bette*', *Nineteenth-Century French Studies*, 20/3–4 (Spring–Summer 1992), 295–304 (298–301).

168 Sand, *Lélia*, 157.

169 Ibid., 158.

170 Ibid., 155.

She contemplated her, stroked her hair and kissed her arm, only to be dismissed by the intimidating gaze of an awakened Lélia.[171]

Pulchérie's behaviour towards Lélia can be described as a 'romantic friendship'. Lillian Faderman and Caroll Smith-Rosenberg have both shown that, in a society where men and women were segregated and where women were not allowed to show their inclination to men, they tended to develop friendships with other women that could be expressed through promises of eternal love and physical manifestations of affection, such as kissing, hugging and sharing the same bed.[172] Smith-Rosenberg has also stressed the importance of family ties between sisters, cousins, sisters-in-law, mothers and daughters.[173] Boundaries between moral and immoral relationships among girls at the time were different from those established in twentieth and twenty-first-centuries Western societies, as sexological precepts had not yet permeated everyday life. Whilst the means of expressing affection for another woman may be similar to the nineteenth-century ones, the modern interpretation of these bonds has changed. What was deemed to be a legitimate female relationship (or romantic friendship) may now be seen as an illegitimate one (or sapphic love).

Although Pulchérie's fondness for her sister seems to correspond to the romantic friendships studied by Faderman and Smith-Rosenberg, Lélia does not judge Pulchérie's sentiment to be innocent. The latter's need to justify her acts when surprised by Lélia highlights her sense of guilt: 'Alors vous ouvrîtes les yeux et votre regard me pénétra d'une *honte* inconnue; je me détournai comme si j'avais fait une action *coupable*. Pourtant, Lélia, aucune pensée *impure* ne s'était même présentée à mon esprit' [Then you opened your eyes and your gaze imbued me with unknown *shame*; I turned away as if I had committed a *guilty* action. However, Lélia, no *impure* thought had even come to my mind].[174] In having Pulchérie use the lexical field of immorality, Sand suggests that her character is aware of the ambiguity of her

171 Ibid., 158.

172 Faderman, *Surpassing the Love of Men*, 157–77; Smith-Rosenberg, *Disorderly Conducts*, 53–76. Likewise, in Gautier's *Mademoiselle de Maupin* (330), Madeleine states that her friendship with female friends resembles passion.

173 Smith-Rosenberg, *Disorderly Conducts*, 62.

174 Sand, *Lélia*, 158 (my emphasis).

actions and feelings. Whereas Pulchérie's declaration of love for Lélia may correspond to non-sexual romantic friendships, the phrases that she employs to describe her sentiments stress their strangeness, as if she were conscious of their 'deviant' nature ('une singulière curiosité' [a peculiar curiosity]; 'un étrange plaisir' [a strange pleasure]; 'cela me troublait étrangement' [that strangely disconcerted me]).[175] She further confesses that this episode was her first lesson in love and, more equivocally, her first sensation of pleasure, even comparing her sister with the man of her previous dream.[176] It can thus be inferred that her 'rêve étrange' [strange dream] was caused by her lying side by side with Lélia.[177]

Lélia and Pulchérie show awareness that the exchange of caresses and love between women might be 'impure', a similar opinion to that of *La Cousine Bette*'s Parisian crowd according to its narrator. This supports Faderman's claim that romantic friendship was deconstructed as lesbianism earlier in France than in Anglo-Saxon countries, by writers such as Gautier and Baudelaire, 'perhaps because the French aesthete writers from the 1830's on delighted in exploring whatever had the potential to astound the bourgeoisie [...], and flaunted exotic images of sex between women in their poetry and prose'.[178] Lélia's severe reaction towards Pulchérie's feelings and the latter's shameful response to her sister's intransigence reflect the growing ambivalence of opinions towards romantic friendship and love between women in the 1830s. The novel shows that the innocence of close relationships between girls has started to be questioned. Although the true nature of these relationships remains the same, they are altered by the external gaze, which both sisters seem to have internalized. 'Respectable' girls are deprived of boys' company and have no one to express their admiration and love for, except themselves – Pulchérie confesses her narcissism – and their female friends, sisters and other female relatives.[179] If even these possibilities are withdrawn, their only options are to become either loveless or 'vicious'. Pulchérie, who became a courtesan, appears to have renounced

175 Ibid., 157.

176 Ibid., 158.

177 Ibid., 156.

178 Faderman, *Surpassing the Love of Men*, 238.

179 Sand, *Lélia*, 156.

her sister's love, whilst choosing the second path. She even encourages Lélia to become either a nun or a courtesan, two traditional female roles outside of the institution of marriage.[180] Lélia defies all traditional categories of female identity in rejecting these two roles.

As such, the sexualities of Lélia, Félicité and Lisbeth may be read as frigidity, nymphomania or lesbianism inside and outside the fictional universe, but, due to their unclassifiable nature and non-conformism, they escape masculine control. Sexually, they 'choisissent les chemins condamnés' [choose forbidden paths], to borrow Beauvoir's words.[181] Men's role during intercourse is rejected or, at least, judged to be limited. This empowerment raises concerns, as shown by the narrators' and characters' attempts to dismiss it through slanders and claims of monstrosity. The anxiety towards masculine women's alternative sexuality reveals the fear of their refusal to conform to social conventions connected to sex, such as childbearing. This will be the focus of the next section.

The Refusal of Motherhood and the Maternal Lover

Before Halberstam's study on female masculinity, Freud had already theorized the existence of the masculine woman. He had defined her as a girl suffering from a masculinity complex, or from a desire to have a penis and to resemble a man, notably after realizing that she was different from her male playmates. The desire to be a man, the sense of inferiority that resulted from the absence of a penis and the resentment towards those who had one were all deemed by Freud to be part of the 'natural' development of women, for which they compensated by wishing to bear a child.[182] In

180 Ibid., 207–08.

181 Beauvoir, *Le deuxième sexe*, I, 479.

182 Freud, 'Three Essays', in *The Freud Reader*, 271; 'Leonardo da Vinci and a Memory of His Childhood', 443–81 (460–61); 'The Dissolution of the Oedipus Complex', 661–66 (662–65); 'Some Psychical Consequences of the Anatomical Distinction Between the Sexes', 670–78 (673–78); 'Anxiety and Instinctual Life', 773–83 (777–79).

contrast, fictional masculine women refuse to submit themselves to the main social conventions that ruled women's lives in the nineteenth century: marriage and motherhood. From a Freudian perspective, it seems as though masculine women are in denial because they refuse to renounce their penis envy and embrace their femininity to become fully mature women. Conversely, I argue that, in the novels analysed here, far from abandoning their maternal function, the protagonists find other ways to exercise it, whilst still asserting their masculinity. Masculine women accordingly create their own gender norms.

The refusal of masculine women to marry is based on their desire to remain independent in a society organized by the Napoleonic Code that can only define them as individuals without rights. For Félicité des Touches, marriage is at first a theoretical concept distant from her everyday concerns. After all, having inherited a tidy sum of money, she does not need the financial support of a husband. What she can see is that being married would equate to renouncing her freedom.[183] The novel does not prove her wrong. Béatrix, entrapped by a husband, a child and a lover, envies Félicité's autonomy.[184] Lisbeth Fischer shares Félicité's reticence regarding marriage, which she sees as a halter, and dreads being mocked for her lack of education and social inferiority.[185]

This rejection of marriage, however, is not final. Lisbeth may well feel real disgust for the animalistic nature of men, but her hatred of men is also partly due to the devoted relative act that she plays to fool her family (when her cousin and niece have to deal with their husbands' infidelity) and partly due to her resentment at not having found a worthy husband.[186] She aspires to becoming Maréchale Hulot by marrying Hector Hulot's brother, the patriarch of the Hulot family. Although she claims otherwise – and claims that her incentive is to help her family – she is not indifferent to the prestige of this position.[187] However, her main motivation lies elsewhere. The domination that, as the matriarch, she could exercise over the members of

183 Balzac, *Béatrix*, 143.
184 Ibid., 193 and 270.
185 Balzac, *La Cousine Bette*, 57.
186 Ibid., 411.
187 Ibid., 263.

the family whom she hates would compensate for the loss of her autonomy and thus allow her to achieve her revenge. Similarly, Félicité contemplates the possibility of marriage, but she is not motivated by love; neither is she fooled by the lot of married women. In search of a partner with whom she can face the fear of ageing and on whom she can exercise a beneficial influence, she considers marrying the journalist Claude Vignon. Above all, she seeks 'un homme supérieur' [a superior man], certainly to other men, but also to herself.[188] That the talented writer does not want to settle for mediocrity – although she entertains this thought – makes sense.[189] It nevertheless suggests that, in having his most intelligent female protagonist consider that the husband must be superior to his wife for the union to be suitable, Balzac rejects complete equality between the sexes: even the greatest woman cannot equal a great man.

Indeed, only Lélia remains opposed to marriage and has no marital projects in the eponymous novel. Sand's character does not express negative opinions towards marriage specifically. However, what Lélia clearly denounces is the romantic script that reduces women's social role to being a lover or a recipient of a man's love: in Lélia's world – and undoubtedly in that of Sand – men define women by their ability to believe in love.[190] Even the chaste Sténio is obsessed with possessing the woman whom he loves; Lélia discerningly tells him: 'Ce n'est point une âme que vous voulez: c'est une femme, n'est-ce pas?' [It is not a soul that you want, it is a woman, isn't it'?].[191] Like Lélia's first and violent lover, he considers her rejection of physical love to be a betrayal, even more so when Lélia tricks him into having sex with Pulchérie, whom he mistakes for the former. Facing his anger, she reminds him that he is responsible for loving her and claims her right not to submit to his fantasy: 'c'est votre faute, il ne fallait pas m'adorer comme une divinité, pour me demander ensuite d'être votre esclave et votre Sulamite' [This is your fault; you should not have worshipped me like a

188 Balzac, *Béatrix*, 156.

189 Ibid., 176.

190 Sand, *Lélia*, 68. See Sivert, 'Lélia and Feminism', 48; Waller, *The Male Malady*, 139 and 161.

191 Sand, *Lélia*, 216.

divinity and then ask me to be your slave and your Shulamite].[192] Through Lélia's rhetorical question, Sand suggests that women's role is more complex: 'Mais quoi! le rôle de la femme se borne-t-il aux emportements de l'amour?' [What! Is woman's role limited to the outbursts of love?].[193] Her female protagonist illustrates the fact that a woman can be friends with a man, be ironic and philosophize, but these abilities are seldom part of the script imposed on women.

The most virulent criticism of marriage is, nonetheless, expressed by Pulchérie, who is not only Lélia's sister but also her doppelgänger despite their physical difference.[194] The courtesan shows the extension of her hatred for marriage by comparing the living conditions of the mother and the prostitute. She first asks an ambivalent question: 'Comparez-vous les travaux, les douleurs, les héroïsmes d'une mère de famille à ceux d'une prostituée?' [Are you comparing the labours, the pains and the heroisms of a mother to those of a prostitute?].[195] Then, she states more overtly: 'être amante, courtisane et mère, trois conditions de la destinée de la femme auxquelles nulle femme n'échappe, soit qu'elle se vende par un marché de prostitution ou par un contrat de mariage' [to be a lover, a courtesan and a mother, three conditions of a woman's destiny from which no woman escapes, whether she sells herself through prostitution or through a marriage certificate].[196] The link between marriage and prostitution (and more generally every relationship between men and women) is also made by feminists of the July Monarchy.[197] This association enables Pulchérie to criticize women's

192 Ibid., 228. See also 230.

193 Ibid., 233.

194 Rogers, 'Psychosexual Identity', 34; Vareille, *Socialité*, 208; Richard B. Grant, 'George Sand's *Lélia* and the Tragedy of Dualism', *Nineteenth-Century French Studies*, 19/4 (Summer 1991), 499–516 (509); Schor, *George Sand and Idealism*, 66; Hatem, 'Scission du cœur', 30. Pulchérie herself claims their likeness when she tells Sténio: 'je suis Lélia elle-même' [I am Lélia herself] (Sand, *Lélia*, 220).

195 Sand, *Lélia*, 152.

196 Ibid., 153.

197 Karen Offen, *European Feminisms 1700–1950: A Political History* (Stanford: Stanford University Press, 2000), 97–98. Beauvoir (*Le deuxième sexe*, II, 247–48) made the same comparison.

lack of freedom and civil rights in the kind of marital institution defined by the Napoleonic Code.

In her condemnation of a woman's lot, Pulchérie does not explicitly distinguish between the wife and the mother. These two figures were often assimilated in nineteenth-century society, as the married woman was expected to become a mother and a 'respectable' woman could not become a mother without being married. The discourse relating to motherhood is extended further in the novels than the discourse about marriage. It is also more ambiguous, as no specific condemnation or praise is pronounced. Before studying the attitude of the masculine woman towards motherhood, I will first discuss the definition of motherhood in the nineteenth century.

In *L'amour en plus* [*Extra Love*], Élisabeth Badinter calls into question the so-called maternal instinct, or maternal love. By this, it is generally understood that a woman 'naturally' aspires to become a mother, that the mother is necessarily attached to her children, that she instinctively knows how to take care of them and that her devotion to them is so intense that it is similar to a sacrifice. Badinter argues that love is a feeling and 'comme tout sentiment, il est incertain, fragile et imparfait' [like any feeling, it is uncertain, fragile and imperfect], in other words, 'contingent'.[198] She convincingly claims that maternal love is an invention of the last third of the eighteenth century, notably developed in Rousseau's *Émile*. Maternal love, it was then claimed, was reputed to be a necessity for both the child and the mother. The latter could only blossom and find a glorifying position in society through her maternal duties. This theory, Badinter argues, was a response to the generally callous attitude towards infants during the seventeenth and eighteenth centuries. French mothers from almost all social milieux, either for economic reasons, to follow a trend, or simply through a lack of interest, abandoned their children to wet nurses and showed indifference to their fate. In so doing, they made them more likely to die in infancy. The purpose of the maternal discourse developed thereafter was thus to reduce the risk of infantile mortality, notably by promoting maternal

198 Élisabeth Badinter, *L'amour en plus: Histoire de l'amour maternel (XVIIe–XXe siècle)* (Paris: Librairie Générale Française, 'Le Livre de Poche', 1982 [Flammarion, 1980]), 29.

breastfeeding and constant care, as well as bringing the custom of swaddling babies to an end. Inadvertently, it developed women's and mothers' guilt and sentiment of abnormality if they did not perceive motherhood as a vocation, refused to have children or felt that they did not correspond to the ideal maternal model.

Early nineteenth-century medical treatises are inscribed in this trend of promoting the maternal instinct analysed by Badinter. Menville de Ponsan is particularly enthusiastic when discussing maternal love to the extent that the style of his essay is sometimes lyrical.[199] He describes maternal love as:

> cette force plus puissante que la douleur et le dégoût, cette force n'est qu'un sentiment animal, un instinct aveugle qui appartient à la plante, à l'insecte, au quadrupède, aux oiseaux, comme à la femme: loi immuable de la nature, loi de conservation, penchant irrésistible auquel nul être sur la terre ne peut se soustraire, auquel la nature a confié la vie![200]
>
> [this force more powerful than pain and disgust, this force is only an animal feeling, a blind instinct which belongs to plants, insects, quadrupeds and birds as well as women: immutable law of nature, law of preservation, irresistible penchant from which no earthly being can escape, to which nature has entrusted life!]

Menville de Ponsan's definition of maternal love asserts its strong bonds with nature, animality and instinctiveness, hence its overwhelming power. Women, like any other female animal, have no choice but to submit to it. This almost 'savage' love inherent to their condition is nonetheless perfected by morals, he claims.[201] Menville de Ponsan depicts women as eager to be pregnant, sometimes to the point of simulating a pregnancy.[202] The young mother recovers astonishingly swiftly from the pains of childbirth to dedicate herself to the pleasures of maternity:

> Ô Rubens! je laisse à ton pinceau le soin de rendre cet état touchant où les dernières impressions d'une douleur qui s'éteint se mêlent encore, dans la femme, à la sérénité

199 Menville de Ponsan, *Histoire médicale*, I, 355. See also extract I, 344 quoted below.
200 Ibid., I, 397.
201 Ibid., I, 401.
202 Ibid., I, 290 and 295. See also Moreau de la Sarthe, *Histoire naturelle*, II, 186.

> de la joie la plus pure; [...] où la crainte, assez naturelle quand on souffre, de perdre le jour, vient faire place au plaisir délicieux de l'avoir donné à un nouvel être![203]
>
> [O Rubens! I leave it to your brush to depict this touching state where the last impressions of a vanishing pain are combined with the serenity of the purest joy in a woman; [...] where the fear, quite natural when she is suffering, of losing her life, yields to the delicious pleasure of having given birth to a new being!]

Menville de Ponsan trivializes labour by seeing in it a mere artistic subject and gives little consideration to parturient mortality, which was still common at that time (see Chapter 1). Only the joys of motherhood and maternal duty count.

Another key step on a mother's journey is breastfeeding. According to Menville de Ponsan, it is not only her duty, but also a guarantee of health for both the mother and the child, as it cures some diseases and prevents others, such as peritonitis, rheumatic fever, breast cancer and, for the baby, vomiting.[204] Above all, it helps the mother to secure the love of her child by creating a special bond between them that rewards her for her sacrifice.[205] In scientific discourse, the love and tenderness of a mother are connected to bodily functions, and therefore to instinct. For instance, the recollection of a beloved baby can trigger lactation.[206]

Furthermore, in the opinion of nineteenth-century scientists, motherhood is the most – and perhaps the only – honourable status for a woman. Consequently, religious vocabulary is frequently used to talk about motherhood.[207] Indeed, Menville de Ponsan observes:

> De toutes les opérations maternelles, l'allaitement est la plus méritoire, parce qu'elle est la seule désintéressée et volontaire; c'est le gage le plus précieux de la tendresse d'une mère; c'est le ministère le plus saint, puisqu'il influe sur le moral comme sur le physique, et que c'est de cette fonction que nous recevons l'influence de nos destinées.[208]

203 Menville de Ponsan, *Histoire médicale*, I, 344. See also Moreau de la Sarthe, *Histoire naturelle*, II, 206.

204 Menville de Ponsan, *Histoire médicale*, I, 369–70 and 381–82.

205 Moreau de la Sarthe, *Histoire naturelle*, II, 211; Menville de Ponsan, *Histoire médicale*, I, 356–57.

206 Moreau de la Sarthe, *Histoire naturelle*, II, 215.

207 Badinter, *L'amour en plus*, 286.

208 Menville de Ponsan, *Histoire médicale*, I, 374–75.

> [Breastfeeding is the most commendable of all the maternal functions, because it is the only one that is disinterested and voluntary; it is the most precious token of a mother's tenderness; it is the holiest ministry, since it influences all things moral as well as physical, and we receive the influence on our fates from this function.]

The alleged selflessness of breastfeeding is, nevertheless, contradicted by the fact that the mother who does not feed her baby herself is often condemned to dreadful diseases – as were the intellectual woman and the unmarried woman.[209] In short, scientific treatises and literary depictions of good mothers perpetuate the ancient worship of the pregnant mother.[210]

In our corpus, the good mother is embodied by Adeline Hulot. She has raised motherhood and wifehood to their highest positions. She is assimilated to a saint and an angel by the narrator and virtually all the characters in *La Cousine Bette*.[211] As such, she incarnates the values of extraordinary devotion and sacrifice promoted by the defenders of maternity. To her daughter, who experiences marital disappointment, she advocates resignation – a trait that is typical of women's conduct according to Beauvoir – and suffering in silence as the guarantees of household peace and a husband's constancy, in the absence of fidelity.[212] Adeline concludes a speech to her daughter with the words 'Courage et silence!', which resonate like a perfect wife's motto.[213] This line of speech is consistent with her belief that softness and submission are a woman's most powerful weapons.[214] Her love goes beyond mere submission; it verges on fanaticism. To Adeline, Baron Hulot is a god; therefore anything that he does is well done, even though it entails ruining and abandoning his family.[215] When Hulot relegates her to a small flat, while providing his mistress with a luxurious one, she accepts

209 Ibid., I, 366–69.

210 Moreau de la Sarthe, *Histoire naturelle*, II, 287–88; Menville de Ponsan, *Histoire médicale*, I, 290–91.

211 For instance, Balzac, *La Cousine Bette*, 51, 108, 363, 429 and 509.

212 Ibid., 286–87. Beauvoir, *Le deuxième sexe*, II, 313.

213 Balzac, *La Cousine Bette*, 288.

214 Ibid., 108–09

215 Ibid., 49, 74 and 108.

her fate with a formula of Christian forbearance: 'Il me veut ainsi: que sa volonté soit faite!' [He wants me like this: his will be done!].[216]

Adeline's words and behaviour illustrate the fact that masochism was a key component in the nineteenth-century discourses that praised maternal love and was therefore deemed to be inherent to women's identity, as Badinter explains:

> Cette théorie du masochisme féminin sert de justification *a posteriori* à l'acceptation de toutes les douleurs et tous les sacrifices. Si la femme est naturellement faite pour souffrir, et que, de surcroît, elle aime cela, il n'y a plus de raison de se gêner.[217]
>
> [This theory of feminine masochism aims at justifying *a posteriori* the acceptance of all pain and all sacrifice. If woman is naturally made for suffering and if, what is more, she loves it, there is no reason to feel awkward.]

It is difficult to disagree with Badinder's analysis when Balzac makes his character say to her unfaithful husband: 'Je suis ta chose, fais de moi tout ce que tu voudras' [I belong to you, do with me what you will].[218] Adeline agrees to be stripped of her free will in order to submit to her husband's whims.

Adeline not only displays excessive religious dedication to her husband, she also acts as a mother to him, as she treats Hector as a spoiled child.[219] The novel makes it clear that Adeline is both an ideal wife and an ideal mother. However, her two natures are increasingly in conflict when the interests of her children become threatened by their father's actions.[220] Consequently, her maternal self ends up being even 'plus forte que la femme' [stronger than the wife], because, if Adeline can forgive Hulot's infidelities, she cannot bear to see her children suffer.[221] Her devotion to her family is so absolute that she is prepared to sacrifice her honour as a respectable wife and mother to become a kept woman in order to fulfil her nurturing duties towards her family. She attempts to prostitute herself to her son's

216 Ibid., 207.
217 Badinter, *L'amour en plus*, 395.
218 Balzac, *La Cousine Bette*, 391.
219 Ibid., 46 and 51.
220 Ibid., 311.
221 Ibid., 53.

father-in-law to obtain two hundred thousand francs that would save her family from ruin and dishonour.

Hortense Steinbock, who physically resembles her mother Adeline, has been told by the latter to accept her lot. She also acts maternally towards her husband Wenceslas, as her mother does. However, she progressively distances herself from Adeline's philosophy. She does not wish to endure the same fate as the latter and decides to leave her unfaithful husband (who, like Hulot, has been seduced by Valérie). The farewell letter that Hortense writes to Wenceslas refers several times to Adeline. She expresses admiration for her mother's devotion, but she admits that this model is out of her range: 'J'ai connu dans son entier le dévouement de ma mère [...]! Elle est héroïque tous les jours depuis vingt-trois ans; mais je ne me sens pas la force de l'imiter' [I am fully aware of the devotion of my mother [...]! She has been heroic, every day, for twenty-three years; but I do not feel I have the strength to imitate her].[222] More than her own mother does, Hortense dissociates the mother from the wife, but she chooses to be a good mother only to her son, and not to her husband. Through the Hulot women's fate, Balzac undermines the claim of philosophers and scientists such as Rousseau and Menville de Ponsan that good mothers were rewarded by the fidelity of their husbands.[223] *La Cousine Bette* ironically shows that such reasoning constitutes a way to encourage women to fulfil their maternal duties, rather than being objective observations.

Unlike Adeline Hulot, but like Saint-Simonian feminists such as Claire Démar, who rejected maternity as a vocation, masculine women do not willingly embrace motherhood.[224] The narrator of *Béatrix* clearly expresses Félicité's absence of maternal instinct: 'elle sentait vivement le prix de l'indépendance et n'éprouvait que du dégoût pour les soins de la maternité'

222 Ibid., 297.

223 Badinter, *L'amour en plus*, 248. For instance, Menville de Ponsan (*Histoire médicale*, I, 371–72) declares: 'La femme qui nourrit est bien plus sûre de l'attachement de son époux, qui est, pour ainsi dire, commandé par le spectacle d'une famille naissante' [The woman who breastfeeds is much more certain of the affection of her husband, who is attracted to the spectacle of an emerging family].

224 Moses, *French Feminism*, 73–74. Some feminists, however, valued the model of the good mother, according to Offen (*European Feminisms*, 99–101).

[she strongly felt the value of independence and she felt nothing but disgust for the cares of motherhood].[225] Contrary to what women should 'normally' feel according to the myth of maternal love, motherhood does not tempt Félicité. Her disgust for maternal care, her awareness of the torments of childbirth and her refusal to believe in the compensatory happiness that a mother is supposed to find in maternal chores should have been subdued by her maternal instinct in the opinion of scientists such as Moreau de la Sarthe and Menville de Ponsan. In depicting her resistance, Balzac suggests that maternal love is not instinctive in every woman. By refusing maternal duties, Félicité appears to break with her female condition. Her position, as Halberstam observes about female masculinity, threatens the division of social roles: 'Presumably, female masculinity threatens the institution of motherhood: [...] if female masculinity is widely approved, then no one will want to take responsibility for the trials and pains of reproduction.'[226] In the novel, Félicité has made her choice: she conceives novels, not children. In a speech that explicitly targets Rousseau, she argues that women should be able to choose whether they want to become mothers.[227]

Félicité, however, does not question the principle according to which maternal instinct is inherent to womanhood. She certainly expresses an unconventional opinion against the supposed innate maternal instinct when she defines what she calls 'les premières malices du mariage' [the first mischiefs of marriage] as 'l'enfant, les couches, et ce trafic de maternité que je n'aime point' [child, childbirth and that business of motherhood that I do not like], but she adds: 'Je ne suis point femme de ce côté-là' [I am not a woman in that way].[228] This therefore suggests that she agrees with the notion of maternal love as determinant in the definition of women and believes that women like herself who lack this love cannot be defined according to the traditional conventions of femininity. This line of reasoning reveals Balzac's ambiguity towards his character, who is at once a model of female achievement and a gender outcast.

225 Balzac, *Béatrix*, 143. See also 'les ennuis de la maternité' [the troubles of motherhood] (171).

226 Halberstam, *Female Masculinity*, 273.

227 Balzac, *Béatrix*, 179.

228 Ibid., 178.

Félicité's perspective on motherhood shows, on the one hand, that maternal love is not intrinsic to the nature of every woman, but, on the other hand, that it is still of relevance for nineteenth-century atypical women. Women like Félicité, Lisbeth and Lélia do display maternal love, not directed towards a child, but rather towards their lovers. Lélia admits feeling like a mother towards Sténio and is depicted several times as acting as a guide and a teacher to the young man; Félicité ensures Calyste's intellectual and musical education that his family had neglected and the young man refers to her as the mother of his intelligence.[229] Calyste, Wenceslas and Sténio, like Octave and Lucien, act like children. The love that masculine women have for them is expressed in a motherly way. Terms of endearment such as 'mon enfant' [my child] and 'cher enfant' [dear child] appear many times in these novels. Calyste confesses to his mother that Félicité has rejected his love because she considers him to be her child. He summarizes her words: 'elle pourrait être ma mère, disait-elle; une femme de quarante ans qui aimait un mineur commettait une espèce d'inceste, elle était incapable d'une pareille dépravation' [she could be my mother, she said; a forty-year-old woman who loved a minor committed a kind of incest, she was incapable of such depravity].[230] The sentiment appears to be shared by Lélia, given that her relationship with the young Sténio is platonic.[231]

Maternal devotion lies in the fact that masculine women put the interest of their lovers ahead of their own. Although besotted with Calyste, Félicité convinces him to fall in love with Béatrix, helps him seduce her and, finally, finds him a fiancée, Sabine de Grandlieu, and bequeaths her fortune to him, thus acting like a fairy godmother.[232] She is proud of his ability to seduce Béatrix although she suffers as a result of it.[233] Similarly, Lisbeth's 'brutale, mais réelle maternité' [brutal but real motherhood] allows Wenceslas to overcome temporarily his 'mollesse morale' [moral softness].[234] The wild, even animal, nature of the old maid is an asset in

229 Sand, *Lélia*, 50–51, 92 and 229; Balzac, *Béatrix*, 132, 164–65, 196 and 278.

230 Balzac, *Béatrix*, 131.

231 Sand, *Lélia*, 233.

232 On Félicité as a fairy, see André Vanoncini, 'Temps et mémoire dans *Béatrix*', *L'Année balzacienne*, 3rd ser., 8 (2007), 275–88 (285–86).

233 Mura, *'Béatrix' ou la logique des contraires*, 226.

234 Balzac, *La Cousine Bette*, 90.

exercising her maternal role. On this subject, Badinter has shown that moralists praised 'savages' and animals as models of motherhood because their maternal instinct was allegedly not corrupted by reason.[235] Under Lisbeth's harsh and animalistic influence, the sculptor can realize his artistic projects: 'Wenceslas, nature rêveuse, avait dépensé tant d'énergie à produire, à s'instruire, à travailler sous la direction despotique de Lisbeth' [Wenceslas, naturally a dreamer, had spent so much energy in producing, educating himself and working under Lisbeth's despotic direction].[236] Although she has little artistic sensibility, she understands the constraints of trade and the necessity to produce frequent mainstream works of art in order to put bread on the table rather than dreaming of a masterpiece.[237] When Wenceslas marries Hortense and is *de facto* away from Lisbeth's influence, he succumbs to idleness and produces no masterpiece. The young man admits his debt towards the Pygmalion-like spinster ('vous m'avez créé' [you have created me]) but cannot repay it in the way that she would like: money or, better, love. As a result, he fears her reaction and begs her: 'ne me détruisez pas' [do not destroy me].[238]

Indeed, Lisbeth displays ambiguous feelings towards Wenceslas. Whilst she acts as a strict but loving mother to him, she is incapable of being as selfless as Félicité in giving him to another woman:

> Elle aimait assez Steinbock pour ne pas l'épouser, et l'aimait trop pour le céder à une autre femme; elle ne savait pas se résigner à n'en être que la mère, et se regardait comme une folle quand elle pensait à l'autre rôle.[239]
>
> [She loved Steinbock enough not to marry him but loved him too much to give him up to another woman; she could not resign herself to being only his mother and considered herself to be insane when thinking about the other role.]

235 Badinter, *L'amour en plus*, 233–45. See Menville de Ponsan's extract quoted earlier.
236 Balzac, *La Cousine Bette*, 255.
237 Ibid., 252.
238 Ibid., 163.
239 Ibid., 102. See also 137.

Accordingly, giving Valérie to him is a huge concession for her[240] but stems from vindictiveness rather than selflessness. She thus allows her malice to take over her maternal side.

Masculine women apparently renounce the enslaving social conventions of marriage and maternity to which a feminine woman such as Adeline Hulot submits without complaint. This liberty allows them to exercise their reputed masculine activities. Their condemnation of marriage and romantic relationships is, overall, more adamant than that of motherhood. Whilst they do not irrevocably escape social conventions, they find unconventional ways to express their masculine, yet nurturing, identity. They display maternal feelings towards a lover, rather than towards a child.

Superiority and Power: Beyond the Angel *vs.* Demon Dichotomy

I introduced this chapter by highlighting the theme of identity in the narratives analysed here. Masculine women are mysteries that narrators, characters and readers alike attempt to decipher. Revealingly, the incipit of Sand's novel is a question that Sténio asks Lélia: 'Qui es-tu?' [Who are you?].[241] Masculine women's identity is connected to the way in which their status is presented in hierarchical terms. In other words, are masculine women superior or inferior beings and is their superiority or inferiority measured in relation to men or women? This section looks at how their complex status is translated into the language of the narratives.

Lélia is the most explicit novel of this corpus regarding the superiority of its heroine. Lélia is an exceptional woman, whose distinction over other women is stated by Sténio in biblical terms that liken her to the Holy Virgin, although he simultaneously highlights her cynicism and scepticism: 'Crois-tu que le Créateur t'ait choisie entre toutes les femmes pour pratiquer la moquerie et le dédain, pour railler les amours sincères, pour

240 Ibid., 164.
241 Sand, *Lélia,* 10.

nier les serments, pour refuser les promesses, pour désespérer la jeunesse crédule et confiante?' [Do you believe that the Creator chose you among all women to mock and despise, to ridicule sincere love, to deny oaths, to reject promises, to drive gullible and trusting youth to despair?].[242] Her superiority over other women is little questioned until a disillusioned Sténio admits in the final part of the novel: 'Lélia elle-même est une femme comme une autre' [Lélia herself is a woman like any other].[243] However, the novel generally claims that Lélia is not only an extraordinary woman, but above all an extraordinary human being. Most characters, including Lélia herself, recognize the superiority of her influence, bodily organization and, above all, intellectual faculties. For Sténio, Lélia is undoubtedly a genius: 'Vous, Lélia, vous êtes plus grande par votre âme et votre génie que tout ce qui existe sur la terre' [You, Lélia, are greater by your soul and genius than everything that exists on earth].[244]

Lélia's intellectual superiority, however, raises suspicion because it borders on atheism; she is accused of being impious and making others doubt God, especially the priest Magnus, to the point of turning them into atheists.[245] Lélia herself admits that, while not being an atheist *per se*, she can momentarily doubt God or not understand Him and that her intelligence takes precedence over her faith, as the former shapes the latter according to her needs.[246] The poet Sténio uses metaphorical language to translate this religious scepticism combined with religious fervour and refers to the angel *vs.* demon dichotomy: 'Tu es un ange ou un démon, mais tu n'es pas une créature humaine' [You are an angel or a demon, but you are not a human creature][247] He expresses the need to label the masculine woman Lélia as either good – that is, generous, tender and submissive – or evil – independent, seductive and assertive. *Lélia* and the other novels studied in this chapter illustrate how restrictive and *de facto* harmful the seemingly

242 Ibid., 223.

243 Ibid., 288.

244 Ibid., 27. See also 13 (Sténio); 65 (Magnus); 163 (Pulchérie); 201, 203, 232 and 267 (Lélia).

245 For instance, ibid., 9, 86, 98 and 285.

246 Ibid., 40, 97, 181, 186 and 196.

247 Ibid., 7.

positive connotations linked to angelic metaphors are. Sand's novel shows that Sténio's error consists in not understanding that Lélia's superiority stems from escaping the constraining and dichotomous labelling that he bestows upon her. This strategy is, however, subtly displayed, which has led some critics to simplify masculine women's ambivalent personality and associate them with one pole of the angel *vs.* demon dichotomy. Lélia, for instance, is listed as a Madonna by Vareille, who interprets the sisters Lélia and Pulchérie as the 'vénérable dichotomie: madone ou putain' [venerable dichotomy: Madonna or whore].[248] Nancy Rogers, on the other hand, lists Sand's heroine as one of the '[m]ore assertive, aggressive women' who die violently, in opposition to the quiet deaths of 'angelic heroines', as if to punish them for their assertiveness.[249] In contrast, Lélia herself does not see her chastity as a sign of purity.[250] As was previously argued, it rather expresses her reaction to the trauma that she suffered at the hands of a man.

As a remarkable *femme auteur*, Félicité des Touches is logically presented, like Lélia, as an exceptional woman, superior not only to women in a social system that values so-called masculine attributes, but also to most human beings, men and women alike.[251] Claude Vignon, for instance, considers her to be the greatest woman; the narrator points out that '[e]lle faisait autorité' [she was an authority], whereas Béatrix is forced to recognize her rival's greatness ('une des plus grandes figures de notre temps' [one of the greatest figures of our time]), her superiority over women, and, above all, her own inferiority.[252] The narrator explains that Félicité surpasses Béatrix not only by her (masculine) talent but also by the nobility of her (feminine) feelings. Madame de Rochefide therefore smugly notices the petty behaviour that likens her great rival to conventional women: 'elle était enchantée de savoir à ce grand écrivain les petitesses de son sexe' [she was delighted to learn that this great writer had the pettiness of her sex].[253]

248 Vareille, *Socialité*, 201.

249 Nancy Rogers, 'The Wasting Away of Romantic Heroines', *Nineteenth-Century French Studies*, 11/3–4 (Spring–Summer 1983) 246–56 (252–53).

250 Sand, *Lélia*, 207.

251 Balzac, *Béatrix*, 153 and 228.

252 Ibid., 224, 153 and 274. See also 280 and 286.

253 Ibid., 257.

The grammatical ambiguity – *écrivain* is masculine but *son sexe* refers to the female sex – presents Félicité as a dual entity in which her masculine writing skills are strictly separated from her sentimental capacity. Balzac thus misogynistically claims that, in matters of the heart, all women, whether great or not, are united in their pettiness.

If the novel almost unambiguously acknowledges Félicité's greatness, it adopts more complex positions regarding her religious sentiment. The characters' opinions must be distinguished from the narrator's depiction. The ultra Catholic and ultra conservative inhabitants of Guérande view her as a diabolical creature and have no shortage of names to describe her: she is perceived as a demon and a witch who has cast a spell on the innocent Calyste.[254] These images influence the reader's interpretation of Félicité, since the gossips present her as a monster before she appears in the novel.[255] Calyste's isolated voice that claims that she is an angel and a saint is lost amongst these negative judgements.[256] The criticism of her narrow-minded detractors is conveyed through traditionally demeaning images towards women in Christian communities, which show their failure to understand Félicité's independence and sexual freedom. Not even her professional activity escapes the disapproval of the Bretons. Abbot Grimont and Fanny du Guénic criticize her artistic ambitions by using disparaging words designating actresses, such as 'baladine' [entertainer] and 'histrionne' [minstrel].[257] By reducing Camille Maupin's creative activities as a writer and a composer to the functions of an untalented executant, they reveal their ignorance of these activities, Parisian culture and the modern literary world. Their understanding of acting may have been determined by performers' status in the seventeenth century, when acting was judged indecent for women, and actors in general were excommunicated by the French Catholic Church.[258] In this case, in an attempt to grasp Félicité's identity, the inhabitants of Guérande reduce the complexity of the masculine woman to

254 For instance, ibid., 120, 122 and 242.

255 Ibid., 134. Van Rossum-Guyon, *Balzac et la littérature réfléchie*, 75–77.

256 Balzac, *Béatrix*, 135 and 335.

257 Ibid., 118 and 121.

258 Prest, *Theatre under Louis XIV*, 15–16.

well-known characteristics taken from the cultural sphere of Christianity and traditional societies.

Where the Bretons are truly confused is regarding her attitude to religion. For them, she is an ungodly woman who neglects her religious duties and spreads impiety in her works.[259] Worse even than her irreligious stance, she is accused of atheism by Abbot Grimont in a free indirect discourse: 'cette monstrueuse créature, qui tenait de la sirène et de l'athée, formait une combinaison immorale de la femme et du philosophe, et manquait à toutes les lois sociales inventées pour contenir ou utiliser les infirmités du beau sexe' [this monstrous creature, who was partly a Siren and partly an atheist, was an immoral combination of woman and philosopher, and she infringed all the social laws that had been invented to contain or use the infirmities of the fair sex].[260] The priest of Guérande first compares Félicité to a traditional mythological female monster to highlight both her alleged impiety and castrating femininity. He then refers to rational, masculine 'monsters' that have been spawned by the Enlightenment: the atheist and the philosopher. These references stress the rationality of the female writer and, while continuing the criticism of her depravity, they nonetheless acknowledge Félicité's exceptional intelligence.[261] However, her cleverness frightens the abbot, because it is seen as an emasculating force that contradicts what he expects of her gender. He accuses her of stepping beyond boundaries by refusing to conform to the social prescriptions imposed on women. Her transgression of gender conventions, more than her seductiveness or atheism, constitute her monstrosity.[262]

Whilst the characters in *Béatrix* and its narrator agree to consider the woman writer a monster, their opinions differ with regard to the definition of a monster. On the one hand, Abbot Grimont, Mademoiselle de Pen-Hoël, the inhabitants of Guérande and even Félicité herself ground their reasoning in the most common definitions of *monster* as an 'animal

259 Balzac, *Béatrix*, 118, 119, 121, 131 and 335.

260 Ibid., 135.

261 Ibid., 118, 121 and 241.

262 Murata ('Sous le signe des monstruosités', 45) points out that it is in *Béatrix* that the word *monstruosité* [monstrosity] has the highest frequency in *La Comédie humaine*.

qui a une conformation contre nature' [animal that is configured against nature] and a 'personne cruelle et dénaturée' [cruel and unnatural person].[263] The Bretons call Félicité a monster because of her supposed immorality, lifestyle and difference, although she attracts genuine, if concealed, fascination, as the long conversations between Calyste's family and Grimont about her demonstrate. On the other hand, the narrator of *Béatrix* justifies his interest in what he calls a secondary character by claiming: 'ne sera-ce pas satisfaire beaucoup de curiosités et justifier l'une de ces monstruosités qui s'élèvent dans l'humanité comme des monuments, et dont la gloire est favorisée par la rareté?' [will it not satisfy much curiosity and justify one of those monstrosities that stand up among humankind like monuments, and whose glory is favoured by rarity?].[264] His observation is based on the etymology of *monster*. The Latin word *monstrum* designates the portent of an event, whether calamitous or marvellous. The monster was originally considered to be a sign sent by the gods or by God to warn people against dangers, to show their/His anger or glory and to create fear or admiration.[265] Accordingly, the narrator's conception of Félicité's monstrosity is not negative. He highlights the extraordinary dimension that is contained in the idea of the monstrous – as *monstrueux* can mean 'prodigieux, excessif' [prodigious, excessive] – and casts new light on Félicité's exceptionality.[266]

But let us return to her alleged lack of religious feeling. At the end of the novel, she decides to become a nun – which makes the priest of Guérande change his opinion about her and compare her to a saint.[267] Apparently sudden and in contradiction to everything that was previously professed about her, Félicité's religious calling is in fact foreshadowed several times. The narrator certainly acknowledges her lack of faith, calling her an 'écrivain incrédule' [non-believing writer], but he also points out her progressive rejection of society and increasing fascination with the mysteries

263 'Monstre', in *Dictionnaire de l'Académie française*, II, 225. References to Félicité's monstrosity: Balzac, *Béatrix*, 116, 156 and 224.

264 Balzac, *Béatrix*, 137.

265 Jean-Louis Fischer, *Monstres: Histoire du corps et de ses défauts* (Paris: Syros-Alternatives, 1991), 47.

266 'Monstrueux', in *Dictionnaire de l'Académie française*, II, 225.

267 Balzac, *Béatrix*, 342 and 362.

of Catholicism.[268] He states, for instance: 'elle trompait ses souffrances en cherchant un sens au mouvement des mondes, et trouvait Dieu dans le sublime désert du ciel. Une fois Dieu reconnu par l'incrédule, il se jette dans le catholicisme absolu, qui, vu comme système, est complet' [she deceived her suffering by seeking a meaning in the motions of the worlds, and she found God in the sublime desert of heaven. Once God is recognized by non-believers, they throw themselves into absolute Catholicism, which, as a system, is complete].[269] Félicité is indeed a complex being who, like Lélia, is capable of intellectual superiority, feminine pettiness, agnosticism and religious fervour, an elusive combination that compels the Bretons to reduce her to a monster.

Whereas Félicité des Touches is wrongfully viewed as an impious and therefore evil woman by the Breton community, Lisbeth Fischer benefits from the opposite prejudice: her family considers her to be the 'bon ange de la famille' [good angel of the family], whereas the omniscient narrator of *La Cousine Bette*, better informed of her intention to harm the Hulots, depicts her as a fundamentally negative character.[270] Balzac's novel borrows narrative elements from fairy tales. Adeline is a modern Cinderella, while Lisbeth seems to be a substitute for evil relatives and ugly ducklings. Hélène Ortali accurately summarizes the plot as 'a contest between Adeline and Lisbeth to determine which can handle men more successfully', the former by virtue, the latter by scheming.[271] Lisbeth temporarily wins by displaying her masculine assets, such as calculation and savagery, before falling victim to her excessive desire for revenge in losing Marshal Hulot before being able to marry him and compelling Victorin Hulot to hire Jacqueline Collin to act against Valérie.

However, the novel is more nuanced as far as her wickedness is concerned. After all, she ultimately has more influence over Wenceslas than either Lélia or Félicité have over their lovers. Whilst promoting their intellectual development, they involuntarily contribute to the young men's

268 Ibid., 256 and 298.

269 Ibid., 281. See also 299.

270 Balzac, *La Cousine Bette*, 168.

271 Hélène Ortali, 'Images of Women in Balzac's *La Cousine Bette*', *Nineteenth-Century French Studies*, 4/3 (Spring 1976), 194–205 (194).

social malaise and misconduct. As Lélia refuses to become his mistress, Sténio falls into a life of debauchery and ends up killing himself. Félicité's sacrifice turns out to be noxious, since Calyste is unhappy with the wife whom she has chosen for him and is seduced by adultery. More importantly, *La Cousine Bette* highlights the fact that Lisbeth's ill-will is fuelled by the contempt to which she is subjected. The poor relative is despised for lacking feminine attributes, for her celibacy and for her financial autonomy obtained through work.[272] She is not only the evil cousin who plots against her generous relatives; she also appears as a victim of, or a scapegoat for, patriarchy.[273] Her female masculinity prevents her from conforming to social conventions, since her lack of beauty impedes both her search for a rich and noble suitor and her climbing of the social ladder. Baron Hulot seems astonished that she can even consider marrying his brother.[274] She must therefore resign herself to being seen as incapable of love, unintelligent and being treated like a servant.[275]

Her strength is to turn her alleged inferiority into an asset. The old maid is a consummate hypocrite. She befriends servants and young people by playing the role of confidante and emissary, she flatters Marshal Hulot by professing Republican opinions and, above all, she poses as the saviour of her family while simultaneously orchestrating its ruin.[276] She reaches the height of hypocrisy when she claims that she is trying to put an end to Baron Hulot's affair with Valérie Marneffe, while his infidelity constitutes a fundamental part of her revenge: 'Je ne veux pas avoir l'air de tremper dans la ruine de ma famille; moi qui ne suis là depuis trois ans que pour l'empêcher' [I do not want to seem involved in the ruin of my family, especially as I have been here for three years in order to prevent it].[277] Lisbeth's death is ambiguous:

272 Balzac, *La Cousine Bette*, 57 and 168.

273 She has been 'immolée' [sacrificed] (Ibid., 55 and 138) for her cousin's happiness: Krumm, 'La Bette noire de Balzac', 259–61. On Lisbeth's ambiguity, see Christopher A. Prendergast, 'Antithesis and Moral Ambiguity in *La Cousine Bette*', *The Modern Language Review*, 68/2 (April 1973), 315–32 (329–32).

274 Balzac, *La Cousine Bette*, 325.

275 Ibid., 25, 58, 63 and 323.

276 Ibid., 59, 168, 210, 262–63, 311 and 371.

277 Ibid., 312.

> Lisbeth, déjà bien malheureuse du bonheur qui luisait sur la famille, ne put soutenir cet événement heureux. Elle empira si bien, qu'elle fut condamnée par Bianchon à mourir une semaine après, vaincue au bout de cette longue lutte marquée pour elle par tant de victoires. Elle garda le secret de sa haine au milieu de l'affreuse agonie d'une phtisie pulmonaire. Elle eut d'ailleurs la satisfaction suprême de voir Adeline, Hortense, Hulot, Victorin, Steinbock, Célestine et leurs enfants tous en larmes autour de son lit, et la regrettant comme l'ange de la famille. [...] – Elle [Adeline] finira par être heureuse! se dit Lisbeth la veille de sa mort en voyant l'espèce de vénération que le baron témoignait à sa femme [...]. Ce sentiment hâta la fin de la cousine Bette, dont le convoi fut mené par toute une famille en larmes.[278]

> [Lisbeth, already made very unhappy by the happiness that shone on the family, could not stand this happy turn of events. Her state worsened so much that Bianchon condemned her to die within a week, defeated at the end of this long fight marked for her by so many victories. She kept her hatred secret through the dreadful agony of a pulmonary tuberculosis. She even had the supreme satisfaction of seeing Adeline, Hortense, Hulot, Victorin, Steinbock, Célestine and their children all in tears at her bedside and mourning her like the angel of the family. [...] – Adeline will end up being happy! Lisbeth thought on the eve of her death when she saw the kind of veneration that the Baron showed his wife [...]. That feeling hastened the death of Cousin Bette, whose funeral procession was led by the whole family in tears.]

Lisbeth's inability to ruin the Hulots' happiness is comprehensibly a source of anger. However, it is unclear whether she finds solace in the fact that she was able to deceive her relatives to the end or whether she is frustrated not to be able to reveal the crucial role that she played in their woes and flaunt her hatred to them.

La Cousine Bette highlights the double standards that determine women's lives in nineteenth-century French society. On the one hand, beautiful, gentle and submissive women like Adeline often manage to obtain what they are supposed to desire because they conform to gender conventions. On the other hand, unconventional masculine women like Lisbeth are unable to achieve what is seen to be their social destiny because their identity questions social norms. They must use masculine means – conspiracy and domination – to endeavour to reach the same result but are often stigmatized for trying.

278 Ibid., 506.

Situated at the intersection of different parameters with opposite poles – gender (masculinity and femininity), religion (angel and demon) and hierarchy (superiority and inferiority) – masculine women create new moral values by subverting them. This combination of diverse factors – and therefore the impossibility of labelling them – is judged to be frightening by other protagonists (Sténio and the Bretons, for instance) and to be a potential sign of inferiority. However, it is arguably masculine women's complexity that determines their superiority over weak young men, ageing beaux and senile old men, as well as over excessively good mothers and vain schemers.

The fact that masculine women's moral or intellectual superiority is controversial shows that what is at stake in these novels is power. What ultimately shocks the other protagonists is not so much the characters' masculine allure, but their alleged will to usurp male prerogatives. Indeed, Halberstam points out: 'Masculinity in this society inevitably conjures up notions of power and legitimacy and privilege.'[279] Conversely, femininity denies access to power. Acknowledging masculine women's superiority is implicitly admitting the possibility that men do not constitute the only pole of power. Masculine women are regarded as threats to the patriarchal organization of society because they have the so-called masculine qualities that are required to exercise power in nineteenth-century society: assertiveness, charisma and authority. If they are given the means, or even take the opportunity, to exercise these in the public sphere, nothing can justify men's unconditional power.

Yet, the full potential of masculine women's power is not exploited in the novels. This restriction may reflect the social context of the July Monarchy. Although 'the cause of women' was characterized by 'a wealth of possibilities and hopes in the 1830s', July Monarchy feminist movements actually had little impact on social organization.[280] In France, feminists' complaints about, and petitions to modify, some articles of the Napoleonic Code did not obtain noteworthy results.[281] More significantly, feminists

279 Halberstam, *Female Masculinity*, 2. See also Bourdieu, *La domination masculine*, 136.

280 Garval, *A Dream of Stone*, 124.

281 Moses, *French Feminism*, 103–05.

were confined within male movements in which feminist claims were only minor issues. Claire Goldberg Moses and Philippe Régnier argue that the feminism within the Saint-Simonian movement as defined by Enfantin was configured from a masculine point of view and existed ideologically rather than concretely, since militant women were increasingly excluded from positions of importance.[282] Similarly, the editor and main author (under female pseudonyms) of the feminist newspaper *La Gazette des femmes* [*The Periodical of Women*] (1836–38) was a man, Charles Frédéric Herbinot de Mauchamps, although he claimed that his mistress was its editor.[283] Even the feminist and socialist Flora Tristan considered that women's emancipation could only take place within the context of a working-class revolution; women alone could not be responsible for their own empowerment.[284] All things considered, although Kimmel and Badinter have argued that the historical causes behind 'crises' of masculinity are grounded in women questioning men's roles,[285] the narratives analysed here suggest that in practice the influence of feminist claims was limited and that headstrong women did not seem sufficiently powerful to undermine even the most fragile men in July Monarchy society.

Accordingly, fiction overtakes society and affirms female masculinity and authority, but only as the fruit of individual and exceptional initiatives. To Sténio's mind, Lélia is an 'exception monstrueuse' [monstrous exception], whereas the narrator of *Béatrix* calls Félicité an 'exception admise' [acceptable exception].[286] The exceptionality and monstrosity of the masculine woman are connected to her 'prodigious' nature: her rapid progress on the path of women's emancipation. Sand disapproved of the feminists of *La Gazette des femmes* because she considered that their claims (women's right to vote or women's eligibility, for instance) were pointless

282 Ibid., 51–59 and 89; Philippe Régnier, 'La réaffirmation et la réinvention saint-simoniennes du masculin: Pour une lecture nouvelle d'un féminisme originel', in *Masculinités en révolution*, 49–76 (esp. 59–61).

283 Moses, *French Feminism*, 98–101.

284 Ibid., 109.

285 Badinter, *XY*, 24–41; Kimmel, 'The Contemporary "Crisis" of Masculinity', 121–53.

286 Sand, *Lélia*, 123; Balzac, *Béatrix*, 153.

as long as women's education was not reformed.[287] Lélia's and Félicité's intellectual faculties prove that they have reached this essential stage in the road towards emancipation. When female education is achieved, social and political liberation is no longer out of reach.

Whilst Beauvoir judges most women to be complicit with men in contributing to their own subjugation,[288] fictional masculine women do not accept women's alleged inferior status. In so doing, they threaten men's political and social superiority. Accordingly, marginal as they may be, they must be forced back into the sphere of male power. As a result, their ascension is hindered by failure, reclusion or death. Lisbeth's revenge fails because of her overeagerness in its execution. It also brings about her failure as a character. As Kris Vassilev points out, she becomes useless in the development of the narrative and her death almost appears as an insignificant event.[289] It contrasts with the deaths of angelic heroines whose medical cause is mysterious, but whose importance is acknowledged by the in-depth descriptions of their agony.[290]

Like Lisbeth, Félicité disappears from the novel when she ceases to be useful to the plot. She is almost absent from the second part of the novel, 'La Lune de miel' [The Honeymoon] (1845), which focuses on Calyste and Sabine's marital problems. After bequeathing her fortune to the young man, Félicité retires to a convent and disavows Camille Maupin's works.[291] Her reclusion and her decision to reject her artistic and intellectual activities have been interpreted differently. Nicole Mozet considers that entering a convent indicates a return to traditional gender roles celebrating motherhood.[292] For Van Rossum-Guyon and Vanoncini, on the other hand, it is

287 Vareille, *Socialité*, 404–06.

288 Beauvoir, *Le deuxième sexe*, I, 328; II, 489.

289 Kris Vassilev, 'Représentation et signification sociale de la vengeance dans un texte réaliste: L'exemple de *La Cousine Bette*', *Romantisme*, 127 (2005), 45–57 (54–55). See also McGuire, 'The Feminine Conspiracy', 295 and 303–04.

290 Germaine de Staël's Corinne, Benjamin Constant's Ellénore and Balzac's Henriette de Mortsauf, for instance: Rogers, 'The Wasting Away of Romantic Heroines', 246–47.

291 Balzac, *Béatrix*, 343.

292 Nicole Mozet, *Balzac au pluriel* (Paris: Presses Universitaires de France, 'Écrivains', 1990), 164–65.

a spiritual choice that corresponds to Balzac's beliefs and is devoid of feminine connotation and conservatism.[293] I see this intellectual abjuration as symptomatic of the patriarchal system that judges women to be neither suitable nor required to exercise power, even when confined to the cultural sphere. A disillusioned Félicité is only too aware that there are few prospects for her after reaching literary glory.

As for Lélia, she appears only in the final chapter of the fifth part of the novel, the previous ones focusing on the male protagonists Trenmor, Magnus and, above all, Sténio. When Lélia finally appears, she experiences the most tragic death of this corpus, strangled by Magnus because he sees the demon in her.[294] It seems as though this so-called demon is the fanatical priest's Catholic interpretation of her assertiveness, autonomy and willpower.

Masculine women's attempt to find agency in a society determined by men, even insecure ones, is revealed to be a challenging project. The combination of masculine and feminine qualities that characterizes them is not fully accepted within the novels. Ultimately, they face repression and restrictions on their independence because, on the one hand, they lay claim to prerogatives that sexual and gender norms deny to women and, on the other, their influence highlights men's weaknesses in a society that was permeated by political upheaval and social instability. Nevertheless, the obstacles placed in the way of their empowerment paradoxically demonstrate that their potential aspirations to power are justified. Masculine women therefore suggest that women who reject the sexual dichotomy could threaten male supremacy in French society.

In conclusion, masculine women appear as agents capable of both causing and preventing young men's malaise in the novels *Lélia*, *La Cousine Bette* and *Béatrix*. Their lifestyle generally reveals social, emotional and financial autonomy that thwarts the norms imposed on women in a society defined by the Napoleonic Code. Their case, however, is presented as exceptional in these novels. They benefit from the advantages that correspond to the claims

293 Van Rossum-Guyon, *Balzac et la littérature réfléchie*, 88; Vanoncini, 'La femme-livre', 164.

294 Sand, *Lélia*, 323. In the 1839 version of *Lélia*, her fate is like Félicité's, as she becomes the abbess of a convent.

of French feminists during the nineteenth century, such as education and economic independence, and they exercise discreet but influential authority over men. Nevertheless, although their masculinity is repeatedly asserted, their femininity is never eradicated. Whilst their refusal of the traditional social roles of wifehood and motherhood contradicts nineteenth-century social conventions for women, masculine women question their own femaleness and express maternal feelings towards their lovers, rather than overtly challenging the validity of these social conventions.

Masculine women eschew patriarchal attempts to reduce their complex identity to the categories defined by sexual and gender norms. The fascination and the unease generated by their unconventional identities and intellectual superiority reveal the controversy that surrounds masculinity. Whilst they are judged to be disturbing, it is not the masculinity of these women that is primarily criticized. Rather, the main concern is their real or assumed claim to power. They experience sexualities at the margins of heteronormativity, they avoid the traditional definitions of women's identities as either angelic or demonic and they advocate autonomy. Their gender ambiguity thus demonstrates the relevance of replacing the relationship of complementarity between men and women by that of equality.

Conclusion

The authors of the narratives analysed in the previous pages of this book reveal the complex position of the nineteenth century in general, and the July Monarchy period in particular, towards gender identities. The period elaborated a strict framework of separation of the sexes. A strong correlation was perceived between sexes and gender roles. Accordingly, gender norms were determined on the grounds of biological sex. A male individual was 'naturally' expected to be manly, which implied physical and moral qualities such as strength, bravery and a sense of duty. A female individual, on the other hand, was deemed to be feminine, hence soft, maternal, changeable and weak. This depiction of women was common in medical texts and legal texts, as well as in the arts and literature.

However, at the same time, it was often acknowledged that the strict model of sexual dichotomy could be fallible. Some individuals did not comply with this benchmark. Scientists and physicians interpreted them as a sign of disorder and labelled them 'abnormal'. A woman who exercised physical and, above all, intellectual activities, who was independent, who practised sapphic love, threatened the model of the passive and devoted wife and mother. Likewise, effeminate and insecure young men such as Lucien de Rubempré and Musset's Octave mirrored the embedded fear that men were becoming more like women, while women were becoming manlier. Similarly, intersex individuals challenged the separation of the sexes by their very existence. They could nonetheless be integrated into the norm if a predominant sex was determined and, above all, if they led a life in accordance with the sexual and gender principles associated with that sex. According to that logic, the homosexual hermaphrodite committed the double offence of being a physical and moral monster.

One of the main interests of the July Monarchy fictional narratives that have been examined in this book is that their authors might refuse to take sexual dichotomy and heteronormativity at face value and go as far as to question them. For instance, they highlighted the difficulty of conforming to the gender expectations of the July Monarchy. The so-called weakness of young men is brought about, they suggest, by expectations which were too high to fulfil and a role model with which it was impossible to compete, that of the Napoleonic soldier. Certainly, some narratives hint at the fact that those who are unable to conform to social and gender norms are condemned to failure, death and a mediocre narrative fate, comprised of conventional marriages and waived ideals. The sexually ambiguous characters Camille, Lucien, Lélia, Lisbeth and Félicité either experience a banal fate or meet a tragic end.

However, the creation of gender-bending protagonists challenges gender constructs in a complex manner. The masculine and yet maternal female characters that include Félicité des Touches, Lisbeth Fischer and Lélia come to mind. Vautrin, the virile homosexual, constitutes another example of this questioning of the gender and sexual norms and creation of alternative models to senile Ancien Régime patriarchs and manly Napoleonic soldiers. Writers even interpreted the gender identity of 'deviant', hermaphroditic, effeminate and mannish characters as an instrument of personal development and social empowerment. Madeleine de Maupin, who defines herself as a member of the third sex and has sexual intercourse with both a man and a woman, is ready to conquer seventeenth-century French society armed with her first-hand knowledge of both sexes.

Although Madeleine is not an intersex individual *per se*, her depiction is in the spirit of the times as it exemplifies the fascination of the July Monarchy for hermaphrodites and androgynes. The complexity of hermaphrodites in literary works constitutes a significant part of this study. It is grounded in their relations to diverse cultural and symbolic domains, such as science, morals and aesthetics. They are, in turn, 'abnormal' human beings scrutinized by scientists and physicians, monsters who defy natural and divine laws and incarnations of ideal beauty. The physical perfection of hermaphrodites plays a crucial part in the narratives studied here. It unites feminine features such as roundness and grace with masculine features such as firmness and strength. More importantly, this beauty symbolically transcends the 'flaws' associated with the opposite sex. The effeminacy of insecure young men and the virility of masculine women can be channelled into promising combinations. Uniting physical and psychological characteristics that allegedly belong to one sex or another allows hermaphroditic characters to influence the gender malaise embedded in the July Monarchy society. Madeleine de Maupin and, to a lesser extent, Vautrin and Félicité des Touches reach the ideal third sex which promotes beauty, tenderness and assertiveness, as well as physical and moral strength. Rising above the so-called incompleteness of hermaphrodites, their androgyny renders them truly complete beings.

Certainly, this depiction of hermaphroditism does not mean that the widely spread concern about sexually ambiguous and gender-bending people is absent from literary texts. The dual nature of hermaphrodites engenders anxiety about their place in a binary society because it compels said society to re-examine the validity of the separation of the sexes. This unease is clear in *Fragoletta*, for example. On several occasions, Camille expresses the ontological angst of not fully belonging to humankind because of their intersexuality. However, through the literary medium, authors addressed some of these concerns. I have shown how the question of the homosexual hermaphrodite was treated in *Fragoletta*, *Sarrasine* and *Mademoiselle de Maupin*, from a clear separation of gender identities that averts explicit homosexuality in the former to an increasing acknowledgement of, and tolerance towards, homosexuality (and especially lesbianism) in the latter. Scientific concerns are thus reduced through the medium of fiction.

Furthermore, the questions of power and its subversion underpin the novels studied in this book. Many young men are depicted as failing to access key influential roles, whereas some women and hermaphroditic characters become increasingly powerful. In this respect, the narratives unkindly depict older characters who built their career and influence during the Empire or the Restoration. Men such as Baron du Guénic, Baron Hulot and his elder brother, who are usually the patriarchs of their families, occupy in the July Monarchy society the positions of authority that are denied to their sons. Their power is nevertheless compromised by their nostalgia, their inability to adapt to present times and even their senility. And yet, the new generation does not challenge the old one because young men seem to fear the consequences of a potential take-over owing to their apparent fragility. Accordingly, gendered connotations of power are often reversed in July Monarchy narratives. Power is shown to be exercised indirectly by those who do not legitimately hold it: women. Lisbeth's animalistic domination, Madeleine's androgynous superiority over d'Albert, Lélia's philosophical thoughts, as well as Camille Maupin's intellectual authority are examples of women's power, usually exercised in the shadows. Generally, this power is not aimed at deposing men; rather, with varying degrees of success, female protagonists support them in their process of empowerment. Noticeably, none of them endeavours to seize power for themselves – there is no feminist revolution on the horizon. This does not mean that social change is absent from our corpus. Vautrin, the homosexual criminal turned Head of Security, is probably the most striking example. Bourgeois order, the narratives suggest, can always be mimicked or even undone, by gender-bending characters.

Compared with July Monarchy writers' ambiguous and sometimes revolutionary positions, where do twenty-first-century Western societies stand on gender identities? The beginning of the twenty-first century has seen some important achievements relating to the awareness of gender and sexual ambiguities, for example, the widespread legalization of same-sex marriage, as well as more visibility and legal recognition for intersex and trans people. Contemporary societies nevertheless remain defined by a binary gender system. It is difficult not to compare the situation described in the books about young men's pervasive malaise regarding their condition in the July Monarchy and the situation in the aftermath of Donald Trump's presidency of the United States of America (2017–21). In 2013, Michael Kimmel published a monograph that somewhat foretold Trump's presidency, or at least the profile of Trump's voters. In *Angry White Men*, the American sociologist describes the anger and the *aggrieved entitlement* of a segment of the American population, which is White, male, heterosexual and lower middle class.[1] These men feel powerless; they feel that they have been robbed of the American Dream that they considered to be their birth right. In retribution, they hold immigrants, women (especially feminists), gay men and other minorities responsible for the loss of their jobs, their social status and

1 Michael S. Kimmel, *Angry White Men: American Masculinity at the End of an Era* (New York: Nation Books, 2017 [2013]).

their economic independence. Kimmel states that their anger is real – they did lose many of their privileges over the last forty years – but misplaced: the responsible parties for the weakening of their privileged status are not the minorities that they criticize (or, in some cases, even attack or kill) but the capitalist system and the corporate elites that they often support, as evidenced by Trump's presidential victory. In this masculine crisis, some of the ingredients of the July Monarchy masculine malaise can be found: a disappointing economic situation, a political power that has not kept its promises and a dashed sense of entitlement. However, where the insecurity of the men of the July Monarchy was expressed as disenchantment, the malaise of these twenty-first-century American men is expressed as rage.

This rage may be fuelled by the fact that, for one of the first times, women's voices are heard. On 5 October 2017, the *New York Times* published an article about allegations of sexual misconduct against the film producer Harvey Weinstein. Ten days later, the actress and activist Alyssa Milano launched the hashtag #MeToo on social media so that women could share their experience of sexual harassment and highlight the omnipresence of the phenomenon. She did not know that the activist Tarana Burke had already founded a MeToo movement in 2006 to support women of colour who were victims of sexual violence (she later corrected her mistake in acknowledging Burke as the original founder of the movement). The hashtag encountered considerable success all around the world (notably under different labels, such as the French version #BalanceTonPorc [grass your pig up]) and led to other initiatives, notably the Time's Up movement founded in 2018 by women in the film industry in order to address the misogynistic working conditions that facilitated gender inequality and sexual harassment.[2] Women (and some men) speaking out not only empowered the victims and survivors of sexual violence, but it also raised awareness of the scale of rampant misogyny, male entitlement and privilege. However, while revealing the fragility of masculine identity, easily threatened by accusations of sexual harassment, #MeToo has also highlighted the permanence of the virile model given the few concrete

2 On the the Weinstein scandal, #MeToo movement and their repercussions, see Laurie Collier Hillstrom, *The #MeToo Movement* (Santa Barbara-Denver: ABC-CLIO, 2018); Karen Boyle, *#MeToo, Weinstein and Feminism* (Cham: Palgrave Macmillan, 2019).

consequences that the movement had on men. For example, the federal judge Brett Kavanaugh and comedians Aziz Ansari and Louis C. K., all accused of sexual misconduct, were able to pursue their careers: being appointed at the Supreme Court and returning to stand-up comedy.[3]

Masculine malaise in France may take different forms. The #MeToo movement has encountered some backlash from men and women alike: it has been accused of blurring the border between sexual violence and awkward flirtation, as evidenced by the open letter published in *Le Monde* on 9 January 2018 and signed by over 100 French women, including the actress Catherine Deneuve, who defend a *liberté d'importuner* [freedom to bother].[4] This backlash is embedded in an ideological shift to the right of the French cultural landscape, which conveys a malaise linked to 'a perceived loss of French identity and the possible loss of the French exception'.[5] In other words, right-wing intellectuals claim that French arts and culture in the broad sense would be threatened by foreign (and especially Anglo-Saxon) cultural models.

One of these conservative intellectuals is the writer Michel Houellebecq. Praised for his supposed ability to depict the *Zeitgeist*, Houellebecq is a controversial author, notably due to his ambiguity: critics cannot agree on whether he sincerely shares the reactionary, antifeminist, or Islamophobic positions that are expressed in his novels or whether he is being ironic. His world is populated with ageing single men characterized by their sexual frustration in a society where sexual interactions follow a model of neoliberal economy paradoxically set up by the sexual liberation of the 1960s

3 Bianca Fileborn and Nickie Phillips, 'From "Me Too" to "Too Far"? Contesting the Boundaries of Sexual Violence in Contemporary Activism', in Bianca Fileborn and Rachel Loney-Howes, eds, *#MeToo and the Politics of Social Change* (Basingstoke: Palgrave Macmillan, 2019), 99–115 (110). Harvey Weinstein constitutes a notable exception since he has been convicted of rape in 2020.

4 Sarah Chiche et al., 'Nous défendons une liberté d'importuner, indispensable à la liberté sexuelle', *Le Monde* (9 January 2018): <https://www.lemonde.fr/idees/article/2018/01/09/nous-defendons-une-liberte-d-importuner-indispensable-a-la-liberte-sexuelle_5239134_3232.html>, accessed 02 August 2021. See also Fileborn and Phillips, 'From "Me Too" to "Too Far"?', 106–08.

5 Michael A. Johnson and Lawrence R. Schehr, 'Editors' Preface: Turns to the Right?', *Yale French Studies*, 116/117 (2009), 1–4 (1).

and 1970s: while masculine power is measured by men's ability to attract sexual partners, Houellebecq's protagonists are condemned to failure.[6]

Éric Zemmour, the highly controversial journalist, polemicist and candidate in the French presidential elections of 2022 associated with far-right intellectuals, offers less subtle and more reactionary criticism of French contemporary society. *Le suicide français* [*The French Suicide*] (2014) denounces the feminization of society as one of the many factors that are supposedly responsible for the decline of French culture. Zemmour attacks not only the disappearance of sexual norms in contemporary France, but also and above all expresses anxiety about 'cette inversion des rôles et des sexes qui obsède notre temps' [this inversion of roles and sexes with which our era is obsessed].[7] The polemicist's discourse is rooted in nostalgia. He believes in a utopian past that certainly precedes the 1970s but that in fact goes back as far as the pre-revolutionary period, since Zemmour sees the French Revolution as a step towards the decline of French culture. None of the ambivalence of the July Monarchy literary narratives examined in this book, which reveal complex views on the feminization of society as both disrupting and empowering, can be found in Zemmour's monolithic position.

The present book has examined the varied sexual and gender ambiguities that were accounted for and critiqued in the first half of the nineteenth century, especially in July Monarchy narratives. The hermaphrodite, the effete man and the masculine woman have proved to be powerful figures to convey the complex position of French society towards sexual dichotomy, gender norms and heteronormativity. In the literary texts analysed here, sexually ambiguous protagonists may certainly echo conventional positions

6 See Douglas Morrey, 'Sex and the Single Male: Houellebecq, Feminism, and Hegemonic Masculinity', *Yale French Studies*, 116/117 (2009), 141–52. See also Douglas Morrey, *Michel Houellebecq: Humanity and Its Aftermath* (Liverpool: Liverpool University Press, 2013), esp. Chapter One: 'Sex and Politics', 13–64; Carol Sweeney, *Michel Houellebecq and the Literature of Despair* (London-New York: Bloomsbury, 2013), esp. Chapter Three: '(Bad) Subjects of Neoliberalism', 69–90 and Chapter Four: 'Liquidating the Sixties', 91–124. The label *incels*, or involuntary celibates, seems appropriate to designate some of Houellebecq's protagonists.

7 Éric Zemmour, *Le suicide français* (Paris: Albin Michel, 2014), 187.

on the sexual dichotomy in a context of increasing awareness of the difficulties inherent to the condition of manhood. However, more often than not, they challenge the rules in order to acknowledge sexual indeterminacy and non-binary roles and offer alternative gender norms. When highlighting the complexity of the depiction of gender identities in works by Balzac, Gautier, Latouche, Musset and Sand, it becomes manifest that these texts occupy a significant position in the worldwide corpus of gender studies.

Bibliography

Primary Sources

Literature

Balzac, Honoré de, *Béatrix*, Michel Lichtlé, ed. (Paris: Librairie Générale Française, 'Le Livre de Poche Classiques', 2012).

—— *Correspondance de H. de Balzac*, 1819–1850, 2 vols (Paris: Calmann Lévy, 1876).

—— *La Cousine Bette*, Roger Pierrot, ed. (Paris: Librairie Générale Française, 'Le Livre de Poche Classiques', 2000).

—— *Illusions perdues*, Maurice Ménard, ed. (Paris: Librairie Générale Française, 'Le Livre de Poche Classiques', 1983).

—— *Le Père Goriot*, Stéphane Vachon, ed. (Paris: Librairie Générale Française, 'Le Livre de Poche Classiques', 1995).

—— *Sarrasine. Gambara. Massimilla Doni*, Pierre Brunel, ed. (Paris: Gallimard, 'Folio Classique', 1995).

—— *Splendeurs et misères des courtisanes*, Patrick Berthier, ed. (Paris: Librairie Générale Française, 'Le Livre de Poche Classiques', 2008).

Byron, George Gordon, *Lord Byron's Cain: Twelve Essays and a Text with Variants and Annotations*, Truman Guy Steffan, ed. (Austin and London: University of Texas Press, 1968).

Chateaubriand, François-René de, *Atala. René. Les aventures du dernier Abencérage*, Jean-Claude Berchet, ed. (Paris: Garnier-Flammarion, 1996).

Duras, Madame de, *Ourika. Édouard. Olivier ou le Secret*, Marie-Bénédicte Diethelm, ed. (Paris: Gallimard, 'Folio Classique', 2007).

Gautier, Théophile, *Mademoiselle de Maupin*, Michel Crouzet, ed. (Paris: Gallimard, 'Folio Classique', 1973).

Latouche, Henri de, *Fragoletta: Naples et Paris en 1799*, 2 vols (Paris: Delloye, 1840).

Musset, Alfred de, *La Confession d'un enfant du siècle*, Sylvain Ledda, ed. (Paris: Garnier-Flammarion, 2010).

Ovid, *Metamorphoses*, A. D. Melville, trans., E. J. Kenney, ed. (Oxford: Oxford University Press, 2008).

Plato, *Lysis. Symposium. Gorgias*, W. R. M. Lamb, ed. and trans. (London and New York: W. Heinemann-Macmillan, 1914).

Rousseau, Jean-Jacques, *Émile ou de l'éducation*, André Charrak, ed. (Paris: Garnier-Flammarion, 2009).
Sand, George, *Lélia*, Pierre Reboul, ed. (Paris: Gallimard, 'Folio Classique', 2003 [Garnier, 1960]).
Stendhal, *Armance*, Armand Hoog, ed. (Paris: Gallimard, 'Folio Classique', 1975).

Science

Cabanis, Pierre-Jean-Georges, *Rapports du physique et du moral de l'homme*, 2 vols (Paris: Crapart, Caille et Ravier, 1802).
Esquirol, Étienne, *Des maladies mentales: Considérées sous les rapports médical, hygiénique et médico-légal*, 2 vols (Paris: Baillière, 1838).
Fodéré, François-Emmanuel, *Traité de médecine légale et d'hygiène publique ou de police de santé*, 6 vols (Paris: Mame, 1813).
Fournier Pescay, François, 'Sodomie', in *Dictionnaire des sciences médicales*, 60 vols (Paris: Panckoucke, 1812–22), LI (1821), 441–48.
Geoffroy Saint-Hilaire, Isidore, *Histoire générale et particulière des anomalies de l'organisation chez l'homme et les animaux, ouvrage comprenant des recherches sur les caractères, la classification, l'influence physiologique et pathologique, les rapports généraux, les lois et les causes des monstruosités, variétés et vices de conformation, ou Traité de tératologie*, 4 vols (Paris: Baillière, 1832–37).
Human, J.-G. (Jean-Louis Michu), *Nosographie des maladies vénériennes ou Étude comparée des divers agents thérapeutiques qui ont été mis en usage pour combattre ce genre d'affections* (Paris: Béchet Jeune, 1838).
Lauvergne, Hubert, *Les forçats considérés sous le rapport physiologique, moral et intellectuel, observés au bagne de Toulon* (Paris: Baillière, 1841).
Mahon, Paul-Augustin-Olivier, *Médecine légale et police médicale*, 3 vols (Paris: Bertrand, 1807).
Marc, Charles-Chrétien-Henri, 'Hermaphrodite', in *Dictionnaire des sciences médicales*, 60 vols (Paris: Panckoucke, 1812–22), XXI (1817), 86–121.
Menville de Ponsan, Charles François, *Histoire médicale et philosophique de la femme considérée dans toutes les époques principales de sa vie avec tous les changements qui surviennent dans son physique et son moral avec l'hygiène applicable à son sexe et toutes les maladies qui peuvent l'atteindre aux différents âges*, 3 vols (Paris: Amyot-Labé, 1845).
Moreau de la Sarthe, Jacques Louis, *Histoire naturelle de la femme suivie d'un traité d'hygiène appliquée à son régime physique et moral aux différentes époques de la vie*, 2 vols (Paris: Duprat, 1803).

Parent-Duchâtelet, Alexandre-Jean-Baptiste, *De la prostitution dans la ville de Paris, considérée sous le rapport de l'hygiène publique, de la morale et de l'administration: Ouvrage appuyé de documents statistiques puisés dans les archives de la Préfecture de police*, 2 vols (Paris: Baillière, 1836).

'Prospectus', *Annales d'hygiène publique et de médecine légale*, 1 (1829), v–viii.

Renauldin, L. J., 'Introduction', in *Dictionnaire des sciences médicales*, 60 vols (Paris: Panckoucke, 1812–22), I (1812), xix–clxix.

Reydellet, 'Pédérastie', in *Dictionnaire des sciences médicales*, 60 vols (Paris: Panckoucke, 1812–22), XL (1819), 37–45.

Salles, Eusèbe-Fr. de, 'Médecine légale', in Jean-Louis Alibert et al., eds, *Encyclopédie des sciences médicales ou Traité général, méthodique et complet des diverses branches de l'art de guérir, 2nd division: Médecine* (Paris: Bureau de l'Encyclopédie, 1835), 1–291.

Tardieu, Ambroise, *Étude médico-légale sur les attentats aux mœurs*, 2nd edn (Paris: Baillière, 1858 [1857]).

Virey, Julien-Joseph, *De la femme, sous ses rapports physiologique, moral et littéraire*, 2nd edn (Paris: Crochard Libraire, 1825 [1823]).

—— 'Frigidité et froideur', in *Dictionnaire des sciences médicales*, 60 vols (Paris: Panckoucke, 1812–22), XVII (1816), 11–28.

—— 'Sexe', in *Dictionnaire des sciences médicales*, LI (1821), 218–44.

Art

Winckelmann, Johann Joachim, *History of the Ancient Art: Among the Greeks*, Giles Henry Lodge, trans. (London: Chapman, 1850).

—— 'Thoughts on the Imitation of the Painting and Sculpture of the Greeks' (1755), in *German Aesthetic and Literary Criticism: Winckelmann, Lessing, Hamann, Herder, Schiller, Goethe*, H. S. Nisbet, ed. and trans. (Cambridge: Cambridge University Press, 1985), 32–54.

Law

Code civil des Français, édition originale et seule officielle (Paris: Imprimerie de la République, 1804).

Secondary Sources

Aguilhon, Maurice, ed., *Les quarante-huitards* (Paris: Gallimard-Julliard, 1975).

Albouy, Pierre, 'Le mythe de l'androgyne dans *Mademoiselle de Maupin*', *Revue d'Histoire Littéraire de la France*, 72/4 (July–August 1972), 600–08.

Allan, Stacie, *Writing the Self, Writing the Nation: Romantic Selfhood in the Works of Germaine de Staël and Claire de Duras* (Oxford: Peter Lang, 2019).

Andrews, Naomi J., 'Utopian Androgyny: Romantic Socialists Confronts Individualism in July Monarchy France', *French Historical Studies*, 26/3 (Summer 2003), 437–57.

Aron, Jean-Paul, and Roger Kempf, *Le pénis et la démoralisation de l'Occident* (Paris: Grasset, 1978).

Badinter, Élisabeth, *L'amour en plus: Histoire de l'amour maternel (XVIIe–XXe siècle)* (Paris: Librairie Générale Française, 'Le Livre de Poche', 1982 [Flammarion, 1980]).

—— *XY: De l'identité masculine* (Paris: Librairie Générale Française, 'Le Livre de Poche', 1994 [Odile Jacob, 1992]).

Barbéris, Pierre, *Balzac et le mal du siècle: Contribution à une physiologie du monde moderne*, 2 vols (Paris: Gallimard, 1970).

—— *Le monde de Balzac* (Paris: Arthaud, 1973).

Barbier, Patrick, *The World of the Castrati: The History of an Extraordinary Operatic Phenomenon*, Margaret Crosland, trans. (London: Souvenir Press, 1996) [*Histoire des Castrats* (Paris: Grasset, 1989)].

Barry, Catherine, 'Camille Maupin: Io to Balzac's Prometheus?', *Nineteenth-Century French Studies*, 20/1–2 (Fall–Winter 1991–92), 44–52.

Barthes, Roland, *S/Z* (Paris: Seuil, 1970).

Beauvoir, Simone de, *Le deuxième sexe*, 2 vols (Paris: Gallimard, 'Idées', 1949).

Bellet, Roger, 'La confession politique des enfants malades du siècle', *Europe: Revue littéraire mensuelle*, 583–84 (November–December 1977), 126–35.

Bertaud, Jean-Paul, 'L'armée et le brevet de virilité', in Alain Corbin, ed., *Histoire de la virilité*, II: *Le triomphe de la virilité: Le XIXe siècle* (Paris: Seuil, 2011), 63–79.

—— 'La virilité militaire', in *Le triomphe de la virilité*, 157–202.

Berthier, Philippe, 'Balzac du côté de Sodome', *L'Année balzacienne*, 1st ser. (1979), 147–77.

Bertrand-Jennings, Chantal, *Un autre mal du siècle: Le romantisme des romancières, 1800–1846* (Toulouse: Presses Universitaires du Mirail, 2005).

Beynon, John, *Masculinities and Culture* (Buckingham and Philadelphia: Open University Press, 2002).

Bordas, Éric, 'Introduction: Comment en parlait-on?', *Romantisme*, 159 (2013), 3–17.
—— '*Sarrasine* de Balzac: Une Poétique du Contresens', *Nineteenth-Century French Studies*, 31/1–2 (Fall–Winter 2002–03), 41–52.
Borowitz, Helen O., 'Balzac's *Sarrasine*: The Sculptor as Narcissus', *Nineteenth-Century French Studies*, 5/3–4 (Spring–Summer 1977), 171–85.
Bouchard, Anne, 'Le masque et le miroir dans *Mademoiselle de Maupin*', *Revue d'Histoire Littéraire de la France*, 72/4 (July–August 1972), 583–99
Bourdieu, Pierre, *La domination masculine* (Paris: Seuil, 'Points/Essais', 2002 [1998]).
—— 'The Link between Literary and Artistic Struggles', Elinor Dorday, trans., in Peter Collier and Robert Lethbridge, eds, *Artistic Relations: Literature and the Visual Arts in Nineteenth-Century France* (New Haven and London: Yale University Press, 1994), 30–39.
Boyle, Karen, *#MeToo, Weinstein and Feminism* (Cham: Palgrave Macmillan, 2019).
Brittan, Arthur, *Masculinity and Power* (Oxford: Blackwell, 1989).
Brix, Michel, 'Nerval et le rêve égyptien', *Romantisme*, 120 (2003), 37–46.
Broglie, Gabriel de, *La Monarchie de Juillet: 1830–1848* (Paris: Fayard, 2011).
Brooks, Peter, 'Balzac: Epistemophilia and the Collapse of the Restoration', *Yale French Studies*, 101 (2001), 119–31.
Burgelin, Claude, 'Le cas Herculine B.', in Jean-Yves Tamet, ed., *Différenciation sexuelle et identités: Clinique, art et littérature*, pref. Pierre Chatelain (Paris: In Press, 2012), 93–100.
Busst, A. J. L., *The Image of the Androgyne in the Nineteenth Century* (London: Routledge & Kegan Paul, 1967) [Offprint from *Romantic Mythologies*, Ian Fletcher, ed., 1–95].
Butler, Judith, *Bodies That Matter: On the Discursive Limits of 'Sex'* (New York and London: Routledge, 1993).
—— 'Doing Justice to Someone: Sex Reassignment and Allegories of Transsexuality', in Susan Stryker and Stephen Whittle, eds, *The Transgender Studies Reader* (New York and London: Routledge, 2006), 183–93.
—— *Gender Trouble: Feminism and the Subversion of Identity* (New York: Routledge, 1990).
Byatt, S., 'The Death of Lucien de Rubempré', in Franco Moretti, ed., *The Novel*, II: *Forms and Themes* (Princeton and Oxford: Princeton University Press, 2006), 389–408.
Castagnès, Gilles, '*La Confession d'un enfant du siècle*: Un romantisme de façade?', in Sylvain Ledda and Frank Lestringant, eds, *Musset: Un romantique né classique* (Toulouse: Presses Universitaires du Mirail, 2009), 87–102.

Cézilly, Frank, 'La monogamie est-elle naturelle?', *Sciences humaines*, 188 (December 2007): <https://www.scienceshumaines.com/la-monogamie-est-elle-naturelle_fr_21607.html>, accessed 05 June 2018.

Chamard-Bergeron, Julia, '"Vous croyez aux amis": L'amitié dans *Illusions perdues*', *L'Année balzacienne*, 3rd ser., 10 (2009), 283–313.

Chase, Cheryl, 'Hermaphrodites with Attitude: Mapping the Emergence of Intersex Political Activism', in Susan Stryker and Stephen Whittle, eds, *The Transgender Studies Reader* (New York and London: Routledge, 2006), 300–14.

Chatelain, Pierre, Yves Morel, and Marc Nicolino, 'Le développement et la différenciation sexuels', in Jean-Yves Tamet, ed., *Différenciation sexuelle et identités: Clinique, art et littérature*, pref. Pierre Chatelain (Paris: In Press, 2012), 31–44.

Chesler, Phyllis, *Women and Madness*, rev. and updated edn (London: Palgrave Macmillan, 2005).

Chiche, Sarah, et al., 'Nous défendons une liberté d'importuner, indispensable à la liberté sexuelle', *Le Monde* (9 January 2018): <https://www.lemonde.fr/idees/article/2018/01/09/nous-defendons-une-liberte-d-importuner-indispensable-a-la-liberte-sexuelle_5239134_3232.html>, accessed 02 August 2021.

Childers, Joseph, and Gary Hentzi, eds, *The Columbia Dictionary of Modern Literary and Cultural Criticism* (New York: Columbia University Press, 1995).

Choquette, Leslie, 'Homosexuals in the City: Representations of Lesbian and Gay Space in Nineteenth-Century Paris', in Jeffrey Merrick and Michael Sibalis, eds, *Homosexuality in French History and Culture* (New York: Harrington Park Press, 2001), 149–67.

Citron, Pierre, 'Interprétation de *Sarrasine*', *L'Année balzacienne*, 1st ser. (1972), 81–95.

—— 'Le rêve asiatique de Balzac', *L'Année balzacienne*, 1st ser. (1968), 303–36.

Clark, Anna, 'The Rhetoric of Masculine Citizenship: Concepts and Representations in Modern Western Political Culture', in Stefan Dudink, Karen Hagemann and Anna Clark, eds, *Representing Masculinity: Male Citizenship in Modern Western Culture* (Houndmills, Basingstoke and New York: Palgrave Macmillan, 2007), 3–22.

Cohen, Margaret, *The Sentimental Education of the Novel* (Princeton: Princeton University Press, 1999).

Coleman, Patrick, *Anger, Gratitude and the Enlightenment Writer* (Oxford: Oxford University Press, 2011).

Collectif Georgette Sand, *Ni vues ni connues: Panthéon, Histoire, mémoire: Où sont les femmes?* (Paris: Hugo Doc, 2017).

Collingham, H. A. C., *The July Monarchy: A Political History of France, 1830–1848* (London and New York: Longman, 1988).

Connell, R. W., *Masculinities*, 2nd edn (Cambridge: Polity Press, 2005 [1995]).

Coquillat, Michelle, *La poétique du mâle* (Paris: Gallimard, 1982).

Corbin, Alain, *Histoire du corps*, II: *De la Révolution à la Grande Guerre* (Paris: Seuil, 'Points', 2005).

—— 'Introduction', in Alain Corbin, ed., *Histoire de la virilité*, II: *Le triomphe de la virilité: Le XIXe siècle* (Paris: Seuil, 2011), 7–11.

—— 'Préface', in Régis Revenin, ed., *Hommes et masculinités de 1789 à nos jours: Contributions à l'histoire du genre et de la sexualité en France* (Paris: Autrement, 2007), 7–11.

Costello, Cary Gabriel, 'Intersex and Trans* Communities: Commonalities and Tensions', in Stefan Horlacher, ed., *Transgender and Intersex: Theoretical, Practical, and Artistic Perspectives* (New York: Palgrave Macmillan, 2016), 83–113.

Counter, Andrew, *The Amorous Restoration: Love, Sex, and Politics in Early Nineteenth-Century France* (Oxford: Oxford University Press, 2016).

Courouve, Claude, *Vocabulaire de l'homosexualité masculine* (Paris: Payot, 1985).

Courtivron, Isabelle de, 'Weak Men and Fatal Women: The Sand Image', in George Stambolian and Elaine Marks, eds, *Homosexualities and French Literature: Cultural Contexts/Critical Texts*, pref. Richard Howard (Ithaca and London: Cornell University Press, 1990 [1979]), 210–27.

Crahay, Géraldine, '"Oh! non, je ne pourrai jamais te le dire… j'aime un homme!": Hermaphrodite et l'esthétisation de l'homosexualité dans *Mademoiselle de Maupin*', *Nineteenth-Century French Studies*, 45/1–2 (Autumn-Winter 2016–17), 17–32.

—— 'Valorising Genitalia: Homosexuality and Hermaphroditism in French Scientific Writings in the Nineteenth Century', *Harts & Minds*, 1/2 (October 2013).

—— '"Voyez les femmes les plus hommasses, ces *viragines* audacieuses": La domestication de la femme masculine dans les traités savants de la première moitié du dix-neuvième siècle', in Sophie Leroy, ed., *Medicine and Maladies: Representing Affliction in Nineteenth-Century France* (Leiden and Boston: Brill-Rodopi, 2018), 19–40.

Creed, Barbara, 'Lesbian Bodies: Tribades, Tomboys and Tarts', in Janet Price and Margrit Shildrick, eds, *Feminist Theory and the Body: A Reader* (Edinburgh: Edinburgh University Press, 1999), 111–24.

Crompton, Louis, *Homosexuality and Civilization* (Cambridge, MA: Harvard University Press, 2003).

Crouzet, Michel, '*Mademoiselle de Maupin* ou l'Éros romantique', *Romantisme*, 8 (1974), 2–21.

—— 'Monstres et merveilles: Poétique de l'Androgyne, à propos de *Fragoletta*', *Romantisme*, 45 (1984), 25–41.

Daston, Lorraine, and Katharine Park, 'The Hermaphrodite and the Orders of Nature: Sexual Ambiguity in Early Modern France', *GLQ*, 1/4 (1995), 419–38.

Davis, Georgiann, *Contesting Intersex: The Dubious Diagnosis* (New York and London: New York University Press, 2015).

Delattre, Geneviève, 'L'imagination balzacienne au travail: La lecture créatrice', *Cahiers de l'Association internationale des études françaises*, 15 (1963), 395–406.

Delcourt, Marie, *Hermaphrodite: Mythes et rites de la Bisexualité dans l'Antiquité classique* (Paris: Presses Universitaires de France, 1958).

DelPlato, Joan, *Multiple Wives, Multiple Pleasures: Representing the Harem, 1800–1875* (London: Associated University Presses, 2002).

Diaz, José-Luis, 'Balzac romancier de la "grande maladie sociale" (1838–1847)', *L'Année balzacienne*, 3rd ser., 17 (2016), 221–40.

—— 'Génération Musset?', *Romantisme*, 147 (2010), 3–12.

—— '"Nous tous, enfants perdus de cet âge critique": La "génération de 1830" par elle-même', *Romantisme*, 147 (2010), 13–28.

Dictionnaire de l'Académie française, 6th edn, 2 vols (Paris: Firmin-Didot frères, 1835).

Didier, Béatrice, 'Le corps féminin dans Lélia', *Revue d'Histoire Littéraire de la France*, 76/4 (July–August 1976), 634–51.

Dreger, Alice Domurat, *Hermaphrodites and the Medical Invention of Sex* (Cambridge, MA, and London: Harvard University Press, 1998).

—— 'Hermaphrodites in Love: The Truth of the Gonads', in Vernon A. Rosario, ed., *Science and Homosexualities* (New York: Routledge, 1997), 46–66.

Dudink, Stefan, Karen Hagemann, and Anna Clark, 'Editors' Preface: Historicizing Male Citizenship', in Stefan Dudink, Karen Hagemann and Anna Clark, eds, *Representing Masculinity: Male Citizenship in Modern Western Culture* (Houndmills, Basingstoke and New York: Palgrave Macmillan, 2007), ix–xv.

Edwards, Tim, *Cultures of Masculinity* (London and New York: Routledge, 2006).

Eliade, Mircea, *Méphistophélès et l'androgyne* (Paris: Gallimard, 'Idées', 1962).

Ender, Evelyne, '"Une femme qui rêve n'est pas tout à fait une femme": Lélia en rupture d'identité', *Nineteenth-Century French Studies*, 29/3–4 (Spring–Summer 2001), 226–46.

Faderman, Lillian, *Surpassing the Love of Men: Romantic Friendship and Love between Women from the Renaissance to the Present* (London: The Women's Press, 1985 [New York: William Morrow and Company, 1981]).

Fausto-Sterling, Anne, *Sex/Gender: Biology in a Social World* (New York and London: Routledge, 2012).

—— *Sexing the Body: Gender Politics and the Construction of Sexuality* (New York: Basic Books, 2000).

Fend, Mechthild, *Les limites de la masculinité: L'androgyne dans l'art et la théorie de l'art en France, 1750–1830*, Jean Torrent, trans. (Paris: La Découverte, 2011)

[*Grenzen der Männlichkeit: Der Androgyn in der französischen Kunst und Kunsttheorie, 1750–1830* (Berlin: Reimer, 2003)].

Fileborn, Bianca, and Nickie Phillips, 'From "Me Too" to "Too Far"? Contesting the Boundaries of Sexual Violence in Contemporary Activism', in Bianca Fileborn and Rachel Loney-Howes, eds, *#MeToo and the Politics of Social Change* (Basingstoke: Palgrave Macmillan, 2019), 99–115.

Finch, Alison, *Women's Writing in Nineteenth-Century France* (Cambridge: Cambridge University Press, 2000).

Fischer, Jean-Louis, *Monstres: Histoire du corps et de ses défauts* (Paris: Syros-Alternatives, 1991).

Foerster, Maxime, *The Politics of Love: Queer Heterosexuality in Nineteenth-Century French Literature* (Durham, NH: University of New Hampshire, 2018).

Forrest, Alan, 'Citizenship and Masculinity: The Revolutionary Citizen-Soldier and His Legacy', in Stefan Dudink, Karen Hagemann and Anna Clark, eds, *Representing Masculinity: Male Citizenship in Modern Western Culture* (Houndmills, Basingstoke and New York: Palgrave Macmillan, 2007), 111–29.

Forth, Christopher E., and Bertrand Taithe, 'Introduction', in Christopher E. Forth and Bertrand Taithe, eds, *French Masculinities: History, Culture and Politics* (Houndsmills, Basingstoke and New York: Palgrave Macmillan, 2007), 1–14.

Foucault, Michel, 'Des espaces autres', in Daniel Defert and François Ewald, eds, *Dits et écrits: 1954–1988*, 4 vols (Paris: Gallimard, 'NRF, Bibliothèque des sciences humaines', 1994), IV, 752–62 [Conférence au Cercle d'études architecturales, 14 March 1967, in *Architecture, Mouvement, Continuité*, 5 (1984), 46–49].

——*Histoire de la sexualité*, I: *La volonté de savoir* (Paris: Gallimard, 1976).

——ed., *Herculine Barbin dite Alexina B.* (Paris: Gallimard, 1978).

Garber, Marjorie, *Vested Interests: Cross-Dressing and Cultural Anxiety* (London: Penguin Books, 1993 [London: Routledge, 1992]).

Garval, Michael D., *A Dream of Stone: Fame, Vision, and Monumentality in Nineteenth-Century French Literary Culture* (Newark, DE: University of Delaware Press, 2004).

Gay, Peter, ed., *The Freud Reader* (London: Vintage Books, 1995).

Gazalé, Olivia, *Le mythe de la virilité: Un piège pour les deux sexes* (Paris: Laffont, 2017).

Gemie, Sharif, 'Balzac and the Moral Crisis of the July Monarchy', *European History Quarterly*, 19 (1989), 469–94.

Genette, Gérard, *Figures III* (Paris: Seuil, 1972).

——*Nouveau discours du récit* (Paris: Seuil, 1983).

——*Palimpsestes: La littérature au second degré* (Paris: Seuil, 1982).

Gilbert, Sandra M., and Susan Gubar, *The Madwoman in the Attic: The Woman Writer and the Nineteenth-Century Literary Imagination*, 2nd edn (New Haven: Yale University Press, 2000 [1979]).

Gildea, Robert, *Children of the Revolution: The French, 1799–1914* (London: Penguin Books, 2009 [London: Allen Lane, 2008]).

Gill, Miranda, *Eccentricity and the Cultural Imagination in Nineteenth-Century Paris* (Oxford: Oxford University Press, 2009).

Gilmore, David, *Manhood in the Making: Cultural Concepts of Masculinity* (New Haven and London: Yale University Press, 1990).

Giraud, Raymond, 'Winckelmann's Part in Gautier's Perception of Classical Beauty', *Yale French Studies*, 38 (1967), 172–82.

Glick, Elisa, 'Sex Positive: Feminism, Queer Theory, and the Politics of Transgression', *Feminist Review*, 64 (Spring 2000), 19–45.

Goruppi, Tiziana, 'Gli eredi de "l'Enfant du siècle": Il mito di Napoleone nella *Confession* di Musset', *Studi Francesi*, 48/1 (January–April 2004), 101–07.

Grant, Richard B., 'George Sand's *Lélia* and the Tragedy of Dualism', *Nineteenth-Century French Studies*, 19/4 (Summer 1991), 499–516.

Grimsley, Ronald, 'Romantic Emotion in Musset's *Confession d'un enfant du siècle*', *Studies in Romanticism*, 9/2 (Spring 1970), 125–42.

Grojnowski, Daniel, ed., *Confessions d'un inverti-né suivies de Confidences et aveux d'un Parisien* (Paris: Corti, 2007).

Groneberg, Michael, 'Hermaphrodite's Voice: Dealing with the Either-Or Attitude in Science, Law, and the Arts', in Stefan Horlacher, ed., *Transgender and Intersex: Theoretical, Practical, and Artistic Perspectives* (New York: Palgrave Macmillan, 2016), 225–52.

Guédron, Martial, ed., *Monstres, merveilles et créatures fantastiques* (Paris: Hazan, 2011).

Guichardet, Jeannine, 'Doublures historiques en scène parisienne', *L'Année balzacienne*, 2nd ser., 5 (1984), 307–25.

Guillet, François, 'Le duel et la défense de l'honneur viril', in Alain Corbin, ed., *Histoire de la virilité*, II: *Le triomphe de la virilité: Le XIXe siècle* (Paris: Seuil, 2011), 83–124.

Gutermann, Déborah, 'Le désir et l'entrave: L'impuissance dans la construction de l'identité masculine romantique: première moitié du XIXe siècle', in Régis Revenin, ed., *Hommes et masculinités de 1789 à nos jours: Contributions à l'histoire du genre et de la sexualité en France* (Paris: Autrement, 2007), 55–74.

Gutterman, David S., 'Postmodernism and the Interrogation of Masculinity', in Stephen M. Whitehead and Frank J. Barrett, eds, *The Masculinities Reader* (Cambridge: Polity Press, 2001), 56–71.

Hahn, Pierre, *Nos ancêtres les pervers: La vie des homosexuels sous le second empire*, pref. Michael Sibalis (Béziers: H&O, 2006 [Paris: Orban, 1979]).

Halberstam, Jack (Judith), *Female Masculinity* (Durham, NC, and London: Duke University Press, 1998).

Harkness, Nigel, *Men of their Words: The Poetics of Masculinity in George Sand's Fiction* (London: Legenda, 2007).

—— 'Resisting Petrification in George Sand's *Lélia* and Balzac's *Sarrasine*', *French Studies*, 59/2 (2005), 159–72.

Hatem, Jad, 'Scission du cœur et désordre du corps dans *Lélia* de Sand', *Romantisme*, 91 (1996), 19–34.

Hazareesingh, Sudhir, *The Legend of Napoleon* (London: Granta Books, 2004).

Heathcote, Owen N., '"Cet être amphibie qui n'est ni homme ni femme": Marginalizing Gender and Gendering the Marginal in Balzac's Camille Maupin', *Nottingham French Studies*, 41/2 (Autumn 2002), 37–46.

Hillstrom, Laurie Collier, *The #MeToo Movement* (Santa Barbara and Denver: ABC-CLIO, 2018).

Holmes, Diana, *French Women's Writing: 1848–1994* (London and Atlantic Highlands: Athlone, 1996).

Horlacher, Stefan, 'Transgender and Intersex: Theoretical, Practical, and Artistic Perspectives', in Stefan Horlacher, ed., *Transgender and Intersex: Theoretical, Practical, and Artistic Perspectives* (New York: Palgrave Macmillan, 2016), 1–27.

Houbre, Gabrielle, 'Dans l'ombre de l'*hermaphrodite*: Hommes et femmes en famille dans la France du XIXe siècle', *CLIO: Histoire, femmes et sociétés*, 34 (2011), 85–104.

——, ed., *Les deux vies d'Abel Barbin, né Adélaïde Herculine (1838–1868)* (Paris: Presses Universitaires de France, 2020).

Intersex Society of North America website <http://www.isna.org>, accessed 02 April 2020.

Jansen, Sebastian, 'Transgender and Intersex: Unavoidable Essentialism and the Normative Struggle for Recognition', in Stefan Horlacher, ed., *Transgender and Intersex: Theoretical, Practical, and Artistic Perspectives* (New York: Palgrave Macmillan, 2016), 115–40.

Johnson, Barbara, *The Critical Difference: Essays in the Contemporary Rhetoric of Reading* (Baltimore and London: The Johns Hopkins University Press, 1980).

Johnson, Michael A., and Lawrence R. Schehr, 'Editors' Preface: Turns to the Right?', *Yale French Studies*, 116/117 (2009), 1–4.

Jourdan, Annie, 'Le triomphe de Lucien de Rubempré: Faux poète, vrai dandy', in Françoise Van Rossum-Guyon, ed., *Balzac: Illusions perdues: 'l'œuvre capitale dans l'œuvre'*, *CRIN*, 18 (1988), 62–77.

Jourde, Pierre, *Littérature monstre: Études sur la modernité littéraire* (Paris: Esprit des péninsules, 2008).

Kerlouégan, François, 'Une relecture de l'androgynie romantique: La quête du genre dans *Fragoletta*, *Sarrasine* et *Mademoiselle de Maupin*', *Insignis*, 1 (May 2010), 36–53.

Kessler, Suzanne, *Lessons from the Intersexed* (New Brunswick and New York: Routledge, 1998).

Kimmel, Michael S., *Angry White Men: American Masculinity at the End of an Era* (New York: Nation Books, 2017 [2013]).

—— 'The Contemporary "Crisis" of Masculinity in Historical Perspective', in Harry Brod, ed., *The Making of Masculinities: The New Men's Studies* (Boston: Allen & Unwin, 1987), 121–53.

—— and Lisa Wade, 'Ask a Feminist: Michael Kimmel and Lisa Wade Discuss Toxic Masculinity', *Signs: Journal of Women in Culture and Society*, 44/1 (2018), 233–54.

Kingston, Ralph, *Bureaucrats and Bourgeois Society: Office Politics and Individual Credit in France 1789–1848* (Houndmills, Basingstoke and New York: Palgrave Macmillan, 2012).

Knight, Diana, 'Skeletons in the Closet: Homosocial Secrets in Balzac's *La Comédie Humaine*', *French Studies*, 57/2 (April 2003), 167–80.

Kolb, Katherine, 'The Tenor of *Sarrasine*', *PMLA*, 120/5 (October 2005), 1560–75.

Krumm, Pascale, 'La Bette noire de Balzac', *Australian Journal of French Studies*, 28/3 (September–December 1991), 254–63.

La Belle, Jenijoy, *Herself Beheld: The Literature of the Looking Glass* (Ithaca and London: Cornell University Press, 1988).

Labouret, Mireille, 'Méphistophélès et l'Androgyne: Les figures du pacte dans *Illusions perdues*', *L'Année balzacienne*, 2nd ser., 17 (1996), 211–30.

Lacan, Jacques, *Écrits* (Paris: Seuil, 1966).

Laforgue, Pierre, *L'éros romantique: Représentations de l'amour en 1830* (Paris: Presses Universitaires de France, 1998).

Laqueur, Thomas, *Making Sex: Body and Gender from the Greeks to Freud* (Cambridge, MA, and London: Harvard University Press, 1990).

—— 'Orgasm, Generation, and the Politics of Reproductive Biology', in Catherine Gallagher and Thomas Laqueur, eds, *The Making of the Modern Body: Sexuality and Society in the Nineteenth Century* (Berkeley, Los Angeles and London: University of California Press, 1987), 1–41.

Lascar, Alex, 'Le suicide dans *La Comédie Humaine*', *L'Année balzacienne*, 3rd ser., 17 (2016), 129–50.

Lehtonen, Maija, *Essai sur la 'Confession d'un enfant du siècle'* (Helsinki: Suomalainen Tiedeakatemia, 1982).

Lejeune, Philippe, 'Autobiographie et homosexualité en France au XIXe siècle', *Romantisme*, 56 (1987), 79–94.

Levin, Susan M., *The Romantic Art of Confession: De Quincey, Musset, Sand, Lamb, Hogg, Frémy, Soulié, Janin* (Columbia, SC: Camden House, 1998).

Linton, Anne E., *Prescribed Fictions: Literary and Medical Representations of Hermaphrodism in Nineteenth-Century France* (doctoral thesis, Yale University, 2011).

Littré, Émile, *Dictionnaire de la langue française*, 2nd edn (1873–77) <http://www.littre.org>, accessed 29 March 2017.

Lloyd, Rosemary, *The Land of Lost Content: Children and Childhood in Nineteenth-Century French Literature* (Oxford: Clarendon Press, 1992).

Lorant, André, *Les Parents Pauvres d'Honoré de Balzac: 'La Cousine Bette'–'Le Cousin Pons': Étude historique et critique*, 2 vols (Genève: Droz, 1967).

Lucas-Dubreton, Jean, *Le culte de Napoléon: 1815–1848* (Paris: Albin Michel, 1960).

Lucey, Michael, *The Misfit of the Family: Balzac and the Social Forms of Sexuality* (Durham, NC, and London: Duke University Press, 2003).

Lyon-Caen, Judith, *La lecture et la vie: Les usages du roman au temps de Balzac*, pref. Alain Corbin (Paris: Tallandier, 2006).

MacInnes, John, *The End of Masculinity: The Confusion of Sexual Genesis and Sexual Difference in Modern Society* (Buckingham and Philadelphia: Open University Press, 1998).

Magraw, Roger, *France 1815–1914: The Bourgeois Century* (London: Fontana, 1983).

Maira, Daniel, and Jean-Marie Roulin, 'Constructions littéraires de la masculinité entre Lumières et Romantisme', in Daniel Maira and Jean-Marie Roulin, eds, *Masculinités en révolution de Rousseau à Balzac* (Saint-Étienne: Publications de l'Université de Saint-Étienne, 2013), 9–28.

Mak, Geertje, *Doubting Sex: Inscriptions, Bodies and Selves in Nineteenth-Century Hermaphrodite Case Histories* (Manchester and New York: Manchester University Press, 2012).

Margadant, Jo Burr, 'Gender, Vice, and the Political Imaginary in Postrevolutionary France: Reinterpreting the Failure of the July Monarchy, 1830–1848', *The American Historical Review*, 104/5 (December 1999), 1461–96.

Martin, Brian Joseph, *Napoleonic Friendship: Military Fraternity, Intimacy, and Sexuality in Nineteenth-Century France* (Durham, NH: University of New Hampshire Press, 2011).

Martin, Xavier, 'Fonction paternelle et Code Napoléon', *Annales historiques de la Révolution française*, 305 (1996), 465–75.

Maugue, Annelise, *L'identité masculine en crise au tournant du siècle: 1871–1914* (Paris: Rivages, 1987).

Mazaleigue-Labaste, Julie, 'De l'amour socratique à l'homosexualité grecque', *Romantisme*, 159 (2013), 35–46.

McGuire, James R., 'The Feminine Conspiracy in Balzac's *La Cousine Bette*', *Nineteenth-Century French Studies*, 20/3–4 (Spring–Summer 1992), 295–304.

Meier, Franziska, 'La "Mort de l'auteur" dans l'écriture autobiographique romantique: À propos du "jeune" François-René de Chateaubriand (*René*) et d'Alfred de Musset (*La Confession d'un enfant du siècle*)', *French Studies*, 67/3 (July 2013), 323–39.

Mellor, Anne K., *Romanticism and Gender* (New York and London: Routledge, 1993).

Mendès-Leite, Rommel, 'Inconstance des sexes et des genres dans les sociétés non occidentales', in Jean-Yves Tamet, ed., *Différenciation sexuelle et identités: Clinique, art et littérature*, pref. Pierre Chatelain (Paris: In Press, 2012), 111–32.

Miller, D. A., 'Balzac's Illusions Lost and Found', *Yale French Studies*, 67 (1984), 164–81.

Miller, John F., and Carole E. Newlands, 'Introduction', in John F. Miller and Carole E. Newlands, eds, *Handbook to the Reception of Ovid* (Hoboken: John Wiley and Sons, 2014), 1–7.

Moi, Toril, *What Is a Woman? And Other Essays* (Oxford: Oxford University Press, 1999).

Molino, Jean, 'Qu'est-ce que le roman historique?', *Revue d'Histoire Littéraire de la France*, 75/2–3 (March–June 1975), 195–234.

Monneyron, Frédéric, *L'androgyne romantique: Du mythe au mythe littéraire* (Grenoble: Ellug, 1994).

Moretti, Franco, 'Homo Palpitans: Balzac's Novels and Urban Personality', in *Signs Taken for Wonders: Essays in the Sociology of Literary Forms*, Susan Fischer, David Forgacs and David Miller, trans. (London and New York: Verso, 1988 [1983]), 109–29.

Morrey, Douglas, *Michel Houellebecq: Humanity and its Aftermath* (Liverpool: Liverpool University Press, 2013).

——'Sex and the Single Male: Houellebecq, Feminism, and Hegemonic Masculinity', *Yale French Studies*, 116/117 (2009), 141–52.

Moscovici, Claudia, 'Gendered Spheres in Balzac's *La Cousine Bette*', *International Journal of Politics, Culture and Society*, 10/3 (1997), 469–96.

Moses, Claire Goldberg, *French Feminism in the Nineteenth Century* (Albany: State University of New York Press, 1984).

Mosher, Donald L., and Mark Sirkin, 'Measuring a Macho Personality Constellation', *Journal of Research in Personality*, 18/2 (1984), 150–63.

Mosse, George L., *The Image of Man: The Creation of Modern Masculinity* (New York and Oxford: Oxford University Press, 1996).

Moulinat, Francis, 'Les Amours grecques: Homosexualité et représentations, du *Léonidas* de Jacques-Louis David (1799–1814) au *Swimming Hole* de Thomas Eakins (1885)', *Romantisme*, 159 (2013), 73–83.

Mouriquand, Pierre, 'S'asseoir dehors', in Jean-Yves Tamet, ed., *Différenciation sexuelle et identités: Clinique, art et littérature*, pref. Pierre Chatelain (Paris: In Press, 2012), 53–64.

Mozet, Nicole, *Balzac au pluriel* (Paris: Presses Universitaires de France, 'Écrivains', 1990).

—— *George Sand: Écrivain de romans* (Saint-Cyr-sur-Loire: Pirot, 1997).

Mura, Aline, *'Béatrix' ou la logique des contraires* (Paris: Champion, 1997).

Murat, Laure, *La loi du genre: Une histoire culturelle du 'troisième sexe'* (Paris: Fayard, 2006).

Murata, Kyoto, 'Sous le signe des monstruosités: La Camille Maupin de Balzac et George Sand', *George Sand Studies*, 26 (2007), 43–54.

Murphy, Margueritte, 'Pure Art, Pure Desire: Changing Definitions of "l'art pour l'art" from Kant to Gautier', *Studies in Romanticism*, 47/2 (Summer 2008), 147–60.

Naginski, Isabelle Hoog, 'Les deux *Lélia*: Une réécriture exemplaire', *Revue des sciences humaines*, 226 (April–June 1992), 65–84.

—— *George Sand: Writing for Her Life* (New Brunswick and London: Rutgers University Press, 1991).

—— '*Lélia*, ou l'héroïne impossible', *Études littéraires*, 35/2–3 (2003), 87–106.

Nesci, Catherine, 'Splendeurs et misères du "grand homme": De la catabase médiévale aux masculinités modernes (Dante, Delacroix, Balzac)', in Daniel Maira and Jean-Marie Roulin, eds, *Masculinités en révolution de Rousseau à Balzac* (Saint-Étienne: Publications de l'Université de Saint-Étienne, 2013), 135–54.

Newton, Esther, 'The Mythic Mannish Woman: Radclyffe Hall and the New Woman', *Signs*, 9/4 (Summer 1984), 557–75.

Noble, Jean Bobby, *Masculinities without Men: Female Masculinity in Twentieth-Century Fictions* (Vancouver and Toronto: University of British Columbia Press, 2004).

Noble, Yvonne, 'Castrati, Balzac, and "BartheS/Z"', *Comparative Drama*, 31/1 (Spring 1997), 28–41.

Le Nouveau Petit Robert: Dictionnaire de la langue française (Paris: Le Robert-SEJER, 2004).

Nye, Robert A., *Masculinity and Male Codes of Honor in Modern France* (Berkeley and Los Angeles: University of California Press, 1998 [Oxford: Oxford University Press, 1993]).

—— 'Sex Difference and Male Homosexuality in French Medical Discourse, 1830–1930', *Bulletin of the History of Medicine*, 63 (1989), 32–51.

Nykrog, Per, '*Illusions perdues* dans ses grandes lignes: Stratégies et tactiques romanesques', in Françoise Van Rossum-Guyon, ed., *Balzac: Illusions perdues: 'l'œuvre capitale dans l'œuvre'*, *CRIN*, 18 (1988), 34–46.

—— 'On Seeing and Nothingness: Balzac's *Sarrasine*', *Romanic Review*, 83/4 (November 1992), 437–44.

O'Donovan, Patrick, 'Avatars of the Artist: Narrative Approaches to the Work of the Painter', in Peter Collier and Robert Lethbridge, eds, *Artistic Relations: Literature and the Visual Arts in Nineteenth-Century France* (New Haven and London: Yale University Press, 1994), 222–36.

Offen, Karen, *European Feminisms 1700–1950: A Political History* (Stanford: Stanford University Press, 2000).

Orr, Mary, *Madame Bovary: Representations of the Masculine* (Bern, Berlin, Brussels, Frankfurt a. M., New York and Vienna: Peter Lang, 1999).

Ortali, Hélène, 'Images of Women in Balzac's *La Cousine Bette*', *Nineteenth-Century French Studies*, 4/3 (Spring 1976), 194–205.

Oswald, Dana, 'Monstrous Gender: Geographies of Ambiguity', in Asa Simon Mittman and Peter J. Dendle, eds, *The Ashgate Research Companion to Monsters and the Monstrous* (Farnham, Surrey and Burlington, VT: Ashgate, 2012), 343–63.

Pavel, Thomas, 'Énergie et illusion: Sur les protagonistes de *Sarrasine*', *French Forum*, 21/3 (September 1996), 301–18.

Péladan, Joséphin, *De l'Androgyne: Théorie plastique* (Puiseaux: Pardès, 'Rebis', 1988 [Paris: Sansot et Cie, 1910]).

Peniston, William A., *Pederasts and Others: Urban Culture and Sexual Identity in Nineteenth-Century Paris* (New York: Harrington Park Press, 2004).

Peponi, Anastasia-Erasmia, 'Fantasizing Lyric: Horace, *Epistles*, 1.19' in Michael Paschalis, ed., *Horace and Greek Lyric Poetry* (Rethymnon: University of Crete, 2002), 19–45.

Petrey, Sandy, 'Speech Acts, and the Realist Differences: *S/Z* versus *Sarrasine*', *PMLA*, 102/2 (March 1987), 153–65.

Picon, Antoine, *Les saint-simoniens: Raison, imaginaire et utopie* (Paris: Belin, 2002).

Piketty, Thomas, *Le capital au XXIe siècle* (Paris: Seuil, 2013).

Pilbeam, Pamela M., *The 1830 Revolution in France* (London: Macmillan, 1991).

Planté, Christine, *La petite sœur de Balzac: Essai sur la femme auteur* (Paris: Seuil, 1989).

Potts, Alex, *Flesh and the Ideal: Winckelmann and the Origins of Art History* (New Haven and London: Yale University Press, 1994).

Praz, Mario, *The Romantic Agony*, Angus Davidson, trans., foreword Frank Kermode, 2nd edn (Oxford: Oxford University Press, 1970 [1933]).

Prendergast, Christopher A., 'Antithesis and Moral Ambiguity in *La Cousine Bette*', *The Modern Language Review*, 68/2 (April 1973), 315–32.

—— 'Towards a Re-assessment of *Béatrix*', *Essays in French Literature*, 9 (1972), 46–62.

Prest, Julia, *Theatre under Louis XIV: Cross-Casting and the Performance of Gender in Drama, Ballet, and Opera* (Houndmills, Basingstoke and New York: Palgrave Macmillan, 2006).

Prosser, Jay, *Second Skins: The Body Narratives of Transsexuality* (New York: Columbia University Press, 1998).

Queffelec, Lise, 'Le lecteur du roman comme lectrice: Stratégies romanesques et stratégies critiques sous la Monarchie de Juillet', *Romantisme*, 53 (1986), 9–21.

Ramacciotti, Valeria, 'L'esilio dell'androgino', *Studi di letteratura francese*, 16 (1990), 188–201.

Rauch, André, *Crise de l'identité masculine: 1789–1914* (Paris: Hachette Littératures, 'Pluriel', 2000).

Reboul, Jean, 'Sarrasine ou la castration personnifiée', *Cahiers pour l'analyse*, 7 (1967), 91–96.

Régnier, Philippe, 'La réaffirmation et la réinvention saint-simoniennes du masculin: Pour une lecture nouvelle d'un féminisme originel', in Daniel Maira and Jean-Marie Roulin, eds, *Masculinités en révolution de Rousseau à Balzac* (Saint-Étienne: Publications de l'Université de Saint-Étienne, 2013), 49–76.

Reid, Martine, 'La confession selon Musset', *Littérature*, 67 (1987), 53–72.

Revenin, Régis, 'Conceptions et théories savantes de l'homosexualité masculine en France, de la monarchie de Juillet à la Première Guerre mondiale', *Revue d'Histoire des Sciences Humaines*, 17/2 (2007), 23–45.

—— 'Homosexualité et virilité', in Alain Corbin, ed., *Histoire de la virilité*, II: *Le triomphe de la virilité: Le XIX*[e] *siècle* (Paris: Seuil, 2011), 369–401.

Rey, Alain, et al., *Dictionnaire historique de la langue française*, 2 vols (Paris: Le Robert, 1992).

Rifelj, Carol, 'Nobody's Perfect: Hair and Sexual Identity in *Mademoiselle de Maupin*', *Dalhousie French Studies*, 69 (Winter 2004), 33–41.

Rivers, Christopher, '*Inintelligibles pour une femme honnête*: Sexuality, Textuality and Knowledge in Diderot's *La Religieuse* and Gautier's *Mademoiselle de Maupin*', *Romance Review*, 86/1 (January 1995), 1–29.

Rizzuto, Anhony, 'Octave in Alfred de Musset's *La Confession d'un enfant du siècle*', *Kentucky Romance Quarterly*, 24/1 (1977), 83–94.

Robb, Graham, *Strangers: Homosexual Love in the Nineteenth Century* (London: Picador, 2003).

Roberts, Mary, *Intimate Outsiders: The Harem in Ottoman and Orientalist Art and Travel Literature* (Durham, NC, and London: Duke University Press, 2007).

Rogers, Nancy, 'Psychosexual Identity and the Erotic Imagination in the Early Novels of George Sand', *Studies in the Literary Imagination*, 12/2 (Fall 1979), 19–35.

—— 'The Wasting Away of Romantic Heroines', *Nineteenth-Century French Studies*, 11/3–4 (Spring–Summer 1983), 246–56.

Rosario, Vernon A., *The Erotic Imagination: French Histories of Perversity* (Oxford and New York: Oxford University Press, 1997).

—— 'Pointy Penises, Fashion Crimes, and Hysterical Mollies: The Pederasts' Inversions', in Jeffrey Merrick and Bryant T. Ragan Jr, eds, *Homosexuality in Modern France* (New York and Oxford: Oxford University Press, 1996), 146–76.

Roulin, Jean-Marie, 'Confusion des sexes, mélange des genres et quête du sens dans *Mademoiselle de Maupin*', *Romantisme*, 103 (1999), 31–40.

Rowland, Antony, Emma Liggins, and Eriks Uskalis, 'Introduction', in Antony Rowland, Emma Liggins and Eriks Uskalis, eds, *Signs of Masculinity: Men in Literature 1700 to the Present* (Amsterdam and Atlanta: Rodopi, 1998), 3–36.

Said, Edward W., *Orientalism: Western Conceptions of the Orient* (London: Penguin Books, 1995 [Routledge & Kegan Paul, 1978]).

Saint-Gérand, Jacques-Philippe, 'Homosexualité des *alphadécédets*: Remarques sur un innommable des dictionnaires conformes, et recours aux excentriques', *Romantisme*, 159 (2013), 19–34.

Saint-Paulien (Maurice-Yvan Sicard), *Napoléon, Balzac et l'Empire de 'la Comédie humaine'* (Paris: Albin Michel, 1979).

Schehr, Lawrence R., 'Homme-diégèse', in Robert Mahieu and Franc Schuerewegen, eds, *Balzac ou la tentation de l'impossible* (Paris: SEDES, 1998), 135–41.

—— *Parts of an Andrology: On Representations of Men's Bodies* (Stanford: Stanford University Press, 1997).

Schor, Naomi, *George Sand and Idealism* (New York: Columbia University Press, 1993).

Schwartz, Regina M., *The Curse of Cain: The Violent Legacy of Monotheism* (Chicago and London: University of Chicago Press, 1997).

Sedgwick, Eve Kosofsky, *Between Men: English Literature and Male Homosocial Desire* (New York: Columbia University Press, 1985).

Segal, Lynne, *Slow Motion: Changing Masculinities, Changing Men* (London: Virago Press, 1990).

Seznec, Jean, 'Diderot et *Sarrasine*', *Diderot Studies*, 4 (1963), 237–45.

Sibalis, Michael, 'The Palais-Royal and the Homosexual Subculture of Nineteenth-Century Paris', in Jeffrey Merrick and Michael Sibalis, eds, *Homosexuality in French History and Culture* (New York: Harrington Park Press, 2001), 117–29.

Sivert, Eileen Boyd, 'Lélia and Feminism', *Yale French Studies*, 62 (1981), 45–66.

Smith, Albert B., 'Gautier's *Mademoiselle de Maupin*: The Quest for Happiness', *Modern Language Quarterly*, 32 (1971), 168–74.

Smith, Anna Marie, 'Hegemony Trouble: The Political Theories of Judith Butler, Ernesto Laclau and Chantal Mouffe', in Jeffrey Weeks, ed., *The Lesser Evil and the Greater Good: The Theory and Politics of Social Diversity* (London: Rivers Oram Press, 1994), 222–34.

Smith, Nigel E., 'Silence, Secrecy, and Scientific Discourse in the Nineteenth Century', in Dominique D. Fisher and Lawrence R. Schehr, eds, *Articulations of Difference: Gender Studies and Writing in French* (Stanford: Stanford University Press, 1997), 83–99.

Smith, Paul, 'Literature and Art', in *French Studies*, 61/1 (January 2007), 1–13.

Smith-Rosenberg, Carroll, *Disorderly Conducts: Visions of Genders in Victorian America* (Oxford and New York: Oxford University Press, 1985).

Sohn, Anne-Marie, *'Sois un homme!': La construction de la masculinité au XIXe siècle* (Paris: Seuil, 2009).

Solomon-Godeau, Abigail, *Male Trouble: A Crisis in Representation* (London: Thames and Hudson, 1997).

Spencer, Michael Clifford, *Charles Fourier* (Boston: Twayne Publishers, 1981).

Stark, Rodney, Daniel P. Doyle, and Jesse Lynn Rushing, 'Beyond Durkheim: Religion and Suicide', *Journal for the Scientific Study of Religion*, 22/2 (June 1983), 120–31.

Storzer, Gerald H., 'The Homosexual Paradigm in Balzac, Gide, and Genet', in George Stambolian and Elaine Marks, eds, *Homosexualities and French Literature: Cultural Contexts/Critical Texts*, pref. Richard Howard (Ithaca and London: Cornell University Press, 1990 [1979]), 186–209.

Sullivan, Nikki, 'The Matter of Gender', in Lisa Downing, Iain Morland, and Nikki Sullivan, eds, *Fuckology: Critical Essays on John Money's Diagnostic Concepts* (Chicago and London: University of Chicago Press, 2015), 19–40.

Sweeney Carol, *Michel Houellebecq and the Literature of Despair* (London and New York: Bloomsbury, 2013).

Tamet, Jean-Yves, Patricia Bretonès, and Claire-Lise Gay, 'Se représenter puis se nommer: Une courte histoire de la nomination de sujets atteints d'anomalie de la différenciation sexuelle', in Jean-Yves Tamet, ed., *Différenciation sexuelle et identités: Clinique, art et littérature*, pref. Pierre Chatelain (Paris: In Press, 2012), 19–27.

Terdiman, Richard, 'Structures of Initiation: On Semiotic Education and Its Contradictions in Balzac', *Yale French Studies*, 63 (1982), 198–226.

Thévenaz, Olivier, 'Sappho au banquet d'Horace', *Camenae*, 12 (June 2012), 1–14: <http://www.paris-sorbonne.fr/IMG/pdf/2-_O-_Thevenaz.pdf>, accessed 29 March 2017.

Thomas, Ruth P., 'The Death of an Ideal: Female Suicides in the Eighteenth-Century French Novel', in Samia I. Spencer, ed., *French Women and the Age of Enlightenment* (Bloomington: Indiana University Press, 1984), 321–31.

Thompson, Victoria, 'Creating Boundaries: Homosexuality and the Changing Social Order in France, 1830–1870', in Jeffrey Merrick and Bryant T. Ragan

Jr, eds, *Homosexuality in Modern France* (New York and Oxford: Oxford University Press, 1996), 102–27.

Tuana, Nancy, *The Less Noble Sex: Scientific, Religious, and Philosophical Conceptions of Woman's Nature* (Bloomington and Indianapolis: Indiana University Press, 1993).

Tulard, Jean, *Le mythe de Napoléon* (Paris: Armand Colin, 1971).

Ussher, Jane, *Women's Madness: Misogyny or Mental Illness?* (New York: Harvester Wheatsheaf, 1991).

Vanoncini, André, 'La femme-livre dans *Béatrix* de Balzac', in Peter Schnyder, ed., *L'homme-livre: Des hommes et des livres, de l'Antiquité au XXe siècle* (Paris: Orizons, chez L'Harmattan, 2007), 153–65.

—— 'Le pacte: Structures et évolutions d'un motif balzacien', *L'Année balzacienne*, 3rd ser., 3 (2002), 279–92.

—— 'Temps et mémoire dans *Béatrix*', *L'Année balzacienne*, 3rd ser., 8 (2007), 275–88.

Van Rossum-Guyon, Françoise, *Balzac et la littérature réfléchie: Discours et autoreprésentations* (Montreal and Paris: Paragraphes, Département d'Études françaises de l'Université de Montréal, 2002).

Vareille, Kristina Wingård, *Socialité, sexualité et les impasses de l'histoire: L'évolution de la thématique sandienne d''Indiana' (1832) à 'Mauprat' (1837)* (Uppsala: Almqvist & Wiksell, 1987).

Vasarri, Fabio, 'I nomi dell'ermafrodito: L'esempio di *Fragoletta*', *Rivista di letterature moderne e comparate*, 44/2 (April–June 1991), 120–29.

Vassilev, Kris, 'Représentation et signification sociale de la vengeance dans un texte réaliste: L'exemple de *La Cousine Bette*', *Romantisme*, 127 (2005), 45–57.

Vinciguerra, Geneviève, 'Trois drames de la paternité frustrée: Goriot, Vautrin, Balzac', *Littératures*, 17 (1970) 3–14.

Waelti-Walters, Jennifer, *Damned Women: Lesbians in French Novels, 1796–1996* (Montreal: McGill-Queen's University Press, 2000).

Waller, Margaret, *The Male Malady: Fictions of Impotence in the French Romantic Novel* (New Brunswick: Rutgers University Press, 1993).

Weil, Kari, *Androgyny and the Denial of Difference* (Charlottesville and London: University Press of Virginia, 1992).

—— 'Romantic Androgyny and its Discontents: The Case of *Mlle de Maupin*', *Romanic Review*, 78/3 (May 1987), 348–58.

Whitehead, Stephen M., and Frank J. Barrett, 'The Sociology of Masculinity', in Stephen M. Whitehead and Frank J. Barrett, eds, *The Masculinities Reader* (Cambridge: Polity Press, 2001), 1–26.

Whittle, Stephen, 'Foreword', in Susan Stryker and Stephen Whittle, eds, *The Transgender Studies Reader* (New York and London: Routledge, 2006), xi–xvi.

Wing, Nathaniel, *Between Genders: Narrating Difference in Early French Modernism* (Newark, DE: University of Delaware Press, 2010).
Yeğenoğlu, Meyda, *Colonial Fantasies: Towards a Feminist Reading of Orientalism* (Cambridge: Cambridge University Press, 1998).
Zemmour, Éric, *Le suicide français* (Paris: Albin Michel, 2014).

Index

French Studies of the Eighteenth and Nineteenth Centuries

Edited by Fiona Cox, Associate Professor of French and Comparative Literature at the University of Exeter

This series publishes the latest research by teachers and researchers working in all the disciplines which constitute French and Francophone studies in this period, in the form of monographs, revised dissertations, collected papers and conference proceedings. Adhering to the highest academic standards, it provides a vehicle for established scholars with specialised research projects but also encourages younger academics who may be publishing for the first time.

The editor takes a broad view of French studies and intends to examine literary and cultural phenomena of the eighteenth and nineteenth centuries, excluding the Romantic movement, against their historical, political and social background in all the French-speaking countries. The editor also welcomes work in comparative studies, and on adaptations, across languages or media.

Volume 1 Malcolm Cook and Annie Jourdan (éds): Journalisme et fiction au 18e siècle.
241 pages. 1999.
ISBN 3-906761-50-9 / US-ISBN 0-8204-4221-6

Volume 2 Paul Rowe: A Mirror on the Rhine? The Nouvelle revue germanique, Strasbourg 1829–1837.
340 pages. 2000.
ISBN 3-906762-39-4 / US-ISBN 0-8204-4233-X

Volume 3 Rachael Langford: Jules Vallès and the Narration of History. Contesting the French Third Republic in the Jacques Vingtras Trilogy.
271 pages. 1999.
ISBN 3-906762-99-8 / US-ISBN 0-8204-4249-6

Volume 4 Malcolm Cook et Marie-Emmanuelle Plagnol-Diéval (éds): Réécritures 1700–1820.
298 pages. 2002.
ISBN 3-906768-28-7

Volume 5 Malcolm Cook et Marie-Emmanuelle Plagnol-Diéval (éds): Anecdotes, Faits-Divers, Contes, Nouvelles 1700–1820.
302 pages. 2000.
ISBN 3-906765-08-3

Volume 6 Timothy Unwin: Textes réfléchissants. Réalisme et réflexivité au dix-neuvième siècle.
217 pages. 2000.
ISBN 3-906758-30-3

Volume 7 Derek Connon and George Evans (eds): Essays on French Comic Drama from the 1640s to the 1780s.
236 pages. 2000.
ISBN 3-906758-49-7 / US-ISBN 0-8204-5071-5

Volume 8 Philip Robinson (éd.): Beaumarchais: homme de lettres, homme de société.
298 pages. 2000, 2002.
ISBN 3-906768-90-2 / US-ISBN 0-8204-5883-X

Volume 9 William Gallois: Zola: The History of Capitalism.
296 pages. 2000.
ISBN 3-906758-60-5 / US-ISBN 0-8204-5077-4

Volume 10 David Kinloch and Gordon Millan (eds): Situating Mallarmé.
191 pages. 2000.
ISBN 3-906766-18-7 / US-ISBN 0-8204-5320-X

Volume 11 Derek Connon: Diderot's Endgames.
333 pages. 2002.
ISBN 3-906768-35-X / US-ISBN 0-8204-5623-3

Volume 12 Robin Howells:
Playing Simplicity. Polemical Stupidity in the Writing of the French Enlightenment.
331 pages. 2002.
ISBN 3-906768-36-8 / US-ISBN 0-8204-5624-1

Volume 13 C. W. Thompson:
Walking and the French Romantics. Rousseau to Sand and Hugo.
160 pages. 2003.
ISBN 3-03910-078-5 / US-ISBN 0-8204-6894-0

Volume 14 Nigel Harkness, Paul Rowe, Tim Unwin and Jennifer Yee (eds):
Visions/Revisions. Essays on Nineteenth-Century French Culture.
330 pages. 2003.
ISBN 3-03910-140-4 / US-ISBN 0-8204-6950-5

Volume 15 Katherine Astbury and Marie-Emmanuelle Plagnol-Diéval (éds):
Le mâle en France 1715–1830. Représentations de la masculinité.
317 pages. 2004.
ISBN 3-03910-177-3 / US-ISBN 0-8204-6987-4

Volume 16 Jelka Samsom:
Individuation and Attachment in the Works of Isabelle de Charrière.
213 pages. 2005.
ISBN 3-03910-187-0 / US-ISBN 0-8204-6997-1

Volume 17 Ceri Crossley:
Consumable Metaphors. Attitudes towards Animals and Vegetarianism in Nineteenth-Century France.
322 pages. 2005.
ISBN 3-03910-190-0 / US-ISBN 0-8204-7175-5

Volume 18 Forthcoming.

Volume 19 Ezequiel Adamovsky:
Euro-Orientalism. Liberal Ideology and the image of Russia in France (c. 1740–1880).
358 pages. 2006.
ISBN 3-03910-516-7 / US-ISBN 0-8204-7552-X

Volume 20 Sarah Capitanio, Lisa Downing, Paul Rowe and Nicholas White (eds):
Currencies. Fiscal Fortunes and Cultural Capital in Nineteenth-Century France.
211 pages. 2005.
ISBN 3-03910-513-2 / US-ISBN 0-8204-7519-X

Volume 21 Anne Betty Weinshenker:
A God or a Bench. Sculpture as a Problematic Art during the Ancien Régime.
379 pages. 2008.
ISBN 978-3-03910-543-4

Volume 22 Malcolm Cook et Marie-Emmanuelle Plagnol-Diéval (éds):
Critiques, Critiques au 18e siècle.
308 pages. 2006.
ISBN 3-03910-537-X

Volume 23 Owen Heathcote:
Balzac and Violence. Representing History, Space, Sexuality and Death in La Comédie humaine.
288 pages. 2009.
ISBN 978-3-03910-551-9

Volume 24 Kirsty Carpenter:
The Novels of Madame de Souza in Social and Political Perspective.
280 pages. 2007.
ISBN 3-03910-898-0 / US-ISBN 0-8204-8018-5

Volume 25 Forthcoming.

Volume 26 James Kearns et Pierre Vaisse (éds):
« Ce Salon à quoi tout se ramène »: Le Salon de peinture et de sculpture, 1791–1890.
180 pages. 2010.
ISBN 978-3-03910-931-9

Volume 27 Mathilde Kang:
Le parcours transatlantique du Journal d'Eugénie de Guérin: un cas de transfert culturel (1850–1950).
215 pages. 2009.
ISBN 978-3-03911-358-3

Volume 28 Andrew Watts:
Preserving the Provinces. Small Town and Countryside in the Work of Honoré de Balzac.
337 pages. 2007.
ISBN 978-3-03910-583-0

Volume 29 Sandrine Léopold:
L'Écriture du regard dans la représentation de la passion amoureuse et du désir. Étude comparative d'œuvres choisies de Madame de Lafayette, Rousseau, Stendhal et Duras.
321 pages. 2009.
ISBN 978-3-03911-542-6

Volume 30 Nigel Harkness et Jacinta Wright (éds):
George Sand: Intertextualité et Polyphonie I. Palimpsestes, Échanges, Réécritures.
359 pages. 2011.
ISBN 978-3-03911-987-5

Volume 31 Nigel Harkness et Jacinta Wright (éds):
George Sand: Intertextualité et Polyphonie II. Voix, Image, Texte.
332 pages. 2011.
ISBN 978-3-03911-988-2

Volume 32 Emma Bielecki:
The Collector in Nineteenth-Century French Literature. Representation, Identity, Knowledge.
242 pages. 2012.
ISBN 978-3-0343-0757-4

Volume 33 James Kearns and Alister Mill (eds):
The Paris Fine Art Salon / Le Salon, 1791–1881.
517 pages. 2015.
ISBN 978-3-0343-1778-8

Volume 34 James Hanrahan and Síofra Pierse (eds):
The Dark Side of Diderot / Le Diderot des ombres.
269 pages. 2016.
ISBN 978-3-0343-1848-8

Volume 35 Rita Oghia-Codsi:
'The Return of the Repressed'.
Uncovering Family Secrets in Zola's Fiction:
An Interpretation of Selected Novels.
332 pages. 2016.
ISBN 978-3-0343-1982-9

Volume 36 Karen Lacey:
A Class Apart.
The Military Man in French and British Fiction, 1740–1789.
247 pages. 2016.
ISBN 978-3-0343-1887-7

Volume 37 Stacie Allan:
Writing the Self, Writing the Nation.
Romantic Selfhood in the Works of Germaine de Staël and Claire de Duras.
240 pages. 2019.
ISBN 978-1-78874-208-5

Volume 38 Helder Mendes Baiao:
Rêves de citoyens. Le républicanisme dans la littérature Suisse romande du XVIIIe siècle.
334 pages. 2021.
ISBN 978-1-78707-648-8

Volume 39 Giada Alessandroni:
Amongst Women. Literary Representations of Female Homosociality in Belle Epoque France, 1880–1914.
244 pages. 2021.
ISBN 978-1-80079-064-3

Volume 40 Géraldine Crahay:
The Hermaphrodite, the Effete and the Butch. Sexual and Gender Ambiguities in Nineteenth-Century French Narratives.
358 pages. 2022.
ISBN 978-1-78997-648-9

Figures